Non-Linear Vibrations

Non-Linear Vibrations

G. Schmidt

A. Tondl

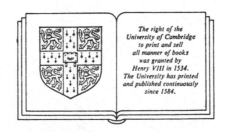

CAMBRIDGE UNIVERSITY PRESS

Cambridge

London New York New Rochelle

Melbourne Sydney

CAMBRIDGE UNIVERSITY PRESS
Cambridge, New York, Melbourne, Madrid, Cape Town, Singapore, São Paulo, Delhi

Cambridge University Press
The Edinburgh Building, Cambridge CB2 8RU, UK

Published in the United States of America by Cambridge University Press, New York

www.cambridge.org
Information on this title: www.cambridge.org/9780521113229

First published 1986
This digitally printed version 2009

A catalogue record for this publication is available from the British Library

Library of Congress Catalogue Card Number: 84-19900

ISBN 978-0-521-26698-7 hardback
ISBN 978-0-521-11322-9 paperback

Preface

It has not been an easy decision to choose the general title of *Non-linear vibrations*. Recently there has been an overwhelming development of the theory and many applications of non-linear vibrations, development reflected in numerous books, many specialist journals and the Equa-Diff conferences and ten International Conferences on Non-linear Oscillations.

In view of these development, we would like this book to be considered not as an attempt to survey the state of the art and extent of current knowledge of non-linear vibrations, but rather as an account of some methods and results in the field of non-linear vibrations we have obtained or encountered in our personal experience. We do not believe in a single, comprehensive theory of non-linear vibrations, not even in the sense of linear vibration theory. On the contrary, we see the specific difficulty — and the attraction — of non-linear vibrations in the non-existence, more than in the existence, of certain rules of order. The nature of our endeavour may perhaps best be conveyed in Rilke's words: ‚Uns überfüllts. Wir ordnens. Es zerfällt. Wir ordnens wieder . . .‘.

We believe that the material presented here can be used for many kinds of vibration problems, although translating it into their terms is no simple task. Various methods and examples are traditionally associated with different notations, and we have not tried to make the notation uniform throughout this book.

The first author is responsible in particular for chapters 5, 6, 10, 11 and 12, the second for chapters 3, 4, 7, 8 and 9.

We wish to thank those who by their interest in our results encouraged, or perhaps seduced, us into writing this book. We especially thank Prof. G. WALLIS, Dr. R. SCHULZ, Prof. W. EBELING, and Dipl.-Ing. D. HAJŠMANOVÁ for many valuable comments. We also thank Cambridge University Press and Akademie-Verlag Berlin for editing this book.

Die Autoren

Contents

Introduction

Finding examples for vibrations of systems which *are linear* is not a trivial matter; there is something in the remark by R. M. ROSENBERG that dividing vibrations into linear and non-linear is like dividing the world into bananas and non-bananas. Nevertheless many important vibration problems must be and *can be treated* in the same way *as linear ones*, and the book *Lineare Schwingungen* by MÜLLER and SCHIEHLEN, for instance, is very helpful for all those grappling with vibration problems.

However, numerous vibration phenomena which are theoretically surprising as well as practically important can only be understood on the basis of non-linear vibration. For instance, the wide field of self-excited, parametric and autoparametric vibration, to which we give special consideration in this book, demands non-linear treatment from the very beginning.

The mathematical theory of non-linear vibrations, the numerical and experimental methods for their evaluation and their many applications in mechanics, in mechanical and civil engineering, in physics, electrotechnology, biology and other sciences, have all developed so rapidly of late that it would require many volumes and many specialists to give a comprehensive picture of non-linear vibration.

The difficulties faced by those engaged in the study of the subject are concerned, on the one hand, with the broad scope of problems and the diversity of systems involved, and on the other hand with the fact that some general laws, such as the principle of superposition and the principle of proportionality, which can be used to advantage in solutions of linear systems, do not apply to non-linear ones. Again analytical or qualitative methods are nearly always approximate and hence even a physical analysis of vibration systems is more difficult if they are non-linear; such methods of solution must be supplemented by digital or analogue computing techniques. On the other hand, numerical methods have led to an understanding of new phenomena in the field of strongly non-linear systems. Similarly, investigations concerning the stability of the solutions are also more difficult because the existence of several steady solutions is a common occurrence in non-linear systems. Consequently, in detailed analysis, it is necessary to investigate stability even for disturbances which can no longer be considered small. Such an investigation of stability in the large is needless in case of linear systems.

Non-linear vibration theory can be divided into three main parts. The first comprises analytical methods dealing with approximate procedures chiefly of steady solutions. Several alternative methods have been developed, based on Poincaré's idea of expressing the solution in the form of an expansion with respect to a small parameter. Their most recent modification is the multiple scale method (see NAYFEH and MOOK (1979)). These procedures, as with the averaging method of Krylov and Bogoljubov, are widely used and thoroughly grounded mathematically. The same can hardly be said of the harmonic balance method, which can nevertheless be a very effective tool when employed by an experienced dynamicist. The integro-differential equation method, less well known than the averaging method although it has some advantages, will be dealt with in detail in this book.

The seond main part of the theory encompasses qualitative methods, stability investigations, bifurcation problems, attractor analysis and determination of the domains of attraction. The analysis of trajectories and singular points in the phase plane (space) is among the first representatives of this group of methods. The determination of the domains of attraction forms an important part of analyses of non-linear systems with several steady stable solutions. So far, this problem has been studied in only a few books. The pioneering work was done by HAYASHI (1964) on systems with one degree of freedom. Using different approaches, the problem of domains of attraction will be discussed in this book without restriction as to the number of degrees of freedom. Investigations of strange attractors and the use of catastrophe theory in solutions of bifurcation and stability problems represent recent contributions to this part of the theory.

The third main part of non-linear vibration theory includes investigations which, using the methods developed in the first two parts, chiefly aim at physical analysis of specific non-linear phenomena and effects or analyses of particular physical systems. The synchronization phenomenon, the effect of a limited exciting energy source and the effect of tuning a system into internal resonance, can be cited as examples of this kind of problem. There are numerous books and monographs which have been devoted to a special class of non-linear problems or to non-linear problems in particular systems, for example in electronic circuits, different machine elements, gyroscopic systems, systems with impacts, self-excited vibrations in machine tools caused by the action of cutting forces, problems of aero- and hydroelasticity, of non-linear control systems, hydraulic systems, astronomy and aerospace engineering, etc.

A survey of the most important books and monographs published during the postwar period, dealing with vibration in non-linear systems which develop a substantial body of non-linear vibration theory, would include the following works. To the oldest studies of this time which attempt to give a general view or to describe some analytical methods, belong those which deal largely with systems with one degree of freedom: books by MINORSKY (1947), McLACHLAN (1950), STOKER (1950), BULGAKOV (1954), PŮST and TONDL (1956), KAUDERER (1958), HAAG and CHALEAT (1960), etc. The treatise by MALKIN (1956) discusses the Poincaré method, that by BOGOLJUBOV and MITROPOL'SKIJ (1963) the asymptotic averaging method of Krylov and Bogoljubov. Of the many publications dealing with this method, the study of MITROPOL'SKIJ (1971) deserves special mention. Books published at a later date for purposes of physical and technical sciences contain numerous examples of analyses used in different fields: BLAQUIERE (1966), EVAN-IWANOWSKI (1976), HAGEDORN (1978), VOJTÁŠEK and JANÁČ (1969)), etc. The book by NAYFEH and MOOK (1979) contains a great many examples and also tackles non-linear continuum systems; it is a very comprehensive work which well represents this class. The newly published third edition of the book by KLOTTER (1980) analyses basic results as well as the applicability of different approximative methods.

The book by ANDRONOV, VITT and CHAIKIN (1959) was the first publication to concentrate on qualitative analysis and topological methods. Of the remaining studies of this kind, mention should be made of the book by BUTENIN (1962) and especially of that by NEJMARK (1972), dealing with mapping. Publications devoted to the theory of stability, which plays an important role in the analyses of non-linear systems, form a special group. Besides the famous work of LJAPUNOV (1950) and the outstanding studies of MALKIN (1952) and ČETAEV (1955), one can name, for example the publications of KARAČAROV and PILJUTNIK (1962) and BOGUSZ (1966) as important contributions in this field. Dynamic stability in connection with parametrically excited vibrations is comprehensively investigated by BOLOTIN (1956). The investigation of the domains of attraction is a relatively new problem, originally tackled by HAYASHI (1964) and then elaborated by NISHIKAWA (1964), UEDA (1968), TONDL (1970a, 1973a, 1973b), ZAKRŽEVSKIJ (1980), GUCKENHEIMER and HOLMES (1983) and others.

This brief survey may be concluded by referring to some books on particular classes of non-linear vibrations and on applications to engineering practice. The first of these is treated in monographs on self-excited systems (CHARKEVIČ (1953), TONDL (1970b, 1979c, 1980b), RUDOWSKI and SZEMPLIŃSKA-STUPNICKA (1977), RUDOWSKI (1979), LANDA (1980), on systems with parametric excitations (McLACHLAN (1947), SCHMIDT (1975), EICHER (1981)), and on non-linear random vibrations (CRANDALL and MARK (1963), BOLOTIN (1979), M. F. DIMENTBERG (1980)). The second — applications of non-linear vibration theory to engineering — has received considerable attention from many authors. A few of the pertinent studies and fields to which they refer include: electronics (CHARKEVIČ (1956), KOTEK and KUBÍK (1962), PHILIPPOW (1963)), machine elements (GRIGORIEV (1961)), rotor dynamics (GROBOV (1961), TONDL (1965, 1974a), MERKER (1981)), systems with impacts (PETERKA (1973, 1981), A. E. KOBRINSKIJ and A. A. KOBRINSKIJ (1973), RAGUL'SKENE (1974), BABICKIJ (1978)), synchronization effects (BLECHMANN (1971, 1981)), mutual effect of self-excitation and of external or parametric excitation (TONDL 1976, 1978)), non-linear damping problems (PISARENKO (1955, 1970), SERGEEV (1969), PANOVKO (1960)), non-linear problems of vibration isolation (RAGUL'SKIS (1963), KOLOVSKIJ (1966), FROLOV and FURMAN (1980), phenomena and problems due to the limited source of energy (KONONENKO (1964)), etc.

Both classes are given equal treatment in STRATONOVICH (1961), a book on self-excited random vibrations in electrotechnology which widely stimulated research work on random vibrations in mechanics too.

A comprehensive handbook on linear and non-linear vibrations in various branches of technology is *Vibracii v technike* (1978ff.) in six volumes.

1. Basic properties and definitions

1.1. Non-linear vibration, non-linear characteristics and basic definitions

As mentioned in the Introduction, vibration of linear systems is only a special case of vibration of non-linear systems. Two important principles, which are valid for linear systems, do not apply to non-linear systems:

(I)   The principle of superposition,
(II)   The principle of proportionality.

In linear systems described by linear differential equations (all terms are functions of dependent variables and their derivatives) the response to the various components of excitation can be added up; if the amplitude of harmonic excitation is increased n times, the amplitudes of steady vibration increase n times. It has been found advantageous to divide linear vibration into two large groups:

(a) Vibration of systems with constant parameters described by linear differential equations with constant coefficients. The term *linear vibration* is usually understood to mean vibration of systems of just this type. The theory of this vibration is essentially complete and will not be discussed in the present book.

(b) Vibration of systems with non-constant parameters described by linear differential equations whose coefficients are generally functions of a dependent variable, chiefly time. An important separate group is formed by systems whose parameters are periodic functions of a dependent variable. Such systems are called rheo-linear and their theory is usually studied together with that of non-linear vibration. One of the reasons for this is the fact that an investigation of stability for small disturbances of steady periodic solutions of non-linear differential equations generally leads to an investigation of stability of the trivial solution of systems of differential equations with periodically variable coefficients.

Using as criteria differential equations which describe non-linear systems and using the character of stationary solutions, non-linear systems can be assigned to the following groups:

(1) Systems governed by homogeneous non-linear differential equations with constant coefficients

$$\ddot{y} + f(\dot{y}, y) = 0 .$$

These systems can be divided into subgroups according to the character of the solution expressing free damped vibration, free undamped vibration, divergent vibration or self-excited vibration of periodic, quasiperiodic or even of chastic nature.

(2) Systems governed by homogenous differential equations whose coefficients are functions of an independent variable (time)

$$\ddot{y} + f(\dot{y}, y, \omega t) = 0 \ .$$

The term consisting of the product of a periodic coefficient and a dependent variable represents the linear parametric excitation. Non-linear parametric excitation is represented by the terms where the dependent variable is replaced by a non-linear function of this variable. When the coefficients are not periodic but random oscillatory functions of time, we speak about stochastic parametric excitation. Moreover, the character of the solutions can be used for the classification into different subgroups.

(3) Systems governed by non-homogeneous differential equations

$$\ddot{y} + f(\dot{y}, y) = F(\omega t)$$

where $F(\omega, t)$ is a periodic or stochastic function of time. Of course, a further criterion is whether or not the homogeneous system is capable of self-excitation.

More general are systems governed by differential equations of the type

$$\ddot{y} + f(\dot{y}, y, \omega t) = F(\omega t)$$

representing combinations of the above groups.

Another classification of non-linear systems can be achieved by estimating how far an analysed system differs from a linear one. It need not be stressed what difficulties can be come across because different points of view, dividing limits and criteria can be used. As an example of one of the different approaches the use of extreme values of non-linear and linear terms in differential equations can be mentioned; as another criterion the characteristic features of a solution can serve. It is very difficult to define the dividing limits because it depends very much on what characteristic features are taken as criteria when comparing a non-linear system with a corresponding linear one. For example, when comparing the vibrations of the systems the vibrations can be estimated as similar although the resonance curves differ substantially from each other (the hysteresis and jump phenomena occur in non-linear systems). As a qualitative difference the occurrence of such a non-linear resonance, typical of non-linear systems only, could be considered — for example, the occurrence of subharmonic resonance. When comparing the vibration of a nonlinear system with that of the corresponding linear undamped system from a broader point of view the vibrations of such systems can be considered to be similar. The vibration at subharmonic resonance of order 1/2 characterized by the dominant half-frequency component and by the excitation frequency component can serve as an example. The first component can be considered to be close to the natural vibration of a linearized undamped system. From the above it can be seen that a certain system can be assigned to different classes when using different criteria.

We principally exclude in this book investigations of the global behaviour of non-linear systems, which cannot be described by approximations starting from the linear system. For the remaining non-linear systems we can distinguish between *strongly non-linear* ones, especially important for applications and mainly dealt with in this book, for which the solutions, although found by approximations starting from the linear solution, in some respect or other substantially — quantitatively or even qualitatively — differ from the linear case, and between *weakly non-linear* ones, for which such a substantial difference does not occur. As an example of such a substantial and qualitative difference a case can be mentioned where, by the action of periodic excitation (external or parametric), the vibration character is non-periodic and stochastic.

Note: The mere non-periodicity of the vibration in self-excited systems with several degrees of freedom cannot be taken as the decisive criterion because a multi-frequency vibration can occur which, though non-periodic, is quasi-periodic in quasi-normal coordinates; the frequencies of the components are close to the natural frequencies of the components of the linearized system vibration. Then the vibration of the non-linear system can be considered to be similar to natural vibration of the linearized undamped system. For that very reason this similarity is utilized in solution methods. These systems cannot be considered strongly non-linear; they should be regarded as weakly non-linear systems.

The various terms (or groups of terms) of non-linear equations which describe the model of an actual system have a physical meaning in mechanical systems (for example, the restoring force of a spring or an elastic element, damping, etc.). The relationship between such quantities and the dependent variables and their derivatives is termed the characteristic (for example, the characteristic of a damper spring). Characteristics cannot in every case be expressed in terms of functions of a single variable. Some of the simpler characteristics which can be expressed as functions of a single variable are discussed below.

Characteristics can be divided into weakly non-linear and strongly non-linear (expressed by functions which differ substantially from linear relationship). Consider first the characteristics of springs which can be expressed in terms of a function of a single variable, for example, the deflection y. If the function expressing the characteristic can be described by the equation

$$f(-y) = -f(y) \tag{1.1, 1}$$

the characteristic is called symmetrical. If $df(y)/dy$ is an increasing function, the characteristic is a *hardening* one; if $df(y)/dy$ is a decreasing function, the characteristic is a *softening* one (Fig. 1.1, 1).

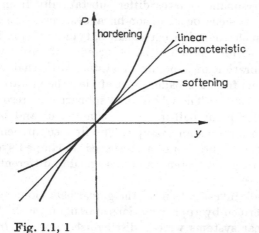

Fig. 1.1, 1

If a spring with a symmetrical non-linear characteristic is acted on by a constant force, the equilibrium position undergoes a change and, relative to the new working point, the characteristic becomes asymmetrical for deflections from the new equilibrium position. Such a case is a common occurrence in mechanical systems — the constant load is produced by the weight of the masses, or by a constant torque in torsional systems. The restoring force of a pendulum (Fig. 1.1, 2) can serve as an

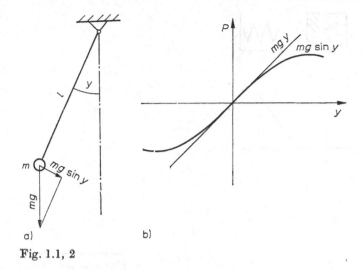

Fig. 1.1, 2

example of a softening symmetrical characteristic. A symmetrical characteristic may not have a monotonic character within the whole range of deflection y; the spring may be softening in one interval and hardening in another. Fig. 1.1, 3 a shows the characteristic for the case when the restoring force is negative in a certain interval of deflections and when several equilibrium positions exist. Fig. 1.1, 3 b shows a system containing a bar loaded by an axial force S which is larger than the Euler buckling force; for transverse deflections the characteristic of the restoring force is as shown in Fig. 1.1, 3 a. A system with a similar characteristic is shown in Fig. 1.1, 3 c.

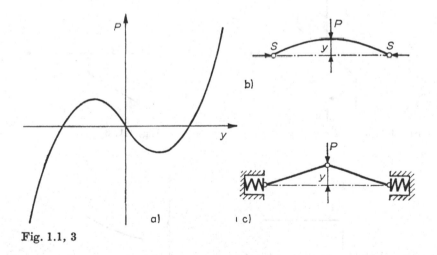

Fig. 1.1, 3

Various broken-line characteristics consisting in part of linear segments form a large group, some examples of which are shown in Figs. 1.1, 4—1.1, 6: Fig. 1.1, 4 — a system with clearance, Fig. 1.1, 5 — a system with stops, Fig. 1.1, 6 — a system with a prestressed (prestress P_0) spring.

Damping which is a function of velocity has similar characteristics. The characteristic of dry friction (Fig. 1.1, 7 b) differs substantially from the usual ones and is often replaced by that of idealized Coulomb friction (Fig. 1.1, 7 a).

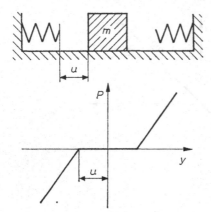

Fig. 1.1, 4

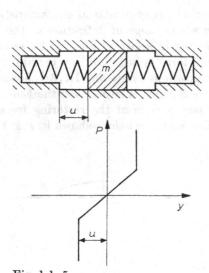

Fig. 1.1, 5

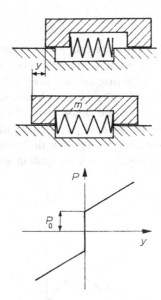

Fig. 1.1, 6

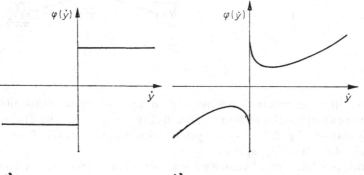

a)   b)

Fig. 1.1, 7

Characteristics cannot always be expressed in terms of a function of a single variable. Both elastic and damping forces are apt to be functions of several variables or to be expressed in terms of vector functions. In cases of this sort it is more to the point to speak of force or damping fields. Fields which can be described by a function whose variable is a two-dimensional vector are very frequently encountered in mechanical systems. If such a field is described by the function

$$F(z, \dot{z}) = [f_1(|z|, |\dot{z}|) + if_2(|z|, |\dot{z}|)] \operatorname{sgn} z \qquad (i = \sqrt{-1}),$$

or by the function

$$F(z, \dot{z}) = [f_1(|z|, |\dot{z}|) + if_2(|z|, |\dot{z}|)] \operatorname{sgn} \dot{z}$$

where z is a two-dimensional vector and $f_1(|z|, |\dot{z}|)$ and $f_2(|z|, |\dot{z}|)$ are scalar functions, the field is central symmetrical. Such a field is characterized by the fact that the absolute value $|F(z, \dot{z})|$ and the radial and tangential components remain unaltered if the absolute values $|z|$ and $|\dot{z}|$ do not vary; the magnitudes of the radial and tangential components do not vary, either. Detailed information concerning two-dimensional fields may be found in a monograph by TONDL (1967c).

Mention should also be made of some of the most important differences between vibrations of linear and non-linear systems. In linear systems, steady vibration produced by external periodic excitation has the same character as the excitation; in non-linear systems this is usually not so. Under harmonic excitation, for example, the steady vibration of a linear system is also harmonic, whereas the response of non-linear systems is either periodic or quasi-periodic. In addition to the basic component, vibration produced by such excitation is apt to contain not only higher but also sub-harmonic components, i.e. the period of the response can be an N-multiple (where N is an integer) of the excitation period. The difference between linear and non-linear systems becomes more marked still when, under periodic or even harmonic excitation, the response is not at all periodic and sometimes not even quasi-periodic. Differences are found to exist in the character of the vibration course as well as in the occurrence of resonant vibration. In linear systems with constant coefficients, resonances occur only for excitation frequencies which are identical with the natural frequencies of the system. Since this stipulation need not be satisfied in non-linear systems, these are likely to be richer in resonances than linear systems.

In linear damped systems without external excitation, steady vibration with finite amplitudes never occurs. Consequently, steady self-excited finite-amplitude vibration is always a property of non-linear systems. Similarly, in the case of linear parametric excitation, finite-amplitude vibration can be obtained only by the action of non-linear terms.

The possible existence of more than one steady solution for equal values of parameters of the system and excitation may be set down as another significant property of non-linear systems.

1.2. Some examples of excited and self-excited systems

Mechanical systems are mostly linear so far as inertia forces are concerned. This applies especially to discrete systems in which concentrated masses can execute only rectilinear motion. In rotating bodies in which the deflections of the gyroscope axis cannot be qualified as small, the gyroscopic effect is expressed by non-linear terms.

Inertial forces assume a non-linear character in the case of reduction of the masses in the mechanism, or are functions of time if a particular member rotates uniformly (for example, in a crank mechanism). Damping is almost always non-linear; linear viscous damping defined by the product of a constant coefficient of damping and the velocity is more or less an idealization, although a very popular one.

The quantity which has the greatest effect of all on the "non-linear" behaviour of discrete systems with concentrated masses is spring non-linearity. A non-linear spring is, for example, a coil steel spring with unequal lead (pitch) whose coils gradually come to rest on one another as the spring deforms. Restoring force non-linearity can also be produced by using several parallel springs of unequal length, some of which come into action only at a definite deflection of the mass. The characteristic of the restoring force thus produced is of the broken-line, in part linear type — Fig. 1.2, 1. Elements whose material has linear elastic properties but which are curved either by the effect of prestress or geometry, have a non-linear character of the restoring force. These elements are examples of geometrical non-linearity in the same way as systems containing a mechanism with one or several components. The simplest system of this kind is a pendulum (Fig. 1.1, 2).

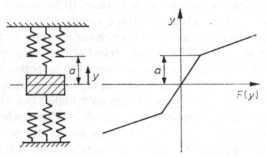

Fig. 1.2, 1

In systems with continuously distributed parameters the system non-linearity can be caused by non-linearity in the boundary conditions. A whole class of these is formed by systems described by linear partial differential equations whose only non-linearity is that introduced by the boundary conditions. A beam supported by non-linear springs, or a shaft rotating in journal bearings whose analysis cannot be restricted to small deflections of the journals in the bearings, serve as examples of such systems.

Non-linear systems having masses which are acted on by periodic forces, described by non-homogeneous non-linear differential equations with constant coefficients, are so common as not to require detailed discussion of particular examples. Attention will therefore be concentrated on systems with parametric excitation or combined external-parametric excitation, which are of less frequent occurrence.

Most vibration textbooks contain the classical example of a parametrically excited system, i.e. vibration of a swing, which can also be modelled by a pendulum having a periodically varying length (Fig. 1.2, 2). Another very popular example is a pendulum whose suspension moves harmonically in the vertical direction (Fig. 1.2, 3), which was first analysed by KLOTTER and KOTOWSKI (1939). Denoting by y the angular deflection of the pendulum, the equation of motion takes the form

$$\Theta\ddot{y} + mgs \sin y + msa\omega^2 \cos \omega t \sin y = 0 \qquad (1.2, 1)$$

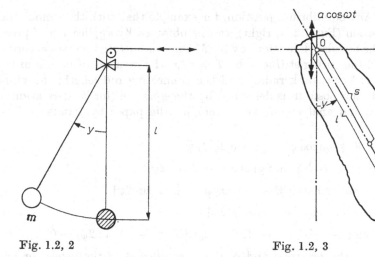

Fig. 1.2, 2   Fig. 1.2, 3

where $\Theta = mk_0^2$ is the moment of inertia about the suspension, m is the mass of the pendulum, $l = k_0^2/s$ is the reduced length of the pendulum, and s is the distance between the centroid and the suspension. For small deflections for which $\sin y \doteq y$, (1.2, 1) turns into the well-known *Mathieu equation*

$$\ddot{y} + \left(\frac{g}{l} + \frac{a\omega^2}{l} \cos \omega t\right) y = 0 \,. \tag{1.2, 2}$$

If the suspension moves in a direction deflected from the vertical by angle α, (1.2, 2) changes to

$$\ddot{y} + \left(\frac{g}{l} + \frac{a\omega^2}{l} \cos \alpha \cos \omega t\right) y = \frac{a\omega^2}{l} \sin \alpha \cos \omega t \tag{1.2, 3}$$

which describes a case of combined external and parametric excitation. For $\alpha = 0°$, the excitation is pure parametric, for $\alpha = 90°$, pure external. This system is an example of kinematic excitation when a periodic motion is specified for a definite point of the system rather than a force acting on the mass. The system is thus made to vibrate by the action of inertial forces.

Fig. 1.2, 4 shows a model of a proposed device for generating energy from tidal motion: a conical buoy carries a crank mechanism whose connecting rod turns the

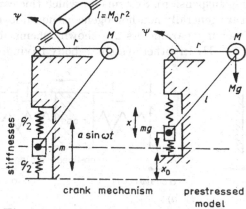

Fig. 1.2, 4

generator rotor. Any equilibrium position, for example that with the connecting rod in horizontal position (Fig. 1.2, 4, right), can be obtained by application of prestress. Denoting by ψ the angular deflection, by m, M the concentrated masses according to Fig. 1.2, 4, by $c/2$ the spring stiffness, by $I = M_0 r^2$ the moment of inertia of the generator, and by r, l the crank radius and the connecting rod length, the vibration about the equilibrium position is described by the equation (for its derivation, a detailed description and analysis of the system see the paper by PARKS and TONDL (1979))

$$r^2[M_0 + M + m(\cos\psi - \tfrac{1}{2}\lambda\sin 2\psi)^2]\,\ddot\psi$$
$$- mr^2\dot\psi^2(\cos\psi - \tfrac{1}{2}\lambda\sin 2\psi)\,(\sin\psi + \lambda\cos 2\psi)$$
$$+ K\dot\psi - ra\omega^2\sin\omega t[(M+m)\cos\psi - \tfrac{1}{2}\lambda m\sin 2\psi]$$
$$+ rg[(M+m)\cos\psi - \tfrac{1}{2}\lambda m\sin 2\psi]$$
$$+ cr\{r[\sin\psi - \tfrac{1}{4}\lambda(1-\cos 2\psi)] - x_0\}\,(\cos\psi - \tfrac{1}{2}\lambda\sin 2\psi) = 0 \qquad (1.2,\,4)$$

where $\lambda = r/l$, x_0 is the prestress and K is the coefficient of the generator load. It may be seen that the excitation of this system is a combination of external and parametric, linear and non-linear, excitations. In consequence the vibration about the equilibrium position grows steadily larger until, when certain conditions are fulfilled, the swinging motion of the crank changes to the desired rotation.

As pointed out by TONDL (1984), combined excitation in the case of kinematic excitation can also be obtained for simpler systems provided the excitation occurs via a non-linear spring. Only for the simplest systems, such as that shown in Fig. 1.2, 5,

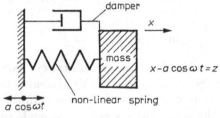

Fig. 1.2, 5

can the equation of motion be converted to a non-linear one with constant coefficients by introducing the relative deflection $z = x - a\cos\omega t$ (x is the absolute deflection and $a\cos\omega t$ is the motion of the spring suspension). Systems for which this cannot be done and whose equations of motion are generally non-homogeneous non-linear differential equations with periodically varying parameters are shown schematically in Fig. 1.2, 6 — (a) system I, (b) — system II. In either system, a body having mass m

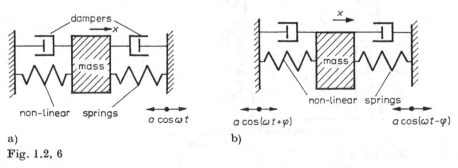

a)

b)

Fig. 1.2, 6

is placed between two non-linear springs and dampers. To simplify, the spring characteristics are assumed to be symmetrical, the non-linearity being expressed by a cubic term; the dampers are linear viscous. In system I the excitation is provided by the motion of the end of one of the springs; in system II both ends move but the motions are generally shifted in phase. The equations describing the two systems will be written for the harmonic motion $a \cos \omega t$. Denoting by x the deflection of the mass, by $\varkappa$ the damping coefficient, by k the coefficient of the linear portion of the restoring force of the spring (assuming the linear terms of the two springs to be identical), and by γ_1 and γ_2 the coefficient of the cubic term of the restoring force of the left-hand and the right-hand spring, respectively, then system I is described by the equation

$$m\ddot{x} + kx + \gamma_1 x^3 + \varkappa\dot{x} + k(x - a \cos \omega t) + \gamma_2(x - a \cos \omega t)^3$$
$$+ \varkappa(\dot{x} + a\omega \sin \omega t) = 0 \tag{1.2, 5}$$

and system II by the equation

$$m\ddot{x} + \varkappa[\dot{x} + a\omega \sin(\omega t + \varphi) + \dot{x} + a\omega \sin(\omega t - \varphi)] + 2kx$$
$$- ak[\cos(\omega t + \varphi) + \cos(\omega t - \varphi)] + \gamma_1[x - a \cos(\omega t + \varphi)]^3$$
$$+ \gamma_2[x - a \cos(\omega t - \varphi)]^3 = 0 . \tag{1.2, 6}$$

A detailed analysis of both systems is presented in Chapter 8.

A frequently discussed example of a parametrically excited system which has been analysed by several authors is a beam subjected to a periodically varying axial load $P = P_0 + S \cos \omega t$ (Fig. 1.2, 7). The lateral vibration of the beam is described by the equation

$$EI\frac{\partial^4 y}{\partial x^4} + (P_0 + S \cos \omega t)\frac{\partial^2 y}{\partial x^2} + m\frac{\partial^2 y}{\partial t^2} = 0 \tag{1.2, 7}$$

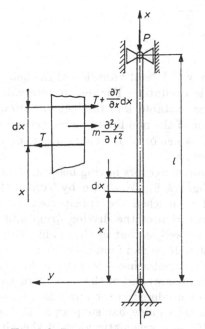

Fig. 1.2, 7

where E is Young's modulus, I the cross-sectional moment of inertia, m the mass per unit length, and $y(x, t)$ the lateral deflection of an element at distance x from the support at time t. A very detailed analysis of this case is contained in the books of BOLOTIN (1956) and SCHMIDT (1975). The torsional vibration of beams under periodically varying boundary conditions is also included in this group of systems.

Another well-known class of systems encompasses rotor assemblies whose shafts — due to the effect of slots — have unequal cross-sectional moments of inertia in two mutually perpendicular directions, or unequal mass moments of inertia. Rotor systems of this kind have been analyzed by DIMENTBERG (1959) and TONDL (1965).

Parametric excitation is also known to occur in torsional vibrating systems containing gear trains, in which both the restoring and the frictional forces during uniform rotation are made to vary periodically by the effect of changes at the point of engagement of two gears. The periodic variation of the forces is one of the main reasons for a deterministic excitation of these systems. They are investigated in Chapter 6.

Several less conventional cases of parametrically excited systems of engineering importance will be dealt with next. Parametric excitation arises, for example, in conveyor belts on which the material being transported is deposited not uniformly but in sizeable amounts at definite time intervals, i.e. the conveyor carries equally spaced heaps of material (Fig. 1.2, 8). As the belt moves at a uniform speed, the distribution

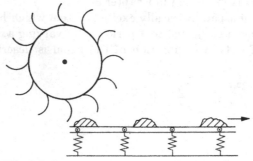

Fig. 1.2, 8

of masses, and thus also the natural frequency of lateral vibration of the belt, undergoes a change which gives rise to parametric excitation. For certain intervals of the belt speed, the equilibrium position becomes unstable, and the belt vibration either grows increasingly larger or, due to the effect of the non-linear terms, becomes stabilized at a finite amplitude. Such vibration is spoken of as parametric resonant and can become very intensive in lightly damped systems.

A similar situation is found to exist in chain conveyors having belts stretched over drums shaped like regular polyhedrons (Fig. 1.2, 9). As shown by TONDL (1967b), these conveyors are parametrically excited even when the transported material is distributed uniformly. Assuming uniform rotation of the driving drum and taking into consideration the stiffness of the belt as well as that of the chains, the author examined the torsional vibration of the tension drum and found the restoring moment per unit torsional deflection to have a periodic — pulsating — waveform (Fig. 1.2, 9b).

If the elasticity of the guide bar (a wooden beam) and of the whole structure supporting the guide bars in their mountings is considered in the analysis, the restoring force for the lateral deflection of the mine cage guide bar support at the point of contact with the guide bar (see Fig. 1.2, 10 — a schematic view of the pit cross-

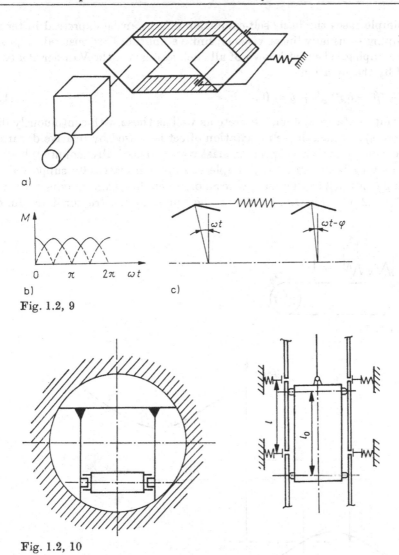

Fig. 1.2, 9

Fig. 1.2, 10

section and of the mine cage in guides) is found to be periodically varying at a constant speed of travel. Details of the analysis of this system may be found in Chapter 8.

In the discussion that follows, at least a few cases of self-excited systems will be taken up and the main origin of self-excitation will be pointed out. One of the causes of self-excitation in mechanical systems is the flow of liquid or gaseous media past an elastic, or a rigid but elastically mounted body, when the hydrodynamic or aerodynamic forces have a destabilizing effect in, at least, a certain region of the system parameters. If a coupling exists between the hydrodynamic or aerodynamic forces and the elastic forces, the self-excitation is termed hydroelastic or aeroelastic, and its topical problems are treated in the separate disciplines of hydroelasticity or aeroelasticity. It is usually very difficult to express forces produced by a flowing medium. Frequently, therefore, artifical models are devised and their parameters adapted to experimental results. In this connection, one often speaks, for example, of bluff body models (DOWELL, 1981).

In fairly simple cases the basic self-excitation effect can be expressed in terms of negative damping — usually linear viscous; finite amplitudes are ensured by progressive positive damping. The best known of all such systems is the Van der Pol oscillator described by the equation

$$y'' - (\beta - \delta y^2)\, y' + y = 0 \,.$$ (1.2, 8)

Another group contains systems (discrete as well as those with continuously distributed parameters) in which the self-excitation effect is caused by relative dry friction arising on the contact surface between an elastic (or an elastically mounted) body and a uniformly moving body. In a very simple example an elastically supported body (c — the spring constant) having mass m rests on an endless band moving with velocity v_0 (see Fig. 1.2, 11). Fig. 1.2, 12 shows the curve of the friction force function

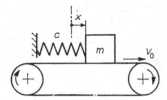

Fig. 1.2, 11

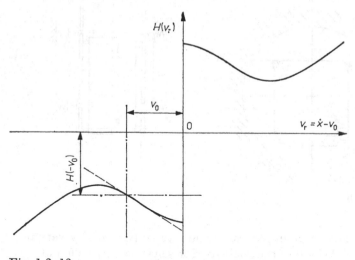

Fig. 1.2, 12

$H(v_r)$ versus the relative velocity $v_r = \dot{x} - v_0$, where x is the deflection of the mass from the equilibrium position (at zero band velocity v_0). The motion of the system is described by the equation

$$m\ddot{x} + cx + H(\dot{x} - v_0) = 0 \,.$$ (1.2, 9)

Expanding function $H(\dot{x} - v_0)$ into a Taylor's series, ignoring terms of orders higher than the third,

$$H(\dot{x} - v_0) = -H(v_0 - \dot{x}) \doteq -H(v_0) + \frac{\partial H}{\partial v_0}\dot{x} - \frac{\partial^2 H}{\partial v_0^2}\dot{x}^2 + \frac{\partial^3 H}{\partial v_0^3}\dot{x}^3$$

and introducing the new coordinate

$$x = x_0 + y \tag{1.2, 10}$$

where $x_0 = H(v_0)/c$, (1.2, 9) takes the form

$$\ddot{y} + \Omega^2 y + h_1 \dot{y} - h_2 \dot{y}^2 + h_3 \dot{y}^3 = 0 \tag{1.2, 11}$$

where

$$\Omega^2 = \frac{c}{m}, \qquad h_1 = \frac{1}{m} \frac{\partial H}{\partial v_0}, \qquad h_2 = \frac{1}{m} \frac{\partial^2 H}{\partial v_0^2}, \qquad h_3 = \frac{1}{m} \frac{\partial^3 H}{\partial v_0^3}.$$

So long as $\partial H/\partial v_0 < 0$, then also $h_1 < 0$, and the equilibrium position $x = x_0$ is unstable; equation (1.2, 11) becomes a system anlogous to the Van der Pol system. As the foregoing discussion reveals, a decrease of function $H(v_r)$, at least in a certain region of the values of the relative velocity, is of substantial importance for the self-excitation effect.

A large group of self-excited systems is formed by rotors in which the self-excitation effect is provided either by the action of forces in journal bearings and forces arising in gap flows (as in the case of labyrinth glands) or by the action of internal damping, i.e. damping which is proportional to the relative velocity of the coordinate system rotating correspondingly to the rotor rotation. Internal damping may also include damping forces coming into existence due to material damping during shaft deformations, or forces of friction arising on the contact surfaces of composite rotors during deformations (such as the friction between the shaft and the disk hub). Denoting by z the vector of the absolute deflection (of the disk, shaft centre, etc.), and by ζ the vector of the deflection in the coordinate system rotating with the angular velocity of rotor rotation ω, the relation between the two vectors may be written as

$$z = \zeta \exp(\mathrm{i}\,\omega t) \tag{1.2, 12}$$

where $z = x + \mathrm{i}y$, $\zeta = \xi + \mathrm{i}\eta$. The relative velocity is then given by

$$\dot{\zeta} = (\dot{z} - \mathrm{i}\omega z) \exp(-\mathrm{i}\omega t). \tag{1.2, 13}$$

Expressing the damping force of internal damping in the form

$$\boldsymbol{P}_\mathrm{r} = -h(|\dot{\zeta}|, |\zeta|) \frac{\dot{\zeta}}{|\dot{\zeta}|} \tag{1.2, 14a}$$

or in the simplest form (linear viscous damping)

$$\boldsymbol{P}_\mathrm{r} = -h\dot{\zeta} \tag{1.2, 14b}$$

then the forces in the absolute (non-rotating) coordinate system are given by the relations

$$\boldsymbol{P}_\mathrm{a} = -h(|\dot{z}|, |z|) \frac{\dot{z} - \mathrm{i}\omega z}{|\dot{z} - \mathrm{i}\omega z|} \tag{1.2, 15a}$$

or

$$\boldsymbol{P}_\mathrm{a} = -h(\dot{z} - \mathrm{i}\omega z). \tag{1.2, 15b}$$

Relations similar to (1.2, 15b) also hold for damping acting on a bladed disk (mounted on a flexible shaft) of an agitator's rotor mixing a viscous liquid in a vessel. Due to the effect of rotor rotation, the liquid rotates in the vessel with a mean velocity not identical with the angular velocity of rotor rotation. Accordingly, in the absolute coordinate system, the damping force has two components: the radial component $(h|\dot{z}|)$ and the tangential component $(-h\omega|z|)$ which is proportional to the angular velocity of rotor rotation ω.

All types of self-excitation forces of rotor systems discussed in the foregoing have one property in common, i.e. that, as the rotor whirls with natural frequency, the vector of the self-excitation force has a tangential component whose direction in a certain interval of rotor rotation agrees with that of the whirling motion and acts in opposition to the external positive damping. In all types of self-excited vibration of rotors, the direction of the whirling motion agrees with that of rotor rotation.

Problems of self-excited vibration of rotors have been in the foreground of interest of many scientists. They have been discussed in numerous papers and comprehensively treated in books (DIMENTBERG (1959), TONDL (1965), KUŠUL (1963)), monographs (TONDL (1961, 1973, 1974)) and lectures (RIEGER (1980)).

1.3.   Basic features of excited systems

An essential distinction should be made between the variously excited systems with regard to their type of excitation (external, parametric, combined). Unlike linear systems, externally excited non-linear systems do not follow the principles of super-position of the solution and of proportionality; sometimes, their response to periodic excitation may differ even qualitatively from the excitation. In a non-linear system excited, for example, harmonically with frequency ω, the response may be periodic having period $2N\pi/\omega$ (N is an integer), the component with frequency ω/N being the dominant one in a particular interval of the excitation frequency. In excited systems, the feature of primary interest, especially from the practical point of view, is the resonant vibration. When dealing with it in connection with non-linear systems, one should alter the conventional "linear" approach to the concept of resonance. In linear systems, resonance is usually defined by a definite value of the excitation frequency (ordinarily by that identical with the natural frequency of the system). In non-linear systems, resonance should be regarded as a phenomenon which brings about a substantial increase of the amplitudes in a certain interval of the excitation frequency. The resonance curve (the amplitude-excitation frequency dependence) of non-linear systems may sometimes have more than one peak typical of the resonance curve of linear systems (Fig. 1.3, 1a). Resonance curves featuring several peaks are often observed in non-linear systems tuned to so-called internal resonance (the ratio of two or several natural frequencies is close to a ratio of small integers) (Fig. 1.3, 1b).

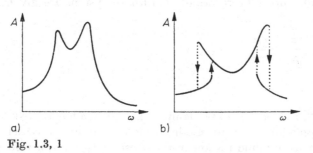

a)   b)

Fig. 1.3, 1

It may be useful to establish a basic classification of resonances. Consider a system governed by the differential equations

$$\ddot{x}_s + \sum_{k=1}^{n} c_{sk}x_k + \varepsilon[h_s(|\dot{x}|, |x|)\,\dot{x} + g_s(x)] = P_s(\omega t) \qquad (s = 1, 2, \dots, n) \qquad (1.3, 1)$$

where c_{sk} are constants, ε is a small parameter. $h_s(|\dot{x}|, |x|)\,\dot{x}$, $g_s(x)$ are analytic functions of $\dot{x}_k$ and x_k, and $P_s(\omega t)$ are periodic or harmonic functions having period $2\pi/\omega$. Functions $h_s(|\dot{x}|, |x|)\,\dot{x}$ representing damping are assumed to be such that no self-excitation can exist. Using the linear transformation

$$x_s = \sum_{k=1}^{n} a_{sk}y_k \qquad (s = 1, 2, \dots, n) \qquad (1.3, 2)$$

where

$$a_{1k} = 1, \qquad a_{jk} = \frac{\Delta_j(\Omega_k)}{\Delta_1(\Omega_k)}$$

and $\Delta_j(\Omega_k)$ is the minor of the determinant of the characteristic equation

$$\det \|c_{sk} - \delta_{sk}\Omega^2\| = 0$$

$$(k, s = 1, 2, \dots, n), \qquad \delta_{ss} = 1, \qquad \delta_{sk} = 0 \quad \text{for} \quad (s \neq k)$$

corresponding to the j-th term of the first row of the determinant for $\Omega = \Omega_k$, system (1.3, 1) can be converted to the quasi-normal form

$$\ddot{y}_s + \Omega_s^2 y + \varepsilon F_s(\dot{y}, y) = q_s(\omega t), \qquad (s = 1, 2, \dots, n). \qquad (1.3, 3)$$

Ω_k are the natural frequencies of the abbreviated system (1.3, 1) (for $\varepsilon = 0$) which are all assumed to be real.

Note: Details which are of use in practical analyses, relating to the coefficients of transformation (1.3, 2) for $s = 2$, may be found in TONDL's book (1965) and monographs (1974a), (1976a).

According to the general theory of quasilinear differential equations of the type of (1.3, 3) (see, for example, MALKIN (1956)) the resonant solutions can be expected to exist in the vicinity of those values of the excitation frequency ω for which

$$r\omega = \sum_{k=1}^{n} N_k \Omega_k \qquad (1.3, 4)$$

where $r = 1, 2, \dots$; $N_k = 0, \pm 1, \pm 2, \dots$

Assuming the excitation to be harmonic or periodic with a dominant first harmonic component, the classification of resonances can be set up as shown in Table 1.3, 1.

If a non-linear system is excited by several periodic components having frequencies $\omega_1, \omega_2, \omega_3, \dots, \omega_M$ whose ratios differ from the ratios of integers, (1.3, 4) can be replaced by the relation

$$\sum_{k=1}^{M} n_k \omega_k = \sum_{j=1}^{n} N_j \Omega_j \qquad (1.3, 5)$$

where n_k, $N_j = 0, \pm 1, \pm 2, \dots$

In systems with only a linear parametric excitation, parametric resonances can be conveniently classified by means of the instability intervals of the linearized system. Let the linearized system converted to the quasinormal form be described by the equations

$$\ddot{y}_s + \Omega_s^2 y_s + \varepsilon \sum_{k=1}^{n} [q_{sk}(\omega t)\,\dot{y}_k + p_{sk}(\omega t)\,y_k] = 0 \qquad (s = 1, 2, \dots, n) \qquad (1.3, 6)$$

Table 1.3, 1

Type of resonance	Excitation frequency close to	Ratio of natural frequencies Ω_j/Ω_k		Note
		differs from	is close to	
(a) Pure main resonance	Ω_j	r/k	—	$j = 1, 2, \dots, n$ $r, k = 1, 2, \dots,$
(b) Subharmonic resonance	$N\Omega_j$			$r \neq k$ $n, N = 2, 3, \dots$
(c) Superharmonic resonance	$\dfrac{1}{n}\Omega_j$			
(d) Subsuperharmonic or subultraharmonic resonance	$\dfrac{N}{n}\Omega_j$			for (d): $\dfrac{N}{n}, \dfrac{n}{N} \neq 1, 2, \dots$
(e) Main internal resonance	Ω_j	—	r/k	
(f) Non-periodic combination resonance	$\dfrac{1}{k}\sum\limits_{j=1}^{n} N_j\Omega_j$	r/k	—	$j = 1, 2, \dots, n$ $N_j = \pm 1, \pm 2, \dots$ $r, k = 1, 2, \dots$ $r \neq k$
(g) Periodic combination resonance		—	r/k	

Note: Superharmonic resonance has been called "ultraharmonic" by many authors.

where q_{sk}, p_{sk} are periodic functions of period $2\pi/\omega$. As the general theory of equations of the type of (1.3, 6) suggests (see, for example, MALKIN (1956)), the intervals of instability of a trivial solution may be expected to occur in the vicinity of such values of ω which are close to

$$\omega_0 = \frac{|\Omega_j \pm \Omega_k|}{N} \qquad (j, k = 1, 2, \dots, n; N = 1, 2, \dots) . \tag{1.3, 7}$$

For $j = k$, the instability intervals are said to be of the first kind and order N; for $j \neq k$, of the second kind and order N. Parametric resonances (the vibration becomes limited due to the action of the non-linear terms) resulting from the instability intervals of the first kind and order N are called parametric resonances of the first kind or simple parametric resonances (for $N = 1$, main parametric resonances); those resulting from the instability intervals of the second kind are simply called combination parametric resonances. The characteristic feature of parametric resonances initiated by the linear parametric excitation is their occurrence at a definite limited interval of exciting frequency. Outside this interval such a system does not vibrate when acted on only by linear parametric excitation.

If an additional non-linear parametric excitation is also present in the system, other resonances than those mentioned above may occur, similarly as they do in externally excited non-linear systems. TONDL (1976 b) showed that of these, subharmonic resonances are of significance in most cases.

The resonances of systems under combined (external and parametric) excitation do not lend themselves readily to classification. The reason for this is the possible interaction of the two types of excitation (they can either combine to increase the vibration amplitude or work in opposition to decrease it) and the difficulty of deciding which of them has the dominant effect. The situation becomes more straightforward if one type prevails over the other, and the dominant excitation is harmonic. The resonances can then be classified according to either Table 1.3, 1 or the instability intervals of parametric resonances. An interesting but less known fact pointed out by several authors is the "non-linear" character of resonances of damped linear systems under combined external and parametric excitation. It should be stressed, however, that systems featuring such resonances must be subjected to damping whose level is such as to ensure complete quenching of all instability intervals associated with the linear parametric excitation (PŮST, TONDL (1956) and, particularly, TONDL, BACKOVÁ (1968), TONDL (1967a) — rotor systems). A system with one degree of freedom, for example, whose harmonic parametric excitation has double the frequency of the harmonic external excitation is likely to have resonance curves of double-peak character (Fig. 1.3, 1(a)) depending on the phase shift between the two excitations.

The findings established above reveal an important feature characteristic of non-linear systems, namely that resonances can be produced at a vibration frequency which is close to one of the natural frequencies of the system or, if the vibration contains several components, at frequencies which are close to the natural frequencies of the system (the frequency of the response being, however, always related, i.e. varying in proportion, to the excitation frequency) by excitation whose frequency may be even far remote from any natural frequency of the system. This cannot happen in linear externally excited systems.

1.4. Basic features of self-excited systems

Self-excited systems described by non-linear homogeneous differential equations with constant coefficients have the characteristic property that there exists, for any initial conditions or only a particular region thereof, a steady vibration which is periodic, quasiperiodic or possibly, non-periodic. This characteristic enables the systems to be divided into two basic groups:

(a) Systems with so-called *soft* self-excitation in which self-excited vibration arises if and only if the equilibrium position is unstable.

(b) Systems with the so-called *hard* self-excitation whose equilibrium is stable only for definite limited disturbances or a definite limited region of the initial conditions in the neighbourhood of the equilibrium position. If the limits are exceeded, the vibration — after a sufficient length of time — converges to a steady self-excited motion.

Consider first a system with one degree of freedom described by a second-order differential equation. An analysis of systems of this kind may conveniently be carried out by mean of phase-plane trajectories which provide a comprehensive picture of the behaviour of the system and stability of steady solutions. The term "phase plane" refers to a plane whose coordinates are deflection and velocity, or, alternatively, the dependent variables obtained in transformation of the original equation to a system of two first-order differential equations. In analyses of self-excited systems the most important of the phase-plane trajectories are the limit cycles. These are closed trajectories which other trajectories in the neighbourhood either approach (a stable limit cycle) or move away from (an unstable limit cycle). The phase-plane analysis becomes particularly simple when dealing with systems having a single equilibrium position, represented by a singular point in the phase plane. In such cases, the limit cycles surround the singular point, the stable cycles alternating with the unstable ones if certain conditions of solution uniqueness are satisfied. As an example of systems with a single equilibrium position consider the system described by a differential equation of the type

$$y'' + y + f(y) + F(|y'|, |y|)\, y' = 0 \qquad (1.4, 1)$$

where $f(y) \neq 0$ if $y \neq 0$. If function $F(|y'|, |y|)$ is a polynomial, the equilibrium position is stable for a positive constant term and unstable for a negative one. If however the function also contains terms of the order $1/n$ $(n > 1)$, the stability of the equilibrium position is decided about by the lowest term (for the largest n). The analysis of phase-plane trajectories is dealt with exhaustively in Chapter 4.

In systems with several degrees of freedom the situation is more complicated. Consider a system described by a set of homogeneous non-linear differential equations whose trivial solution represents a single equilibrium position; let all the non-linear terms be of higher degree than the linear terms, and let the characteristic equation of the linearized system contain n pairs (n is the number of degrees of freedom) of complex roots. Further, let k pairs have real positive parts $\alpha_k \pm i\Omega_k$ where $\alpha_k > 0$. Assuming that the non-linear terms of the equations expressing damping are all positive, i.e. the damping is positive progressive, the equilibrium position is unstable with respect to k natural modes of the system; this is a necessary (but not sufficient) condition for the system to oscillate in r $(k \geqq r > 1)$ modes, the frequency of one vibration mode being close to the frequency Ω_k. When the system vibrates in only one mode of frequency Ω_k, the vibration is termed *single-frequency self-excited vibration*. Assuming that the characteristic equation of the linearized system has k roots, the steady vibration can have as many as k components of different frequencies. Such vibration is called *multi-frequency* (two-, three-, ..., k-frequency) *self-excited vibration*. As the value of k increases, the number of possible multi-frequency vibrations grows larger. Since the ratios of the frequencies are generally different from the ratios of small integers, multi-frequency vibrations have the distinctive quality of being generally non-periodic. On the other hand, vibrations expressed in terms of quasinormal coordinates are found to be largely quasi-periodic. For further details on the subject refer to a monograph by TONDL (1970b).

Analysis of systems with several degrees of freedom, especially those capable of oscillating with multi-frequency vibrations, is difficult for a number of reasons, one of them being the complications which are apt to arise in connection with the graphical representation and interpretation of the trajectories in the *phase space*.

Even less amenable to analysis are self-excited systems which are subjected to additional external or parametric excitation. At resonance, i.e. when the external or parametric excitation is at resonance with respect to the abbreviated (undamped linearized) system, a phenomenon called *vibration synchronization* occurs. The term refers to a situation in which the component of the excited vibration fully synchronizes with the corresponding component of the self-excited vibration, and the frequency is controlled by the self-excitation frequency. Details are given in Chapter 5.

1.5.   Stability

In nearly all stability investigations of deterministic systems, stability is understood in the sense of Ljapunov as uniform continuity: A solution $y(\tau)$ of the vibrational differential equation

$$y'' + \lambda y = \Phi(y, y', \tau) \qquad\qquad (1.5, 1)$$

with Φ being continuous in y, y', τ is *stable* if for given positive ε any (not necessarily periodic) solution[1])

$$Y(\tau) = y(\tau) + z(\tau)$$

of (1.5, 1) satisfies the condition

$$|z(\tau)|, \ |z'(\tau)| < \varepsilon$$

for all $\tau > \tau_0$ if

$$|z(\tau_0)|, \ |z'(\tau_0)| < \eta$$

holds with a positive $\eta = \eta(\varepsilon)$, in other words, if the solutions keep a given maximum distance for all time when they have a certain positive maximum distance for an initial time $\tau = \tau_0$. A solution is *unstable* if this condition is not fulfilled. A solution is *asymptotically stable* if additionally

$$\lim_{\tau \to \infty} \{|z(\tau)|, \ |z'(\tau)|\} = 0$$

holds, that is, if every solution $Y(\tau)$ tends to $y(\tau)$ for $\tau \to \infty$.

If we subtract the differential equations (1.5, 1) written down for Y and y we get the *variational equation*

$$z'' + \lambda z = \Phi(y + z, y' + z', \tau) - \Phi(y, y', \tau)$$

which can be expressed (under the assumption of continuous partial derivatives) in the form

$$z'' + \lambda z = u(\tau) \, z + v(\tau) \, z' + \Psi(z, z', \tau) \qquad\qquad (1.5, 2)$$

where

$$\frac{\Psi(z, z', \tau)}{\text{Max} \, \{|z|, |z'|\}} \to 0 \quad \text{as} \quad \text{Max} \, \{|z|, |z'|\} \to 0 \, .$$

Neglecting the non-linear expression Ψ we get the *linear variational equation*

$$z'' + \lambda z = u(\tau) \, z + v(\tau) \, z' \, . \qquad\qquad (1.5, 3)$$

The stability of a solution $y(\tau)$ can be determined by inserting this solution in the variational equations (1.5, 2) or (1.5, 3). If we use the linear variational equations (1.5, 3), we get *infinitesimal stability*. In general infinitesimal stability also determines the stability of the non-linear equations (1.5, 2). This is denoted *stability in the large* or *practical stability* (compare for instance CESARI (1963), MALKIN (1952)).

The differential equation

$$z'' = V(\tau) \, z + v(\tau) \, z' \qquad\qquad (1.5, 4)$$

---

[1]) We assume that the solution $Y(\tau)$ as well as $y(\tau)$ exist for every $\tau \geqq \tau_0$, compare Cesari (1963), p. 4.

whose coefficients are real continuous periodic functions with period 2π,

$$V(\tau + 2\pi) = V(\tau) , \qquad v(\tau + 2\pi) = v(\tau) \tag{1.5, 5}$$

has, following for instance HORN (1948), a fundamental set of two linearly independent solutions z_1, z_2, and any solution of (1.5, 4) can be written as a linear combination of z_1 and z_2. In particular, the functions

$$\bar{z}_1(\tau) = z_1(\tau + 2\pi) , \qquad \bar{z}_2(\tau) = z_2(\tau + 2\pi) , \tag{1.5, 6}$$

which arise from the solutions z_1, z_2 if τ is replaced by $\tau + 2\pi$, are solutions of (1.5, 4) because of (1.5, 5) and are therefore linear combinations of z_1 and z_2,

$$\bar{z}_1 = c_{11}z_1 + c_{12}z_2 , \qquad \bar{z}_2 = c_{21}z_1 + c_{22}z_2 . \tag{1.5, 7}$$

They are also linearly independent and therefore a fundamental set of solutions because a linear dependence for every τ and (1.5, 6) would yield the linear dependence of z_1, z_2, hence

$$\begin{vmatrix} c_{11} & c_{12} \\ c_{21} & c_{22} \end{vmatrix} = 0 . \tag{1.5, 8}$$

The solutions of (1.5, 4) are in general not periodic, but there exists at least one solution ζ which only multiplies with an (in general) complex constant k if τ is replaced by $\tau + 2\pi$:

$$\bar{\zeta} = k\zeta . \tag{1.5, 9}$$

To show this, we write ζ as a linear combination of z_1, z_2,

$$\zeta = l_1z_1 + l_2z_2 . \tag{1.5, 10}$$

Substituting in (1.5, 9) gives

$$l_1\bar{z}_1 + l_2\bar{z}_2 = k(l_1z_1 + l_2z_2) ,$$

which is, because of (1.5, 7), a linear relation between z_1 and z_2, the coefficients of which are zero because z_1, z_2 are linearly independent:

$$\left. \begin{array}{l} (c_{11} - k) l_1 + c_{21}l_2 = 0 , \\ c_{12}l_1 + (c_{22} - k) l_2 = 0 . \end{array} \right\} \tag{1.5, 11}$$

Because both quantities l_1, l_2 do not vanish simultaneously, the equation

$$\begin{vmatrix} c_{11} - k & c_{21} \\ c_{12} & c_{22} - k \end{vmatrix} = 0 \tag{1.5, 12}$$

holds; this is called the *characteristic equation*. The roots of this equation

$$k_{1, 2} = \frac{c_{11} + c_{22}}{2} \pm \frac{1}{2}\sqrt{(c_{11} - c_{22})^2 + 4c_{12}c_{21}} \tag{1.5, 13}$$

are non-zero because of (1.5, 8).

If the roots k_1, k_2 are unequal, (1.5, 11) yields two pairs of values l_1, l_2 which lead by means of (1.5, 10) to two solutions ζ_1, ζ_2 for which

$$\bar{\zeta}_1 = k_1\zeta_1 , \qquad \bar{\zeta}_2 = k_2\zeta_2 \tag{1.5, 14}$$

hold. These two solutions are linearly independent and therefore constitute a fundamental set of solutions because a linear relation

$$\varkappa_1 \zeta_1 + \varkappa_2 \zeta_2 = 0 \qquad (1.5, 15)$$

with non-vanishing coefficients would, if τ were replaced by $\tau + 2\pi$, lead to $k_1 \varkappa_1 \zeta_1 + k_2 \varkappa_2 \zeta_2 = 0$ because of (1.5, 14), thence to $k_1 = k_2$ because of (1.5, 15).

Setting

$$\zeta_1 = e^{\varrho_1 \tau} Z_1(\tau), \qquad \zeta_2 = e^{\varrho_2 \tau} Z_2(\tau), \qquad (1.5, 16)$$

substituting in (1.5, 14), choosing ϱ_1, ϱ_2 such that

$$e^{2\pi\varrho_1} = k_1, \qquad e^{2\pi\varrho_2} = k_2, \qquad (1.5, 17)$$

and substituting in (1.5, 14), we find

$$Z_1(\tau + 2\pi) = Z_1(\tau), \qquad Z_2(\tau + 2\pi) = Z_2(\tau),$$

i.e., the functions Z_1, Z_2 are periodic with period 2π. The quantities ϱ_1, ϱ_2 defined by (1.5, 17) are called *characteristic exponents*, they, and the functions Z_1, Z_2, are in general non-real. The imaginary parts of the characteristic exponents are chosen in the interval

$$-\tfrac{1}{2} \leqq \operatorname{Im}\{\varrho_1\}, \operatorname{Im}\{\varrho_2\} < \tfrac{1}{2} \qquad (1.5, 18)$$

which is possible because $e^{2\pi i} = 1$.

If the roots k_1, k_2 are equal, they are (because of (1.5, 13)) equal to $(c_{11} + c_{22})/2$ and therefore real. In the case $c_{12} = c_{21} = 0$, the equations (1.5, 7) are already of the form (1.5, 14). If on the other hand at least one of these quantities, c_{21} say, is not zero, a solution of (1.5, 11) is $l_1 = c_{21}, l_2 = k_1 - c_{11}$, and the corresponding solution (1.5,10),

$$\zeta_1 = c_{21} z_1 + (k_1 - c_{11}) z_2, \qquad (1.5, 19)$$

yields, when τ is replaced by $\tau + 2\pi$,

$$\bar{\zeta}_1 = k_1 \zeta_1. \qquad (1.5, 20)$$

As now a second solution of this kind does not exist, we choose as second solution $\zeta_2 = z_2$. Because of (1.5, 7), (1.5, 19),

$$\bar{\zeta}_2 = \bar{z}_2 = c_{21} z_1 + c_{22} z_2 = \zeta_1 + (c_{11} + c_{22} - k_1) \zeta_2,$$

that is

$$\bar{\zeta}_2 = k_1 \zeta_2 + \zeta_1 \qquad (1.5, 21)$$

holds. We choose

$$\zeta_1 = e^{\varrho_1 \tau} Z_1(\tau), \qquad \zeta_2 = e^{\varrho_1 \tau} \left[Z_2(\tau) + \frac{\tau}{2\pi k_1} Z_1(\tau) \right].$$

Substituting in (1.5, 20), (1.5, 21) and choosing

$$e^{2\pi\varrho_1} = k_1$$

shows that the (now real) functions ζ_1, ζ_2 are periodic with period 2π. Thus we have proved the following theorem.

Floquet theorem. The differential equation (1.5, 4) with continuous and periodic coefficients with period 2π has a fundamental set of solutions of the form (1.5, 16) or the form

$$\zeta_1 = e^{\varrho_1 \tau} Z_1(\tau) , \qquad \zeta_2 = e^{\varrho_1 \tau} [Z_2(\tau) + K\tau Z_1(\tau)] \qquad (1.5, 22)$$

with in general complex constants ϱ_1, ϱ_2 satisfying (1.5, 18), continuous and in general complex periodic functions Z_1, Z_2 with period 2π and constant K which can be zero.

A result of the form (1.5, 16) or (1.5, 22) of the solutions is the following corollary.

Corollary. If the real parts of the characteristic exponents are negative, all solutions tend to zero for $\tau \to \infty$, in other words, we have asymptotic stability. If on the other hand the real part of at least one characteristic exponent is positive, solutions exist which are unbounded for $\tau \to \infty$, that is, we have instability.

For a stability investigation in the *critical cases* when characteristic exponents with real parts equal to zero, but no characteristic exponent with a positive real part exist, compare MALKIN (1952).

The method based on these results and on the knowledge of solutions of the vibrational differential equations is called *Ljapunov's first method*. With this method the stability will be determined in Section 2.4 and Chapter 5. Extensions to systems of differential equations are given in Chapter 6.

A method based not on knowledge of solutions of the differential equation but on the construction of Liapunov functions with certain properties is known as the *second method of Ljapunov*, compare for instance MALKIN (1952).

If several steady stable solutions exist, infinitesimal stability is often called *local stability*. When the solution is asymptotically stable, only one steady solution exists and divergent vibrations are absent, we speak of *absolute stability*.

In connection with the periodic motions of planets, another definition of stability has been introduced, *orbital stability*. A periodic motion is orbitally stable if this motion and disturbed periodic motions correspond with (closed) phase curves in the y, $\dot{y}$ plane which differ slightly if the disturbances are small. For an orbitally stable motion, the period can be slightly different from the periods of the disturbed motions so that after a long time the distance of the solutions increases infinitely. This shows that an orbitally stable solution can be unstable in the sense of Ljapunov, but an orbitally unstable solution is always unstable in the sense of Ljapunov. Very often these two forms of stability coincide (compare for instance STOKER (1950), KLOTTER (1980)).

As an example of a system, which is orbitally stable and unstable in the sense of Ljapunov, consider the free motion of an undamped pendulum. The motion in the phase plane is represented by closed trajectories. If two sets of slightly different initial conditions are applied, the trajectories lie close together. At a certain time, however, the distance between the corresponding points is no longer small because the period of free vibration of a pendulum depends on its initial deflection.

It is only possible to mention recent investigations of non-periodic, irregularly oscillating behaviour of deterministic systems the trajectories of which are very sensitive to small variations of the initial conditions and which is termed *chaotic behaviour*, also interpreted as a transient motion of infinite duration. Connected with such a behaviour are attractors (limit sets in the phase space against which every solution tends after long time) which are neither a point nor a limit cycle nor a surface and which are called *strange attractors*. A general view of these investigations can be got from YUNG-CHEN LU (1976), TROGER (1982, 1984), HOLMES (1980), HAKEN (1982), POPP (1982), SPARROW (1982), GUCKENHEIMER and HOLMES (1983), SCHMIDT (1986). In this book, strange attractor problems are not treated; attractors are always understood in the sense of simple attractors.

2.   Methods of solution

2.1.   The harmonic balance method

This method is based on a very simple idea and that is why it may seem less rigorous mathematically and not exact enough in approach. However, when combined with experience and sound engineering judgment on the part of the analyst it can be turned into a very efficient tool for solving non-linear problems, sometimes even excelling procedures which are grounded more thoroughly in mathematics. It may, for example be usefully employed in solution of resonant vibration of systems excited by a harmonic force, in which one or two harmonic components can be assumed to predominate in the response. In problems of excited vibration, the solution of a particular type is always restricted to a definite interval of the excitation frequency. Proceeding from an analysis involving considerations of the symmetry of the characteristics, the type of resonance and the form of the response, the engineer assumes a certain form of the solution, substitutes it in the equations of motion and compares the coefficients of the same harmonic components; from this last step the method derives its name.

To illustrate the application of the method, consider a Duffing system described by the equation

$$m\ddot{x} + h\dot{x} + cx + \varepsilon x^3 = P \cos \omega t. \tag{2.1, 1}$$

where m is the mass of the system, h is the damping coefficient, c is the stiffness of the linearized spring, and ε is the coefficient of the non-linear term of the restoring force. It is usually advantageous to convert the equation to the dimensionless form. Denoting by $P/c = x_0$ the static deflection of the linearized system produced by the action of force P, writing

$$\frac{x}{x_0} = y, \qquad \sqrt{\frac{c}{m}} = \omega_0, \qquad \frac{\omega}{\omega_0} = \eta, \qquad \varkappa = \frac{h/m}{\omega_0}, \qquad \frac{\varepsilon x_0^2}{c} = \gamma$$

and using the time transformation

$$\omega_0 t = \tau \tag{2.1, 2}$$

(2.1, 1) becomes

$$y'' + \varkappa y' + y + \gamma y^3 = \cos \eta \delta . \tag{2.1, 3}$$

The task now is to find the solution in the main resonance, i.e. $\eta \sim 1$. Since the characteristics of the damping and restoring forces are both symmetrical and the excitation is harmonic without a constant term, this solution can be approximated (for $\gamma \ll 1$) by

$$y = a \cos \eta \tau + b \sin \eta \tau = A \cos (\eta \tau - \varphi) \tag{2.1, 4}$$

in which the following relations hold:

$$A \cos \varphi = a , \qquad A \sin \varphi = b , \qquad A^2 = a^2 + b^2 . \qquad (2.1, 5)$$

Substituting (2.1, 4) in (2.1, 3) and comparing the coefficients of $\cos \eta\tau$ and $\sin \eta\tau$ gives the following algebraic equations for determining a and b, or possibly, A and φ:

$$(1 + \tfrac{3}{4} \gamma A^2 - \eta^2) \, a + \varkappa\eta b = 1 ,$$
$$-\varkappa\eta a + (1 + \tfrac{3}{4} \gamma A^2 - \eta^2) \, b = 0 . \qquad (2.1, 6)$$

Using (2.1, 5), equations (2.1, 6) after rearrangement (the first multiplied by $\cos \varphi$ is added to the second multiplied by $\sin \varphi$; the first multiplied by $\sin \varphi$ is added to the second multiplied by $-\cos \varphi$) take the form

$$(1 + \tfrac{3}{4} \gamma A^2) \, A = \cos \varphi ,$$
$$\varkappa\eta A = \sin \varphi \qquad (2.1, 7)$$

from which we get the equations

$$[(1 + \tfrac{3}{4} \gamma A^2 - \eta^2)^2 + \varkappa^2\eta^2] \, A^2 = 1 , \qquad (2.1, 8)$$

$$\tan \varphi = \varkappa\eta / (1 + \tfrac{3}{4} \gamma A^2 - \eta^2) . \qquad (2.1, 9)$$

Equation (2.1, 8) readily yields the function $\eta(A)$

$$(\eta^2)_{1, 2} = 1 + \frac{3}{4} \gamma A^2 - \frac{1}{2} \varkappa^2 \mp \left[\frac{1}{A^2} - \left(1 + \frac{3}{4} \gamma A^2 \right) \varkappa^2 + \frac{1}{4} \varkappa^4 \right]^{1/2} \qquad (2.1, 10)$$

and hence also the inverse function $A(\eta)$. Equation (2.1, 9) is used to determine φ as a function of η.

Equations (2.1, 7) can also be obtained directly. Shifting the time origin by φ/η, (2.1, 3) becomes

$$y'' + \varkappa y' + y + \gamma y^3 = \cos (\eta\tau + \varphi) \qquad (2.1, 3\,\text{a})$$

whose solution can be approximated by

$$y = A \cos \eta\tau . \qquad (2.1, 4\,\text{a})$$

The stability of the solution can be examined by means of the procedure used in the case of small disturbances, i.e. by substituting the approximate solution for the exact one. In the example being considered this involves substituting $y = A \cos \eta\tau + \xi$ for y in (2.1, 3a). The variational equation thus obtained, i.e.

$$\xi'' + \varkappa\xi' + \xi + \tfrac{3}{2} \gamma A^2 (1 + \cos 2\eta\tau) \, \xi = 0 \qquad (2.1, 11)$$

is the well-known Mathieu equation. The boundary of the instability interval of the first order is approximately determined as follows.

The solution on the boundary of the instability region which is represented by the Mathieu function of the first order is replaced by a harmonic function, i.e. approximated by

$$\xi = u \cos \eta\tau + v \sin \eta\tau . \qquad (2.1, 12)$$

Substituting (2.1, 12) in f2.1, 11) gives the equations

$$(1 + \tfrac{9}{4} \gamma A^2 - \eta^2) \, u + \varkappa\eta v = 0 ,$$
$$-\varkappa\eta u + (1 + \tfrac{3}{4} \gamma A^2 - \eta^2) \, v = 0 .$$

The condition of non-trivial solution ($u \neq 0$, $v \neq 0$) leads to the equation

$$(1 + \tfrac{3}{2}\,\gamma A^2 - \eta^2)^2 + (\varkappa\eta)^2 - (\tfrac{3}{4}\,\gamma A^2)^2 = 0$$

and in turn, to the equation for obtaining the function $\eta(A)$:

$$(\eta^2)_{1,2} = 1 + \tfrac{3}{2}\,\gamma A^2 - \tfrac{1}{2}\,\varkappa^2 \mp [(\tfrac{3}{4}\,\gamma A^2)^2 - (1 + \tfrac{3}{2}\,\gamma A^2)\,\varkappa^2 + \tfrac{1}{4}\,\varkappa^4]^{1/2}.$$

$$(2.1, 13)$$

In the (A, η)-plane, this curve forms the boundary between the stable and the unstable solutions.

Fig. 2.1, 1 shows the $A(\eta)$ curve drawn for the case of $\varkappa = 0.05$ and $\gamma = 10^{-2}$ (heavy full line — stable solution, dashed line — unstable solution) as well as the boundary of the solution stability calculated from (2.1, 13) (light solid line). Since subsequent chapters contain a number of diverse examples solved with the aid of this method, it is unnecessary to continue here with illustrations of its applicability. However, mention should be made at this point of the suitability of the method for dealing with the steady vibration of self-excited systems.

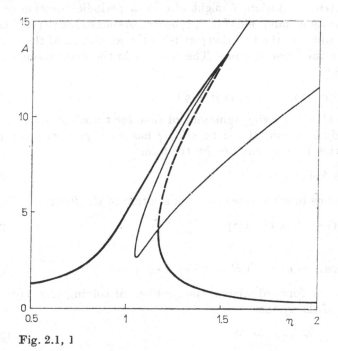

Fig. 2.1, 1

If the system to be examined has one degree of freedom and the solution can be approximated by a single harmonic component, the procedure is even simpler than in the previous case. Since the phase shift between excitation and response need not be taken into account, the solution can be approximated by the simple equation

$$y = A \cos \Omega\tau.$$

$$(2.1, 14)$$

Substituting this solution in the equation of motion and comparing the coefficients of $\cos \Omega\tau$ and $\sin \Omega\tau$ results in two algebraic equations which readily yield the unknown amplitude A and the unknown frequency Ω.

Chapter 9 will show an expedient procedure for solving more complicated self-excited systems with several degrees of freedom. In the case that the solution involved is of single-frequency type, no transformation to the quasi-normal form (such as that required for multi-frequency solutions) is necessary and the equations can be processed directly. If, moreover, the single-frequency vibration is close to a harmonic motion, the harmonic balance method yields very accurate results.

2.2.   The Van der Pol method

As in the harmonic balance method, the solution is approximated by a definite form; in the vicinity of the stationary solution, the coefficients of the various terms of this form are assumed to be functions slowly varying with time. To illustrate this, consider a system governed by the differential equation (already transformed to the dimensionless form)

$$y'' + \varkappa y' + f(y', y) = \cos \eta \tau . \qquad (2.2, 1)$$

In the case of parametric excitation, f might also be a periodic function of time $(f = f(y', y, \eta\tau))$, and the right-hand side of (2.2, 1) absent. Assume that the coefficients $\varkappa$ and $\partial f/\partial y'$ are small, and that the function f satisfies the conditions of the resonant vibration being close to the harmonic one. The solution in the main resonance can then be sought in the form

$$y = a \cos \eta\tau + b \sin \eta\tau = A \cos (\eta\tau - \varphi) \qquad (2.2, 2)$$

where a, b or A, φ are slowly varying functions of time for transient vibration and constants for the steady solution. If the function f has an asymmetric non-linear characteristic, the solution is approximated by the form

$$y = a \cos \eta\tau + b \sin \eta\tau + Y . \qquad (2.2, 3)$$

Note: It is advantageous in some cases to use a solution of the form

$$y = U \exp (i\eta\tau) + V \exp (-i\eta\tau) \qquad (2.2, 4)$$

where

$$U + V = A \cos \varphi = a , \qquad i(U - V) = A \sin \varphi = b . \qquad (2.2, 5)$$

On the assumptions put forward above, the problem of solving (2.2, 1) may be changed to an analysis of the system

$$a' = F_1(a, b) , \qquad b' = F_2(a, b) \qquad (2.2, 6)$$

or (in case of an asymmetric characteristic) the system

$$a' = F_1(a, b, Y) , \qquad b' = F_2(a, b, Y) , \qquad F_3(a, b, Y) = 0 . \qquad (2.2, 7)$$

The stationary solution is represented by the singular points

$$F_1(a, b) = 0 , \qquad F_2(a, b) = 0 \qquad (2.2, 8)$$

or

$$F_1(a, b, Y) = 0 , \qquad F_2(a, b, Y) = 0 , \qquad F_3(a, b, Y) = 0 . \qquad (2.2, 9)$$

Since the procedure for obtaining solutions in the phase plane is discussed separately in Chapter 4, no analysis of system (2.2, 6) or (2.2, 7) will be presented at this point.

Suffice it to say that for the same form of approximation, the steady solution is identical with that obtained by the harmonic balance method. For the Duffing system, for example, equations (2.2, 8) are identical with (2.1, 6).

Before turning to an examination of the *stability* of steady vibration *for small disturbances*, a general note on the process of solution will be in order. Frequently, especially when trying to find solutions in secondary resonances (or when the excitation is periodic but non-harmonic), approximation (2.2, 2) or (2.2, 3) is not satisfactory and additional harmonic components, for example, the terms $c \cos 2\eta\tau$, $d \sin 2\eta\tau$, etc., should also be considered. The literature contains numerous references to the case of a system with n degrees of freedom defined by n second-order differential equations whose solution is approximated by a form having K terms with slowly varying coefficients ($K > 2n$); the problem is reduced to an analysis of a system of K differential equations of the first order. Any solution of this new system is thus defined by a greater number of initial conditions than that of the original system, and this is physically inadmissible (for further details refer to TONDL (1970a)). Some authors claim that the initial conditions of the original system form a subset of the set of initial conditions of the new system of K first-order differential equations. A procedure which appears more correct physically consists of choosing only $2n$ time-variable coefficients and assuming the remaining $K - 2n$ coefficients to be constant. This leads to $2n$ differential equations of the first order and to $K - 2n$ algebraic equations.

The stability of the steady solution for small disturbances is determined using the characteristic equation of the linearized perturbation system

$$
\begin{vmatrix}
\dfrac{\partial F_1}{\partial a} - \lambda & \dfrac{\partial F_1}{\partial b} \\[2mm]
\dfrac{\partial F_2}{\partial a} & \dfrac{\partial F_2}{\partial b} - \lambda
\end{vmatrix} = 0 ,
\tag{2.2, 10}
$$

or

$$
\begin{vmatrix}
\dfrac{\partial F_1}{\partial a} - \lambda & \dfrac{\partial F_1}{\partial b} & \dfrac{\partial F_1}{\partial Y} \\[2mm]
\dfrac{\partial F_2}{\partial a} & \dfrac{\partial F_2}{\partial b} - \lambda & \dfrac{\partial F_2}{\partial Y} \\[2mm]
\dfrac{\partial F_3}{\partial a} & \dfrac{\partial F_3}{\partial b} & \dfrac{\partial F_3}{\partial Y} - \lambda
\end{vmatrix} = 0 ,
\tag{2.2, 11}
$$

i.e. on the basis of an examination of the stability of singular points (see Chapter 4 for the types and stability of the singular points).

As shown in Appendix I of TONDL's monograph (1970), the vertical tangent rule applies to system (2.2, 6) and others, i.e. the points of the $A(\eta)$ curve at which the tangent is vertical form the boundary between the stable and the unstable solutions.

Proof: Recalling the relations

$$
a = A \cos \varphi , \qquad b = A \sin \varphi , \qquad A = (a^2 + b^2)^{1/2}
$$

the slope of the tangent to the $A(\eta)$ curve is defined by the equation

$$
\frac{dA}{d\eta} = \frac{\partial A}{\partial a} \frac{da}{d\eta} + \frac{\partial A}{\partial b} \frac{db}{d\eta}
\tag{2.2, 12}
$$

where

$$\frac{\partial A}{\partial a} = a(a^2 + b^2)^{-1/2}, \qquad \frac{\partial A}{\partial b} = b(a^2 + b^2)^{-1/2}.$$
(2.2, 13)

Eliminating the derivatives $da/d\eta$ and $db/d\eta$ from the system

$$\frac{dF_1}{d\eta} = \frac{\partial F_1}{\partial a}\frac{da}{d\eta} + \frac{\partial F_1}{\partial b}\frac{db}{d\eta} + \frac{\partial F_1}{\partial \eta} = 0, \qquad \frac{dF_2}{d\eta} = \frac{\partial F_2}{\partial a}\frac{da}{d\eta} + \frac{\partial F_2}{\partial b}\frac{db}{d\eta} + \frac{\partial F_2}{\partial \eta} = 0$$
(2.2, 14)

and using the notation

$$\Delta_a = \begin{vmatrix} -\dfrac{\partial F_1}{\partial \eta} & \dfrac{\partial F_1}{\partial b} \\[2mm] -\dfrac{\partial F_2}{\partial \eta} & \dfrac{\partial F_2}{\partial b} \end{vmatrix}, \qquad \Delta_b = \begin{vmatrix} \dfrac{\partial F_1}{\partial a} & -\dfrac{\partial F_1}{\partial \eta} \\[2mm] \dfrac{\partial F_2}{\partial a} & -\dfrac{\partial F_2}{\partial \eta} \end{vmatrix}, \qquad \Delta = \begin{vmatrix} \dfrac{\partial F_1}{\partial a} & \dfrac{\partial F_1}{\partial b} \\[2mm] \dfrac{\partial F_2}{\partial a} & \dfrac{\partial F_2}{\partial b} \end{vmatrix}$$
(2.2, 15)

leads to

$$\frac{da}{d\eta} = \frac{\Delta_a}{\Delta}, \qquad \frac{db}{d\eta} = \frac{\Delta_b}{\Delta}.$$
(2.2, 16)

Substituting (2.2, 16) and (2.2, 13) in (2.2, 12) gives

$$\frac{dA}{d\eta} = [\Delta(a^2 + b^2)^{1/2}]^{-1}(a\Delta_a + b\Delta_b).$$
(2.2, 17)

The characteristic equation (2.2, 10) can have the form

$$\lambda^2 + \alpha_1\lambda + \alpha_2 = 0,$$

where

$$\alpha_1 = -\left(\frac{\partial F_1}{\partial a} + \frac{\partial F_2}{\partial b}\right), \qquad \alpha_2 = \Delta.$$
(2.2, 18)

In systems with positive damping, the first condition of stability

$$\alpha_1 = -\left(\frac{\partial F_1}{\partial a} + \frac{\partial F_2}{\partial b}\right) > 0$$
(2.2, 19)

is always satisfied and it is only the sign of the term Δ that is decisive; consequently, the limit of stability is at

$$\alpha_2 = \Delta = 0.$$
(2.2, 20)

At this point, $dA/d\eta$ increases beyond all bounds (see (2.2, 17) provided that (unlike in an extraordinary case) the term $a\Delta_a + b\Delta_b$ is not simultaneously equal to zero). It follows from the above that:

The necessary and sufficient condition for applying the criterion of vertical tangents to the resonant amplitude curve of the system (2.2, 1) is

$$\operatorname{sgn}\left(\frac{\partial F_1}{\partial a} + \frac{\partial F_2}{\partial b}\right) = -1.$$

As shown in Chapter 8, the vertical tangent rule is not a sufficient criterion in some cases as it provides only partial information about the system's stability. In the example presented there, owing to the softening characteristic of the restoring force of the spring, divergent vibration is likely to exist next to the steady one in some domain of the parameters, and the slope of the characteristic can even become negative for larger deflections.

2.3.   The integral equation method

The method widely used in this book is the method of non-linear integro-differential equations (in short: integral equation method).

Vibrations of various kinds can be modelled by a differential equation

$$y'' + \lambda y = \Phi \tag{2.3, 1}$$

where dashes denote derivatives with respect to a dimensionless time τ and

$$\Phi = \Phi(y, y', \varepsilon_p; \tau) = \Phi[\tau]$$

is given by a power series in y, y' and parameters ε_p $(p = 1, 2, \ldots, P)$ which may also depend on τ.

Restricting ourselves for simplicity to continuous solutions with continuous first and second derivatives, we can prove the following *theorem* (compare SCHMIDT (1961, 1975)): Every periodic solution of the differential equation (2.3, 1) is a solution of the integro-differential equation

$$y(\tau) = \int\limits_0^{2\pi} G(\tau, \sigma)\, \Phi[\sigma]\, \mathrm{d}\sigma + \delta_\lambda^{n^2}(r \cos n\tau + s \sin n\tau) \tag{2.3, 2}$$

where

$$G(\tau, \sigma) = \frac{1}{\pi}\left[\frac{1}{2\lambda} + \sum\limits_{\nu=1}^{\infty} \frac{\cos \nu(\tau - \sigma)}{\vartheta\lambda - \nu^2}\right]$$

is the corresponding generalized Green's function,

$$\delta_\lambda^{n^2} = \begin{cases} 1 & \text{for} \quad \lambda = n^2, \quad n \text{ being an integer,} \\ 0 & \text{otherwise} \end{cases}$$

is the Kronecker symbol, and

$$\vartheta = \vartheta_\lambda^{n^2} = 1 - \delta_\lambda^{n^2}$$

so that the denominator does not vanish. The bifurcation parameters r, s appearing in the resonance case

$$\lambda = n^2$$

are to be determined by the bifurcation or periodicity equations

$$r = \frac{1}{\pi} \int\limits_0^{2\pi} y(\tau) \cos n\tau \, \mathrm{d}\tau\,, \qquad s = \frac{1}{\pi} \int\limits_0^{2\pi} y(\tau) \sin n\tau \, \mathrm{d}\tau \tag{2.3, 3}$$

which are equivalent to

$$\int\limits_0^{2\pi} \Phi[\tau] \cos n\tau \, \mathrm{d}\tau = \int\limits_0^{2\pi} \Phi[\tau] \sin n\tau \, \mathrm{d}\tau = 0\,. \tag{2.3, 4}$$

The solutions of (2.3, 2) and (2.3, 3) respectively (2.3, 4) can be found by the method of successive approximations based upon the following equations for the approximate solutions $y_j(\tau)$, $j = 1, 2, 3, \ldots$,

$$y_j(\tau) = \int_0^{2\pi} G(\tau, \sigma) \, \Phi_{j-1}[\sigma] \, d\sigma + \delta_\lambda^{n^2}(r \cos n\tau + s \sin n\tau)$$

where

$$\Phi_0[\tau] = \Phi(0, 0, \varepsilon_p; \tau),$$

$$\Phi_j[\tau] = \Phi(y_j, y_j', \varepsilon_p; \tau), \qquad j = 1, 2, 3, \ldots$$

The convergence of the method can be proved under the assumption that the parameters ε_p are less than certain upper bounds. In most cases, the actual evaluation of the radius of convergence is a very complicated problem, and one has to content oneself with a verification of the solution for special values by direct numerical integration of the differential equation. A second derivative y'' in Φ can be handled in the same way, for the proof of convergence it can be substituted successively by means of the differential equation.

As a simple example for the integral equation method, we once more consider the *Duffing equation* (2.1, 1). We introduce a dimensionless time by

$$\tau = \omega t$$

— derivatives with respect to τ are denoted by dashes — instead of (2.1, 2) and write ω with a fixed frequency ω_0 and the small frequency variation α in the form

$$\omega = \omega_0(1 + \alpha).$$

As in Section 2.1 we relate the deflection x to the static deflection of the linear system by setting

$$y = \frac{c}{P} x,$$

assume the main resonance $\omega_0^2 = c/m$ and use the abbreviations

$$\varkappa = \frac{h}{m\omega_0}, \qquad \gamma = \frac{P^2 \varepsilon}{c^3}.$$

Instead of (2.1, 3) we get

$$y'' + y = -(2 + \alpha) \alpha y'' - (1 + \alpha) \varkappa y' - \gamma y^3 + \cos \tau \equiv \Phi. \tag{2.3, 5}$$

Substituting the first approximation

$$y_1 = r \cos \tau + s \sin \tau$$

into (2.3, 4) and taking into consideration only the first power of the small coefficients α, $\varkappa$ yields

$$\left(\frac{3\gamma}{4} A^2 - 2\alpha\right) r + \varkappa s = 1, \qquad -\varkappa r + \left(\frac{3\gamma}{4} A^2 - 2\alpha\right) s = 0$$

instead of (2.1, 6). Elimination of r respectively s in the first equation by help of the second equation gives for the resonance amplitude

$$A = \sqrt{r^2 + s^2}$$

the formula

$$\left[\left(\frac{3\gamma}{4} A^2 - 2\alpha\right)^2 + \varkappa^2\right] A^2 = 1 \tag{2.3, 6}$$

instead of (2.1, 8) which can be written

$$2\alpha = \frac{3\gamma}{4} A^2 \pm \sqrt{\frac{1}{A^2} - \varkappa^2} \tag{2.3, 7}$$

instead of (2.1, 10)

Formulae (2.1, 8), (2.1, 10) seem to be a better approximation than (2.3, 6), (2.3, 7). If we had also taken into consideration the second power of α, $\varkappa$ in first approximation, we would have got as easily the formula

$$\left[\left(\frac{3\gamma}{4} A^2 - 2\alpha - 2\alpha^2\right)^2 + (1 + \alpha)^2 \varkappa^2\right] A^2 = 1 \tag{2.3, 8}$$

which exactly corresponds with (2.1, 8) because $\eta = 1 + \alpha$. But the second approximation terms

$$y_2 = -2\alpha(r \cos \tau + s \sin \tau) + \varkappa(s \cos \tau - r \sin \tau)$$

yield the additional terms

$$\Phi_2 = (-4\alpha^2 + \varkappa^2)(r \cos \tau + s \sin \tau) + 4\alpha\varkappa(s \cos \tau - r \sin \tau)$$

of second order in α, $\varkappa$, that is the complete second order amplitude formula

$$\left[\left(\frac{3\gamma}{4} A^2 - 2\alpha + \alpha^2 - \varkappa^2\right)^2 + (1 - 3\alpha)^2 \varkappa^2\right] A^2 = 1 . \tag{2.3, 9}$$

It shows that the partial consideration of second order terms by a one-step method in (2.1, 8) respectively in (2.3, 8) can not determine the real second order corrections which are given by (2.3, 9).

The main advantages of the integral equation method are:

1. The use of small parameters leads to solutions to every degree of accuracy needed.

2. The use of *several independent* small parameters saves unnecessary evaluations (in our example the evaluation of the complete second approximation).

3. The simple and clear mechanism of finding the successive approximations enables us to estimate their influence without explicitly evaluating them.

4. The method also permits us to investigate different multiple resonances (Chapters 6 and 12), stability (Section 2.4, Chapter 5), and systems with many degrees of freedom where it is often necessary to adapt the method to the special problem in order to confine oneself to a reasonable number of evaluations (Chapters 6 and 12), as well as narrow-band random excitations (Chapters 10 and 12).

5. The method is especially suitable for combination with computer algebra methods in order to transfer extensive analytical evaluations susceptible to errors to a computer. Hints on computer algebra methods are given in Sections 6.9 and 12.7.

2.4.   Stability conditions

The integral equation method together with the Floquet theorem of Section 1.5 leads
to conditions for stability. To find them, we consider the linear variational equations
(1.5, 3)

$$z'' + \lambda z = u(\tau)\, z + v(\tau)\, z' \,. \tag{2.4, 1}$$

The coefficient functions $u(\tau)$, $v(\tau)$ can be assumed to be real, continuous, periodic
with period 2π and sufficiently small in modulus, with the absolutely and uniformly
convergent Fourier series

$$u(\tau) = u_0 + 2 \sum_{j=1}^{\infty} (u_j \cos j\tau + U_j \sin j\tau)\,,$$

$$v(\tau) = v_0 + 2 \sum_{j=1}^{\infty} (v_j \cos j\tau + V_j \sin j\tau)\,.$$

Corresponding to the Floquet theorem, we write the solution of (2.4, 1) in the form

$$z = e^{\varrho\tau}\, Z\,,$$

where ϱ is the characteristic exponent, and get the differential equation

$$\left. \begin{aligned} Z'' + \lambda Z &= \Phi\,, \\ \Phi &= [u(\tau) + \varrho v(\tau) - \varrho^2]\, Z + [v(\tau) - 2\varrho]\, Z' \end{aligned} \right\} \tag{2.4, 2}$$

for the function $Z(\tau)$ which is periodic with period 2π and in general complex.

The integral equation method of Section 2.3 is applicable to (2.4, 2) which can be
shown by applying it to the real and the imaginary part of (2.4, 2) and then again
combining the real and imaginary parts (compare SCHMIDT (1969 b, 1975)). The first
approximation is

$$Z_1(\tau) = \delta_\lambda^{n^2}(R \cos n\tau + S \sin n\tau)\,,$$

hence for the resonance case $\lambda = n^2$,

$$\Phi_1 = (u_0 + v_0\varrho - \varrho^2)\,(R \cos n\tau + S \sin n\tau)$$

$$+ \sum_{j=1}^{\infty} [n_j R \cos (n-j)\,\tau + u_j S \sin (n-j)\,\tau + u_j R \cos (n+j)\,\tau + u_j S \sin (n+j)\,\tau$$

$$+ U_j S \cos (n-j)\,\tau - U_j R \sin (n-j)\,\tau - U_j S \cos (n+j)\,\tau + U_j R \sin (n+j)\,\tau]$$

$$+ \varrho \sum_{j=1}^{\infty} [v_j R \cos (n-j)\,\tau + v_j S \sin (n-j)\,\tau + v_j R \cos (n+j)\,\tau + v_j S \sin (u+j)\,\tau$$

$$+ V_j S \cos (n-j)\,\tau - V_j R \sin (n-j)\,\tau - V_j S \cos (n+j)\,\tau + V_j R \sin (n+j)\,\tau]$$

$$+ n(v_0 - 2\varrho)\,(S \cos n\tau - R \sin n\tau)$$

$$+ n \sum_{j=1}^{\infty} [v_j S \cos (n-j)\,\tau - v_j R \sin (n-j)\,\tau + v_j S \cos (n+j)\,\tau - v_j R \sin (n+j)\,\tau$$

$$- V_j R \cos (n-j)\,\tau - V_j S \sin (n-j)\,\tau + V_j R \cos (n+j)\,\tau + V_j S \sin (n+j)\,\tau]\,.$$

The periodicity conditions (2.3, 4) now read

$$(u_0 + v_0\varrho - \varrho^2)\binom{R}{S} + (u_{2n} + v_{2n}\varrho - nV_{2n})\binom{R}{-S}$$

$$+ (U_{2n} + V_{2n}\varrho + nv_{2n})\binom{S}{R} + n(v_0 - 2\varrho)\binom{S}{-R} = 0\,,$$

the determinant condition for non-vanishing (in general complex) solutions R, S yields

$$(u_0 + v_0\varrho - \varrho^2)^2 - (u_{2n} + v_{2n}\varrho - nV_{2n})^2$$
$$-(U_{2n} + V_{2n}\varrho + nv_{2n})^2 + n^2(v_0 - 2\varrho)^2 = 0 . \qquad (2.4, 3)$$

The characteristic exponent ϱ is of no higher order of magnitude than the Fourier coefficients appearing because otherwise approximately the equation

$$\varrho^4 + 4n^2\varrho^2 = 0$$

that is $\varrho^2 = -4n^2$ would hold, in contradiction to (1.5, 18). Therefore the terms $v_0\varrho$, $v_{2n}\varrho$, $V_{2n}\varrho$ and ϱ^2 in (2.4, 3) can be omitted in the same way as products of the Fourier coefficients have not been taken into consideration in the above approximation, and (2.4, 3) yields

$$2n\varrho = nv_0 \pm \sqrt{(u_{2n} - nV_{2n})^2 + (U_{2n} + nv_{2n})^2 - u_0^2} .$$

Because of the Corollary in Section 1.5, we have *asymptotic stability if and only if the inequalities*

$$v_0 < 0 \qquad (2.4, 4)$$

and

$$n^2v_0^2 > (u_{2n} - nV_{2n})^2 + (U_{2n} + nv_{2n})^2 - u_0^2 \qquad (2.4, 5)$$

hold, but certainly instability if at least one of these inequalities holds with the opposite sign.

As an example we investigate the stability of the *Duffing equation* (2.3, 5). The corresponding linear variational equation is in the approximation used for the solution (2.3, 7)

$$z'' + z = 2\alpha z - \varkappa z' - 3\gamma y_1^2 z .$$

Using

$$2y_1^2 = A^2 + (r^2 - s^2) \cos 2\tau + 2rs \sin 2\tau ,$$

we get

$$u_0 = 2\alpha - \frac{3\gamma}{2} A^2 , \qquad v_0 = -\varkappa ,$$

$$u_2 = -\frac{3\gamma}{4} (r^2 - s^2) , \qquad U_2 = -\frac{3\gamma}{2} rs , \qquad v_2 = V_2 = 0 .$$

The first stability condition (2.4, 4) holds with $\varkappa > 0$, the second stability condition (2.4, 5) requires that

$$\varkappa^2 > \frac{9\gamma^2}{16} (r^2 - s^2)^2 + \frac{9\gamma^2}{4} r^2s^2 - \left(2\alpha - \frac{3\gamma}{2} A^2\right)^2 = \frac{9\gamma^2}{16} A^4 - \left(2\alpha - \frac{3\gamma}{2} A^2\right)^2$$

or, after inserting (2.3, 7),

$$\mp \frac{3\gamma}{2} A^2 \sqrt{\frac{1}{A^2} - \varkappa^2} + \frac{1}{A^2} > 0 .$$

Differentiation of (2.3, 7) yields

$$2A \sqrt{\frac{1}{A^2} - \varkappa^2} \frac{d\alpha}{dA} = \frac{3\gamma}{2} A^2 \sqrt{\frac{1}{A^2} - \varkappa^2} \mp \frac{1}{A^2} ,$$

a comparison with (2.4, 6) shows that the solution A of (2.3, 7) with the upper sign (the upper branch of the resonance curve) is asymptotically stable if $d\alpha/dA < 0$, the lower solution if $d\alpha/dA > 0$ whereas the opposite sign causes instability. The points of the resonance curve with vertical tangent correspond to the boundary points of the instability region.

2.5.   The averaging method

The well-known averaging method of Krylov and Bogoljubov (compare BOGOLJUBOV and MITROPOL'SKI (1963), KLOTTER (1980)) will be introduced here by means of the Duffing equation (2.3, 5) which we write in the form, up to small terms of higher order of magnitude,

$$y'' + y = 2\alpha y - \varkappa y' - \gamma y^3 + \cos \tau .$$
$$(2.5, 1)$$

In connection with the solution

$$y = a_0 \cos (\tau + \vartheta_0) \qquad (a_0,\ \vartheta_0 \text{ constants})$$

for vanishing right-hand side, a solution of (2.5, 1) is sought for in the form

$$y = a(\tau) \cos [\tau + \vartheta(\tau)] = a(\tau) \cos \varphi(\tau)$$
$$(2.5, 2)$$

with the slowly varying functions $a = a(\tau)$ and $\vartheta = \vartheta(\tau)$ for which a condition can be fixed:

$$a' \cos \varphi - a\vartheta' \sin \varphi = 0 .$$
$$(2.5, 3)$$

Differentiation yields

$$y' = a' \cos \varphi - a(1 + \vartheta') \sin \varphi = -a \sin \varphi$$
$$(2.5, 4)$$

from (2.5, 3), hence

$$y'' = -a' \sin \varphi - a(1 + \vartheta') \cos \varphi$$

so that the differential equation (2.5, 1) reads

$$-a' \sin \varphi - a\vartheta' \cos \varphi = 2\alpha y - \varkappa y' - \gamma y^3 + \cos \tau .$$

Solving this equation and (2.5, 3) and substituting (2.5, 2), (2.5, 4) in the right-hand sides gives the equations in standard form

$$\left. \begin{array}{l} a' = (-2\alpha a \cos \varphi - \varkappa a \sin \varphi + \gamma a^3 \cos^3 \varphi - \cos \tau) \sin \varphi , \\ a\vartheta' = (-2\alpha a \cos \varphi - \varkappa a \sin \varphi + \gamma a^3 \cos^3 \varphi - \cos \tau) \cos \varphi , \end{array} \right\}$$
$$(2.5, 5)$$

two differential equations of first order for a, ϑ instead of the differential equation of second order (2.5, 1).

Both the right-hand side of (2.5, 1) and the right-hand sides of (2.5, 5) are small in comparison with the left-hand sides. The basic idea of the averaging method consists in averaging the terms on the right-hand side containing φ over one period (the slowly varying term $\cos \tau$ being assumed as constant during this period):

$$a' = -\frac{\varkappa}{2} a - \frac{1}{2} \sin \vartheta , \qquad a\vartheta' = -\alpha a + \frac{3\gamma}{8} a^3 - \frac{1}{2} \cos \vartheta .$$

For a stationary solution, $a' = \vartheta' = 0$, there follows the amplitude formula

$$2\alpha = \frac{3\gamma}{4}a^2 \pm \sqrt{\frac{1}{a^2} - \varkappa^2}$$

which is the same as (2.3, 7), and

$$\tan \vartheta = \frac{\varkappa}{2\alpha - \dfrac{3\gamma}{4}a^2}.$$

The advantage of the averaging method lies in the possibility of evaluating not only stationary, that is periodic solutions but also non-stationary ones such as transition processes or random vibrations. The drawback lies in a comparatively complicated way of evaluating higher approximations and of estimating the accuracy. We will use this method, also in higher approximation, in Section 10.1 and in Chapters 11 and 12 in connection with random vibrations.

3. Auxiliary curves for analysis of non-linear systems

3.1. Characteristic features of auxiliary curves, particularly the backbone curves and the limit envelopes

In harmonically excited systems, especially in systems with one degree of freedom, the specific characteristics of the auxiliary curves have been found to be very useful for preliminary qualitative analyses of stationary vibration as well as for identification of the various elements (e.g. damping) of the system being examined on the basis of experimental results. In the former case they enable the analyst to make a prompt estimate of the basic properties of the system and of the effect of various parameters on its behaviour, in the latter, to identify the specific properties of the system and in turn to formulate a suitable analytic expression of the forces acting in a particular element for the purpose of a mathematical model.

Let us first consider the characteristics of the so-called *backbone* (or skeleton) *curve*, and of the curves connecting the points at which $\sin \varphi = $ const (φ is the phase angle between response and excitation). The limiting case of the latter curves ($\sin \varphi = \pm 1$; the minus sign has no meaning except in special cases stated farther on) is the so-called *limit envelope* (this term was proposed by Tondl (1973d)). The backbone curves have been used in routine analyses for a long time; the application of the second type of curves, the limit envelope and the $\sin \varphi = $ const curves, is less common. The limit envelope was used for the first time by Abramson (1954) and later applied to qualitative analyses by Kolovskij (1966). Assuming linear viscous damping, Novák (1963) employed the $\sin \varphi = $ const curves for the construction of the backbone curve from the measured vibration amplitude-excitation frequency dependence. Other possible uses of the properties of these curves have been investigated and reported by Tondl (1971b), (1973d), (1975c), (1979a), (1980c). Tondl was the first to propose the use of the limit envelopes for the identification of damping and to extend the field of application of these curves to parametrically excited systems (1975c), higher-order systems (1974b), (1977a) and to systems with several degrees of freedom (1978a).

To facilitate the subsequent explanation, the basic properties of the auxiliary curves will be described using the example of a single-degree-of-freedom system having mass m excited by a harmonic force whose amplitude is either constant (P) or proportional to the square of the excitation frequency (inertial excitation — $\varepsilon m_0 \omega^2$). To simplify matters, it is assumed that the damping and the restoring force have both symmetric characteristics and can, therefore, be expressed in terms of the functions $c[1 + f(|y|)] y$, $\varkappa(|\dot{y}|, |y|) \operatorname{sgn} \dot{y}$ where y is the deflection and $\dot{y}$ the velocity of the mass. A system of this kind is described by the equation

$$m\ddot{y} + \varkappa(|\dot{y}|, |y|) \operatorname{sgn} \dot{y} + c[1 + f(|y|)] y = \begin{cases} P \cos \omega t \\ \varepsilon m_0 \omega^2 \cos \omega t \end{cases}. \qquad (3.1, 1)$$

Using the harmonic balance method (described in Chapter 2) and introducing the time shift $t_0 = \varphi/\omega$, i.e. substituting, in (3.1, 1) $\cos(\omega t + \varphi)$ for $\cos \omega t$, and approximating the stationary resonance solution by

$$y = A \cos \omega t \qquad (3.1, 2)$$

the following equations are obtained for determining A and φ:

$$A[c + cF(A) - m\omega^2] = \begin{cases} P \cos \varphi \\ \varepsilon\, m_0\omega^2 \cos \varphi \end{cases}, \qquad (3.1, 3)$$

$$K(\omega A, A) = \begin{cases} P \sin \varphi \\ \varepsilon m_0\omega^2 \sin \varphi \end{cases} \qquad (3.1, 4)$$

where

$$F(A) = \frac{\omega}{\pi} \int\limits_0^{2\pi/\omega} f(|A \cos \omega t|) \cos^2 \omega t \, dt \ ,$$

$$K(\omega A, A) = \frac{\omega}{\pi} \int\limits_0^{2\pi/\omega} \varkappa(|-\omega A \sin \omega t|, |A \cos \omega t|) \frac{\sin^2 \omega t}{|\sin \omega t|} \, dt \ .$$

The backbone curve is defined by the equation obtained from (3.1, 3) in which $\cos \varphi = 0$, i.e. $\varphi = \pi/2$, viz.

$$c[1 + F(A)] - m\omega^2 = 0 \ ;$$

this equation can be given the form of the dependence of frequency on vibration amplitude A (which is identical with the dependence of the natural frequency of the undamped vibration on the initial deflection at zero initial velocity), viz.

$$\omega = \left\{ [1 + F(A)] \frac{c}{m} \right\}^{1/2} = \Omega(A) \ . \qquad (3.1, 5)$$

As shown schematically in Fig. 3.1, 1, the various points of the resonance curve $A(\omega)$ can also be obtained as the points of intersection of the curve defined by (3.1, 4) in which $\sin \varphi = \mathrm{const} = S$

$$K(\omega A, A) = \begin{cases} PS \\ \varepsilon m_0\omega^2 S \end{cases} \qquad (3.1, 6)$$

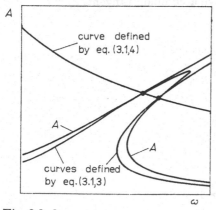

A

curve defined
by eq. (3.1,4)

A

A

curves defined
by eq. (3.1,3)

ω

Fig. 3.1, 1

and the pair of curves defined by (3.1, 3) in which $\cos \varphi = \pm C = \pm \sqrt{1 - S^2}$

$$A[c + cF(A) - m\omega^2] = \begin{cases} \pm PC \\ \pm \varepsilon m_0 \omega^2 C \end{cases}.$$ (3.1, 7)

If

$$K(\omega A, A) \gtreqless 0$$ (3.1, 8)

for any A, then

$$0 \leq \sin \varphi \leq 1 \; ;$$ (3.1, 9a)

if the function $K(\omega A, A)$ assumes negative values in a particular interval of A,

$$-1 \leq \sin \varphi \leq 1 \; .$$ (3.1, 9b)

For the extreme values, i.e. $\sin \varphi = 1$ or $\sin \varphi = \pm 1$, equation (3.1, 6) becomes

$$K(\omega A, A) = \begin{cases} P \\ \varepsilon m_0 \omega^2 \end{cases}$$ (3.1, 10a)

or (if (3.1, 8) does not apply to all A),

$$K(\omega A, A) = \begin{cases} \pm P \\ \pm \varepsilon m_0 \omega^2 \end{cases}.$$ (3.1, 10b)

These equations define the limit envelope.

The limit envelope divides the (A, ω) plane into two regions: one for which

$$K(\omega, A, A) \leq \begin{cases} P \\ \varepsilon m_0 \omega^2 \end{cases}$$ (3.1, 11a)

or

$$\left. \begin{array}{l} -P \\ -\varepsilon m_0 \omega^2 \end{array} \right\} \leq K(\omega A, A) \leq \begin{cases} P \\ \varepsilon m_0 \omega^2 \end{cases}$$ (3.1, 11b)

the other, in which inequalities (3.1, 11a, b) are not satisfied.

The response curve $A(\omega)$ can lie only in the first region. The limit envelope thus forms the envelope of all possible response curves which touch it at the points at wich $\varphi = \pm \pi/2$. These points are the points of intersection of the limit envelope $A_L(\omega)$ and the backbone curve $A_S(\omega)$ (Fig. 3.1, 2) (Subscripts L and S are used to distinguish these curves from the resonance curve $A(\omega)$). At these points the limit envelope and the response curve $A(\omega)$ have a common tangent.

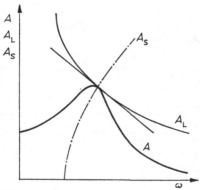

Fig. 3.1, 2

The following fundamental characteristics of the backbone curve and the limit envelope can be deduced from the equations defining these curves:

The position of the backbone curve is affected by the linear term of the restoring force of the spring and by the magnitude of the mass. The form (curvature) is influenced by the non-linear term of the restoring force characteristic. If this characteristic is linear, the backbone curve is a straight line parallel to the axis of amplitude A, i.e. the natural frequency is independent of initial deflection or velocity. The backbone curve is not altered by change of damping.

The limit envelope is affected only by damping and the excitation amplitude. It is invariant to changes of the mass, i.e. all the resonance curves obtained for different masses m touch the same limit envelope (Fig. 3.1, 3), as well as to changes of the spring characteristic — both its linear and non-linear terms (Fig. 3.1, 4).

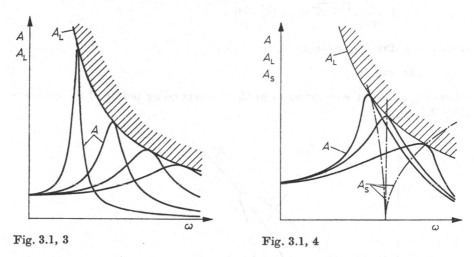

Fig. 3.1, 3   Fig. 3.1, 4

Another feature of interest is the relation between the points of intersection of the curve defined by (3.1, 6) and the backbone and the resonance curves. For the curve defined by (3.1, 6) we can write the $\omega - A$ dependence for a certain value of $\sin \varphi$ = const in the form

$$\omega = \omega^*(A\,;\varphi)\;; \tag{3.1, 12a}$$

the reciprocal to function (3.1, 12a) is

$$A = A^*(\omega\,;\varphi)\,. \tag{3.1, 12b}$$

The point of intersection of the above curve and the backbone curve has the ω-coordinate denoted by Ω_φ which is defined by the equation

$$\Omega_\varphi^2 = \frac{c}{m}\,[1 + F(A)]\,. \tag{3.1, 13}$$

Accordingly, for the alternative with a constant excitation amplitude, equation (3.1, 7) takes the form

$$A(\Omega_\varphi^2 - \omega^2)\,m = \pm P|C| = \pm P\sqrt{1 - S^2}\,. \tag{3.1, 14}$$

Equation (3.1, 14) is satisfied by both φ = const and $\pi - \varphi$.

Consider first the case of viscous (generally non-linear) damping ($\varkappa(|\dot{y}|)\,\dot{y}$ where $\varkappa(|\dot{y}|) > 0$). For a constant excitation amplitude, (3.1, 6) has the form

$$\omega A K(\omega A) = PS \tag{3.1, 15}$$

from which it follows that

$$\omega A = \text{const} = k_0(\varphi) \tag{3.1, 16}$$

i.e. the curves for $\sin\varphi = \text{const}$ and thus also the limit envelope are rectangular hyperbolas. The points of the resonance curve $A(\omega)$ are obtained by establishing the points of intersection of this curve for a given φ, with the curves defined by (3.1, 14). Substituting $A = k_0/\omega$ in (3.1, 14) and some rearrangement gives

$$\omega^2 \pm \frac{P}{mk_0}\sqrt{1 - S^2}\,\omega - \Omega_\varphi^2 = 0\ ;$$

from this follows the relation

$$\Omega_\varphi^2 = \omega_{\mathrm{I}}\omega_{\mathrm{II}} \tag{3.1, 17}$$

where ω_{I}, ω_{II} are the coordinates of the corresponding points of the resonance curve (Fig. 3.1, 5).

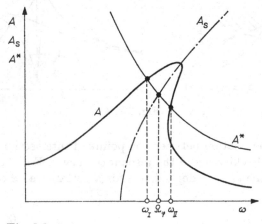

Fig. 3.1, 5

Consider now the case of damping defined by the function $\varkappa(|y|)\,\text{sgn}\,\dot{y}$ where $\varkappa(|y|) > 0$. In this case, (3.1, 15) becomes

$$K(A) = PS\ . \tag{3.1, 18}$$

The ($\sin\varphi = \text{const}$)-curves are straight lines parallel to the ω axis, i.e. $A = \text{const}$. It follows then from (3.1, 4) that

$$\Omega_\varphi^2 - \omega^2 = \pm H_0 \tag{3.1, 19}$$

where $H_0 = \dfrac{P}{Am}\sqrt{1 - S^2} = \text{const}$ for $\sin\varphi = \text{const}$. Equation (3.1, 19) leads to

$$\Omega_\varphi^2 = \tfrac{1}{2}\left(\omega_{\mathrm{I}}^2 + \omega_{\mathrm{II}}^2\right) \tag{3.1, 20}$$

where ω_I, ω_{II} are again the ω coordinates of the points of intersection of the $A^*(\omega)$ curve and the resonance curve $A(\omega)$. Relations applicable to other types of damping can be derived in a similar way.

For the specified types of damping, Table 3.1, 1 shows the equations applicable to the case of a constant excitation amplitude, Table 3.1, 2 those used in connection with an excitation amplitude proportional to ω^2.

Table 3.1, 1

Type of damping	Applicable equation
$\varkappa(|\dot{y}|)\,\dot{y}$	(3.1, 17)
$\varkappa(|y|)\,\mathrm{sgn}\,\dot{y}$	(3.1, 20)

Table 3.1, 2

Type of damping	Applicable equation
$\varkappa_0\dot{y}$	(3.1, 17)
$\varkappa_0|\dot{y}|\,\dot{y}$	(3.1, 17)
$\varkappa_0|y|\,\mathrm{sgn}\,\dot{y}$	(3.1, 20)
$\varkappa_0 = \mathrm{const}$	

It has been assumed in the foregoing that the damping and the restoring force characteristics are both symmetric, and that the system is not loaded with a constant force. An analysis will now be made of a system described by the differential equation

$$m\ddot{y} + f(y)\,y + h(y) + [\varkappa(\dot{y}, y) + \gamma(\dot{y}, y)]\,\dot{y} = \begin{cases} P_0 + P\cos\omega t \\ P_0 + m_0\varepsilon\omega^2\cos\omega t \end{cases} \tag{3.1, 21}$$

where m is the mass, P_0 the constant load; the term $\bigl(f(y)\,y + h(y)\bigr)$ represents the restoring force, i.e. the symmetric and the non-symmetric part of the characteristic because the functions $f(y)$ and $h(y)$ satisfy the relations

$$f(-y) = f(y)\,, \qquad h(-y) = h(y) \tag{3.1, 22}$$

Similarly, the functions $\varkappa(\dot{y}, y)\,\dot{y}$ and $\gamma(\dot{y}, y)\,\dot{y}$ represent the symmetric and the non-symmetric terms of the damping and satisfy the relations

$$\varkappa(-\dot{y}, y) = \varkappa(\dot{y}, -y) = \varkappa(\dot{y}, y)\,, \tag{3.1, 23}$$

$$\gamma(-\dot{y}, y) = -\gamma(\dot{y}, y)\,, \qquad \gamma(\dot{y}, -y) = \gamma(\dot{y}, y)\,. \tag{3.1, 24}$$

Assuming (3.1, 21) to be quasi-linear, the stationary solution can be approximated by

$$y = Y + A\cos(\omega t - \varphi)\,. \tag{3.1, 25}$$

Just as in the preceding case, the introduction of the time shift $t_0 = \varphi/\omega$ where φ is the phase shift makes it possible to approximate the stationary solution by

$$y = Y + A\cos\omega t \tag{3.1, 25a}$$

(the term $\cos \omega t$ in (3.1, 21) having been replaced by $\cos (\omega t + \varphi)$). Using the harmonic balance method, the following equations are obtained for determining Y, A and φ:

$$F_0(Y, A)\, Y + H(Y, A) + G(Y, A, \omega A)\, \omega A = P_0 \,, \qquad (3.1, 26)$$

$$[F(Y, A) - m\omega^2]\, A = \begin{cases} P \cos \varphi \\ m_0 \varepsilon \omega^2 \cos \varphi \end{cases}, \qquad (3.1, 27)$$

$$\omega A K(Y, A, \omega A) = \begin{cases} P \sin \varphi \\ m_0 \varepsilon \omega^2 \sin \varphi \end{cases} \qquad (3.1, 28)$$

where

$$Y F_0(Y, A) = \frac{\omega}{2\pi} \int\limits_0^{2\pi/\omega} f(AY + A \cos \omega t)\,(Y + A \cos \omega t)\, dt \,,$$

$$H(Y, A) = \frac{\omega}{2\pi} \int\limits_0^{2\pi/\omega} h(Y + A \cos \omega t)\, dt \,,$$

$$G(Y, A, \omega A) = -\frac{\omega}{2\pi} \int\limits_0^{2\pi/\omega} \gamma(-\omega A \sin \omega t,\, Y + A \cos \omega t) \sin \omega t\, dt \,,$$

$$Y F(Y, A) = \frac{\omega}{\pi} \int\limits_0^{2\pi/\omega} f(Y + A \cos \omega t)\,(Y + A \cos \omega t) \cos^2 \omega t\, dt \,,$$

$$K(Y, A, \omega A) = \frac{\omega}{\pi} \int\limits_0^{2\pi/\omega} \varkappa(-\omega A \sin \omega t,\, Y + A \cos \omega t) \sin^2 \omega t\, dt \,.$$

In view of (3.1, 22) to (3.1, 24), the following equations will apply:

$$\int\limits_0^{2\pi/\omega} f(Y + A \cos \omega t) \cos \omega t \sin \omega t\, dt = 0 \,,$$

$$\int\limits_0^{2\pi/\omega} h(Y + A \cos \omega t) \cos \omega t\, dt = 0 \,, \qquad \int\limits_0^{2\pi/\omega} h(Y + A \cos \omega t) \sin \omega t\, dt = 0 \,,$$

$$\int\limits_0^{2\pi/\omega} \varkappa(-\omega A \sin \omega t,\, Y + A \cos \omega t) \sin \omega t\, dt = 0 \,,$$

$$\int\limits_0^{2\pi/\omega} \varkappa(-\omega A \sin \omega t,\, Y + A \cos \omega t) \sin \omega t \cos \omega t\, dt = 0 \,,$$

$$\int\limits_0^{2\pi/\omega} \gamma(-\omega A \sin \omega t,\, Y + A \cos \omega t) \sin^2 \omega t\, dt = 0 \,,$$

$$\int\limits_0^{2\pi/\omega} \gamma(-\omega A \sin \omega t,\, Y + A \cos \omega t) \sin \omega t \cos \omega t\, dt = 0$$

as well as

$$F_0(-Y, A) = F_0(Y, -A) = F_0(Y, A) \,, \tag{3.1, 29}$$

$$H(-Y, A) = H(Y, -A) = H(Y, A) \,, \tag{3.1, 30}$$

$$\left. \begin{aligned} G(-Y, A, \omega A) &= G(Y, -A, \omega A) = G(Y, A, \omega A) \,, \\ G(Y, A, -\omega A) &= -G(Y, A, \omega A) \,, \end{aligned} \right\} \tag{3.1, 31}$$

$$K(-Y, -A, \omega A) = K(Y, A, -\omega A) = K(Y, A, \omega A) \,. \tag{3.1, 32}$$

With $\cos \varphi = 0$ substituted in (3.1, 27), the backbone curve is defined by the equations

$$\left. \begin{aligned} F_0(Y_{\rm S}, A_{\rm S}) \, Y_{\rm S} + H(Y_{\rm S}, A_{\rm S}) + G(Y_{\rm S}, A_{\rm S}, \omega A_{\rm S}) \, \omega A_{\rm S} &= P_0 \,, \\ F(Y_{\rm S}, A_{\rm S}) &= m\omega^2 \end{aligned} \right\} \tag{3.1, 33}$$

and the limit envelope by the equations

$$F_0(Y_{\rm L}, A_{\rm L}) \, Y_{\rm L} + H(Y_{\rm L}, A_{\rm L}) + G(Y_{\rm L}, A_{\rm L}, \omega A_{\rm L}) \, \omega A_{\rm L} = P_0 \,,$$

$$\omega A_{\rm L} K(Y_{\rm L}, A_{\rm L}, \omega A_{\rm L}) = \begin{cases} P \\ m_0 \varepsilon \omega^2 \end{cases} \tag{3.1, 34}$$

(subscripts L, S refer to the backbone curve and the limit envelope, respectively). If $K(Y_{\rm L}, A_{\rm L}, \omega A_{\rm L}) > 0$ does not apply in some ranges of $Y_{\rm L}, A_{\rm L}$, the minus sign must also be considered in the right-hand side of the last of equations (3.1, 34).

As in the case of systems with symmetric characteristics of the restoring force and damping, the limit envelope is not altered by changes of the mass. It is, however, in general no longer independent of changes in the restoring force non-linearity.

Condition (3.1, 32) implies the validity of the equation

$$(A_{\rm L})_P = -(A_{\rm L})_{-P} \quad \text{or} \quad (A_{\rm L})_\varepsilon = -(A_{\rm L})_{-\varepsilon} \,. \tag{3.1, 35}$$

On the strength of this one can deduce the following:

Upon determining the two extreme values of the deflection y ($[y]_{\max}$, $[y]_{\min}$) as functions of the excitation frequency for various masses, and drawing the corresponding limit envelopes ($[y]_{\rm L\,max}$, $[y]_{\rm L\,min}$) (Fig. 3.1, 6), one finds that, within the scope of the present approximate approach, the following equations apply:

$$A_{\rm L} = \tfrac{1}{2} \left([y]_{\rm L\,max} - [y]_{\rm L\,min} \right) , \tag{3.1, 36}$$

$$Y_{\rm L} = \tfrac{1}{2} \left([y]_{\rm L\,max} + [y]_{\rm L\,min} \right) . \tag{3.1, 37}$$

If function $\varkappa(\dot{y}, y)$ is not a function of the deflection or if it depends solely on the deflection amplitude ($\varkappa(\dot{y}, \text{amp } y)$) function K is only a function of $A_{\rm L}$ and the limit envelope depends neither on the constant load nor on the restoring force non-linearity. Consequently, the course of the limit envelope $A_{\rm L}(\omega)$ can be used for identification of the symmetric component of damping (see Section 3.6 for further details). Identification of the non-symmetric component of damping is more complicated by far. The considerations which follow should aid in obtaining information concerning the presence or absence of the non-symmetric component of damping on the basis of experimental results.

The simplest of all is the case of zero constant load and a symmetric restoring force characteristic, i.e. $P_0 = 0$ and $h(y) = 0$, when the first of equations (3.1, 34) becomes

$$F_0(Y_{\rm L}, A_{\rm L}) \, A_{\rm L} + G(Y_{\rm L}, A_{\rm L}, \omega A_{\rm L}) = 0 \,.$$

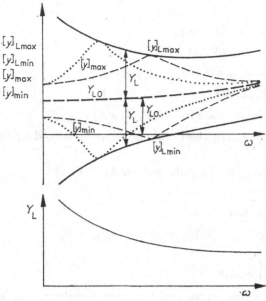

Fig. 3.1, 6

If, according to experimental results, $Y_L = 0$ in the whole range of ω, it must necessarily be

$$G(Y_L, A_L, \omega A_L) = 0$$

and hence also

$$\gamma(\dot{y}, y) = 0 \ .$$

If for a constant load P_0 and $-P_0$

$$(Y_L)_{P_0} = -(Y_L)_{-P_0} \qquad\qquad (3.1, 38)$$

in the whole range of ω, then, as (3.1, 29) and (3.1, 31) imply,

$$H(Y_L, A_L) + G(Y_L, A_L, \omega A_L) \equiv 0 \qquad\qquad (3.1, 39)$$

for all ω. This identity cannot be satisfied in the whole range of ω except for $H(Y_L, A_L) = 0$ and $G(Y_L, A_L, \omega A_L) = 0$. This means that $h(y) = 0$, $\gamma(\dot{y}, y) = 0$. Since it is easy to determine whether or not $h(y) = 0$, equation (3.1, 38) is a reliable guide to establishing the presence (or absence) of damping asymmetry.

It may be of interest, in concluding the discussion, to set down some of the equations applicable to a simple system harmonically excited by a force with a constant amplitude for which

$$f(y) = c = \text{const}\ , \qquad h(y) = 0\ , \qquad P_0 = 0\ , \qquad \varkappa = \varkappa(\dot{y})\ , \qquad \gamma = \gamma(\dot{y})\ .$$
$$(3.1, 40)$$

The limit envelope of such a system is described by the equations

$$c Y_L + G(\omega A_L)\, \omega A_L = 0\ , \qquad \omega A_L K(\omega A_L) = P\ . \qquad\qquad (3.1, 41)$$

Considering the accuracy of the approximate solution (3.1, 25), one can write

$$V_L = \omega A_L \qquad\qquad (3.1, 42)$$

where V_L is the limit envelope of the velocity resonance curves. On the strength of (3.1, 42), equations (3.1, 41) can be written in the form

$$cY_L + G(V_L) V_L = 0 , \qquad V_L K(V_L) = P .$$ \hfill (3.1, 41 a)

As the second equation implies,

$$V_L = \Phi(P)$$ \hfill (3.1, 43)

and the $A_L(\omega)$ curve is a rectangular hyperbola. For a definite value of P, V_L is independent of ω; consequently, Y_L is also independent of ω and constant. Different values of Y_L, however, are obtained for different values of P.

In the case of the excitation amplitude proportional to ω^2, (3.1, 43) no longer applies and Y_L as well as V_L depend on ω.

What remains is to ascertain how well the equations derived above on the basis of the approximate solution, describe the real behaviour of the system. So far as the limit envelopes are concerned, this problem was investigated by SVAČINA and FIALA (1980) who, for a number of systems, compared the limit envelope determined analytically by application of the theory presented above, with the limit envelope of a set of resonance curves obtained by analogue modelling for a gradually varying mass. They found the quantitative difference between the analytic and the real limit envelope to be very small for systems with symmetric characteristics of the restoring force and damping (a Duffing system with hard and soft characteristics of the restoring force was solved for different types of damping), and somewhat larger for systems with non-symmetric characteristics, especially those with non-symmetric damping. The fact that no qualitative difference exists between the approximate analytic and the analogue computer results, is very important from the point of view of other applications of the limit envelopes, particularly that of damping identification.

3.2. Use of auxiliary curves for preliminary analysis

It has been found expedient in investigations of various systems to obtain first the backbone curve and the limit envelope, and to form, on their basis, a rough idea of the system's behaviour before undertaking a detailed analysis. Thus, for example, knowledge of the course of the limit envelope and the backbone curve enables the analyst to estimate the height of the resonant peak and the form of the resonance curve and to determine whether the phase angle for the maximum amplitude is greater or smaller than 90°. If the limit envelope is a decreasing function of increasing frequency ω, the amplitude reaches its maximum value for $\varphi < \pi/2$; in the case of a softening spring characteristic, this maximum is larger than for a hardening spring. The reverse happens when the limit envelope is an increasing function of increasing frequency ω. It is, therefore, incorrect and naive to assume that the resonance amplitude will always be reduced by the use of a non-linear spring.

The backbone curve and the limit envelope of non-linear systems are sometimes apt to have more than one point of intersection, and a resonance curve which is no longer a simple continuous line but consists of several branches. A case of this sort is shown in Fig. 3.2, 1 (the limit envelope is an increasing function of excitation frequency ω and the backbone curve changes twice its sense of curvature). Systems in which the backbone curve is not unique in a certain interval of the excitation frequency, include those in which the spring characteristic changes from softening to

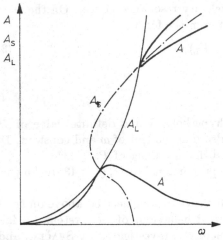

Fig. 3.2, 1

hardening as the deflection increases, and those with a hardening spring characteristic additionally loaded with a constant force. A limit envelope which is not unique in the whole range of the excitation frequency and hence has more than one points of intersection with the backbone curve, belongs, for example, to systems for which the condition $K(\omega A, A) > 0$ is not satisfied for all A. These systems, however, are counted in the class of self-excited systems in which the interaction of self-excitation and parametric excitation must be taken in consideration. They will be treated in greater detail in Chapter 5.

The courses of the limit envelope and the backbone curve or the equation describing them make it possible to decide whether or not a specified damping is capable of ensuring a limited resonance amplitude. As an example, consider the Duffing system with a progressive characteristic, excited by a harmonic force having an amplitude proportional to the square of the excitation frequency and with linear viscous damping. This system is governed by the differential equation

$$\ddot{y} + \varkappa\dot{y} + y + \varepsilon y^3 = \omega^2 \cos \omega t \qquad (3.2, 1)$$

where $\varepsilon > 0$, $\varkappa > 0$. In this case the limit envelope is a straight line passing through the origin of the coordinates, described by the equation

$$A_{\mathrm{L}} = \frac{\omega}{\varkappa} ; \qquad (3.2, 2)$$

the backbone curve is defined by the equation

$$\omega^2 = 1 + \tfrac{3}{4}\varepsilon A_S^2 . \qquad (3.2, 3)$$

The coordinate A of the point of intersection of the two curves is given by the equation

$$A^2 = (\varkappa^2 - \tfrac{3}{4}\varepsilon)^{-1} ;$$

as this equation implies, the condition of existence of a point of intersection and in turn, a finite value of amplitude A is the fulfilment of the inequality

$$\varkappa^2 > \tfrac{3}{4}\varepsilon . \qquad (3.2, 4)$$

As another example, consider a system with dry friction and damping defined by the function $\varkappa|y|$ sgn $\dot{y}$. This system is defined by the differential equation

$$\ddot{y} + y + (\vartheta + \varkappa\,|y|)\,\text{sgn}\,\dot{y} = P\cos\omega t \qquad (3.2,\,5)$$

and its limit envelope, by the equation

$$A_{\mathrm{L}} = \frac{\dfrac{\pi}{2}P - 2\vartheta}{\varkappa}. \qquad (3.2,\,6)$$

So long as $\dfrac{\pi}{2}P - 2\vartheta > 0$, the limit envelope is a straight line parallel to the ω axis, i.e. independent of frequency. It follows from this fact that for a damping which is independent of frequency and a constant-amplitude excitation, the maximum resonance amplitude is independent of the spring characteristic as well as the magnitude of the mass; the phase angle at resonance is always $\varphi = \pi/2$. For $\varkappa \to 0$, $\lim\limits_{\varkappa \to 0} A_{\mathrm{L}} = \infty$, i.e. dry friction alone cannot ensure a limited amplitude at resonance so long as the condition

$$\frac{\pi}{4}P > \vartheta \qquad (3.2,\,7)$$

is satisfied. Failure to satisfy this inequality means that the excitation force amplitude is so small as to be unable to make the system vibrate even in resonance.

3.3. Use of auxiliary curves for preliminary analysis of parametrically excited systems

This section will show that the backbone curve and the limit envelope are also very convenient tools of analysis of parametrically excited systems. The way in which these curves are applied to such systems will be illustrated by way of the following example:

Consider a system with a periodically variable stiffness described by a non-linear differential equation of the Mathieu type, viz.

$$m\ddot{y} + \varkappa(|\dot{y}|,\,|y|)\,\dot{y} + [c(1 + \mu\cos 2\omega t) + f(y)]\,y = 0 \qquad (3.3,\,1)$$

where m is the mass of the system, $\varkappa(|\dot{y}|,\,|y|)\,\dot{y}$ — the damping force, c — the mean value of the restoring force stiffness, and the function $f(y)\,y$ — the non-linearity of the restoring force. To simplify, assume that $f(-y) = f(y)$. For convenience, (3.3, 1) is usually rearranged so as to make the coefficients dimensionless. As in Section 3.1, no such rearrangement will be made here in order to show clearly the effect of the various parameters of the system.

The solution is sought in the *main resonance* (in the interval of instability of the first-order trivial solution), i.e. in the vicinity of $\omega = \sqrt{c/m}$. Introducing the shift of the origin of time, φ/ω, the term $\cos 2\omega\,t$ is replaced by $\cos 2(\omega t - \varphi)$ and the solution approximated by

$$y = A\cos\omega t\,. \qquad (3.2,\,2)$$

Using the approach outlined in Section 3.1, the following equations are obtained for the determination of A and φ:

$$A[-m\omega^2 + c(1 + \tfrac{1}{2}\mu \cos 2\varphi) + F(A)] = 0 \,,$$

$$A[-\omega K(\omega A, A) + \tfrac{1}{2}\mu c \sin 2\varphi)] = 0 \; ; \tag{3.3, 3}$$

functions $F(A)$ and $K(\omega, A, A)$ are derived from $f(y)$ and $\varkappa(|\dot{y}|, |y|)$. In a certain interval of ω, the equations can, in addition to a trivial solution, have a non-trivial solution given by the equations

$$m\omega^2 - c - F(A) = \tfrac{1}{2}\mu c \cos 2\varphi \,, \tag{3.3, 4}$$

$$\omega K(\omega, A, A) = \tfrac{1}{2}\mu c \sin 2\varphi \,. \tag{3.3, 5}$$

On eliminating φ (by squaring and adding (3.3, 4) and (3.3, 5)) one obtains the following equation for determining the dependence $A = A(\omega)$:

$$(\omega^2 - \Omega^2)^2 + \left[\frac{\omega}{m} K(\omega A, A)\right]^2 = \left(\frac{1}{2}\mu\omega_0^2\right)^2 \; ; \tag{3.3, 6}$$

in the above

$$\omega_0^2 = \frac{c}{m} \,, \qquad \Omega^2 = \frac{1}{m}[c + F(A)]$$

and φ is obtained from the equation

$$\tan \varphi = \frac{\dfrac{\omega}{m} K(\omega A, A)}{\dfrac{1}{2}\mu\omega_0^2 - \Omega^2 + \omega^2} \,. \tag{3.3, 7}$$

As (3.3, 4) and (3.3, 5) also imply, in the interval $(0, 2\pi)$ two values of the angle φ (i.e. φ and $\pi + \varphi$) belong to every value of A. For $\cos 2\varphi = 0$ (or $\mu = 0$), (3.3, 4) yields the following equation for the determination of the backbone curve $A_\mathrm{S} = A_\mathrm{S}(\omega)$:

$$\omega = \left\{\frac{1}{m}[c + F(A_\mathrm{S})]\right\}^{1/2} \,. \tag{3.3, 8}$$

Equation (3.3, 8) is wholly analogous to that for the backbone curve of systems excited by an external force.

The limit envelope is obtained from (3.3, 5) by putting $\sin 2\varphi = 1$ (on the assumption that $K(\omega A, A) \geqq 0$ for any ω, A); this is the extreme value of the right-hand side of (3.3, 5), viz.

$$\omega K(\omega A, A) = \tfrac{1}{2}\mu c \,. \tag{3.3, 9}$$

Equation (3.3, 9) makes it possible to determine the dependence $A_\mathrm{L} = A_\mathrm{L}(\omega)$. The resonance curve can only lie on one side of the limit envelope. As in systems excited by an external force, the limit envelope is independent of the mass of the system and the spring non-linearity but depends (unlike in those systems) on the stiffness c and the coefficient μ, which is the amplitude of the parametric excitation.

The various points of the resonance curve can also be obtained as the points of intersection of a set of curves described by (3.3, 4) and (3.3, 5) for a particular φ. The right-hand side of (3.3, 5) is the same for both φ and $\pi/2 - \varphi$. In (3.3, 4), on the other hand, the sign of $\cos 2\varphi$ changes if φ is replaced by $\pi/2 - \varphi$. For specified values of φ and $\pi/2 - \varphi$, the corresponding points of the resonance curve can thus be obtained as the points of intersection of the curves

$$\omega^2 = \Omega^2 \pm \tfrac{1}{2}\,\mu\omega_0^2 \cos 2\varphi \qquad\qquad (3.3,\,10)$$

and the curve

$$\frac{\omega}{m}\,K(\omega, A, A) = \frac{1}{2}\,\mu\omega_0^2 \sin 2\varphi \;. \qquad\qquad (3.3,\,11)$$

Denoting by Ω_φ the coordinate ω of the point of the backbone curve which also lies on the curve defined by (3.3, 11), one can write the equation

$$\omega^2 = \Omega_\varphi^2 \pm \tfrac{1}{2}\,\mu\omega_0^2 \cos 2\varphi$$

from which it follows that

$$\Omega_\varphi^2 = \tfrac{1}{2}\,(\omega_I^2 + \omega_{II}^2)\;. \qquad\qquad (3.3,\,12)$$

As in section 3.1, ω_I, ω_{II} denote the coordinates ω of the points of intersection of the limit envelope and the resonance curve. Fig. 3.3, 1 shows schematically the dependence of the amplitude A in parametric resonance on ω (heavy solid lines — the stable solution, dashed lines — the unstable solution); the backbone curve is drawn in dot-and-dash lines, the curve defined by (3.3, 11) for a particular value $\sin 2\psi = $ const, in light solid lines.

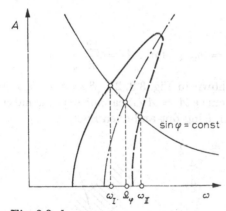

Fig. 3.3, 1

Equation (3.3, 12) is identical with (3.1, 20) and, so far as parametric excitation is concerned, applies to a broad class of damping.

As an example, consider a system which (after rearrangement to the dimensionless form of the linearized part of the equation) is described by the following differential equation:

$$\ddot{y} + \delta y^2 \dot{y} + \vartheta\,\mathrm{sgn}\,\dot{y} + [1 + \mu \cos 2(\omega t - \varphi)]\,y + \gamma y^3 = 0\;. \qquad\qquad (3.3,\,13)$$

Substituting the approximate solution (3.3, 2) and comparing the coefficients of $\cos \omega t$ and $\sin \omega t$ results in the following equations for determining A and φ:

$$\omega^2 - (1 + \tfrac{3}{4}\,\gamma A^2) = \tfrac{1}{2}\,\mu \cos 2\varphi\;, \qquad\qquad (3.3,\,14)$$

$$\frac{1}{4}\,\delta\omega A^2 + \frac{4}{\pi}\,\vartheta\,\frac{1}{A} = \frac{1}{2}\,\mu \sin 2\varphi\;. \qquad\qquad (3.3,\,15)$$

The amplitude in parametric resonance, A, as a function of ω is obtained from the equation

$$(\omega^2 - \Omega^2)^2 + \left(\frac{1}{4}\delta\omega A^2 + \frac{4}{\pi}\vartheta\frac{1}{A}\right)^2 = \frac{1}{4}\mu^2 \tag{3.3, 16}$$

where

$$\Omega^2 = 1 + \tfrac{3}{4}\gamma A^2 .$$

In the case of $\vartheta = 0$, equation (3.3, 16) is biquadratic in ω; hence the dependence of ω on A and in turn $A = A(\omega)$ are readily determined. The dependence $\varphi = \varphi(\omega)$ is obtained from the equation

$$\tan\varphi = \left[\frac{1}{4}\delta\omega A^2 + \frac{4}{\pi}\frac{\vartheta}{A}\right]\left(\frac{1}{2}\mu - \Omega^2 + \omega^2\right)^{-1}, \tag{3.3, 17}$$

the backbone curve from the equation

$$\omega = \Omega = (1 + \tfrac{3}{4}\gamma A_{\rm S}^2)^{1/2} \tag{3.3, 18}$$

and the limit envelope from the equation

$$\omega = \frac{4}{\delta A_{\rm L}^2}\left(\frac{1}{2}\mu - \frac{4}{\pi}\frac{\vartheta}{A_{\rm L}}\right). \tag{3.3, 19}$$

Since

$$\lim_{A_{\rm L}\to 0}(\omega) = \infty , \qquad \lim_{A_{\rm L}\to\infty}(\omega) = 0 ,$$

$$\omega = 0 \quad \text{for} \quad A_{\rm L} = \frac{8}{\pi}\frac{\vartheta}{\mu} , \qquad \omega = \omega_{\max} \quad \text{for} \quad A_{\rm L} = \frac{12}{\pi}\frac{\vartheta}{\mu}$$

the limit envelope looks like the curve shown in Fig. 3.3, 2a ($\vartheta \neq 0$) or that in Fig. 3.3, 2b ($\vartheta = 0$). The region in which the curve $A = A(\omega)$ cannot exist, is hatched in. The backbone curves for $\gamma > 0$ are drawn using dot-and-dash lines.

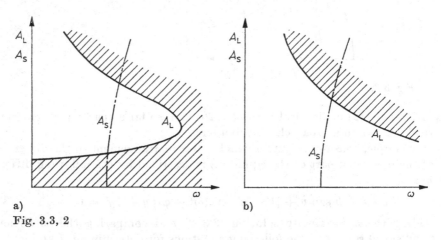

a)   b)

Fig. 3.3, 2

The course of the auxiliary curves enables the analyst to draw some interesting conclusions. For systems without dry friction ($\vartheta = 0$), provided that there exists an interval of instability of the trivial solution, the first-order parametric resonance always arises spontaneously because the resonance curve $A(\omega)$ must start from the ω axis.

For systems with dry friction ($\vartheta \neq 0$), the trivial solution is stable in the whole range of ω and parametric resonance can arise if there exists a point of intersection of the backbone curve and the limit envelope; the process, however, is not spontaneous but takes place only on application of definite initial conditions. If the backbone curve does not intersect the limit envelope but lies whole outside the region of the possible points of the resonance curve (in the hatched region in Fig. 3.3, 2a), parametric vibration cannot arise at all. The course of the limit envelope also provides information as to which spring characteristic (softening or hardening) is the more advantageous for the system in question. The maximum amplitude of the system being discussed here is smaller for $\gamma > 0$ than for $\gamma < 0$. Fig. 3.3, 3a shows the curves $A(\omega)$, $A_L(\omega)$, $A_S(\omega)$ for $\vartheta = 0$, $\delta = 0.01$, $\mu = 0.2$ and $\gamma = 0.01$, Fig. 3.3, 3b, the curve $\varphi(\omega)$. Figs. 3.3, 4a, b show these curves for the same value of δ, μ and γ but for $\vartheta = 0.1$. The sections of the curves $A(\omega)$ and $\varphi(\omega)$ to which corresponds an unstable solution (for small disturbances) are drawn in dashed lines. As seen in the figures, the dependence $A = A(\omega)$ forms a closed curve. If non-trivial solutions of A exist for a definite ω,

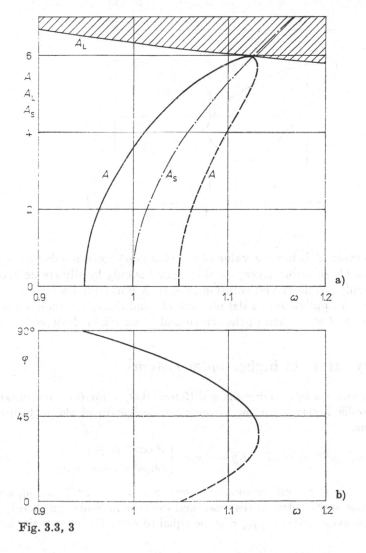

Fig. 3.3, 3

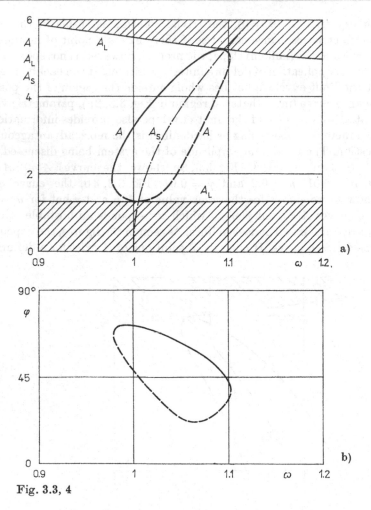

Fig. 3.3, 4

then one value (to every A belongs a value of φ and $\pi + \varphi$) corresponds to a stable, the other to an unstable solution. Every stable solution is only locally stable because there exists the locally stable trivial solution $A = 0$. A non-trivial solution cannot be obtained except on application of definite initial conditions, i.e. there exist, for certain ω the domains of attraction of the trivial and non-trivial solutions.

3.4. Auxiliary curves of higher-order systems

Consider a system governed by a higher-order differential equation (i.e. one containing derivatives of an order higher than the second; the derivative of the n-th order is denoted by $y^{(n)}$), viz.

$$f_{2n}y^{(2n)} + f_{2n-1}y^{(2n-1)} + \ldots + f_1\dot{y} + f_0y = \begin{cases} P \cos{(\omega t + \varphi)} \\ \varepsilon m_0 \omega^2 \cos{(\omega t + \varphi)} \end{cases}, \qquad (3.4, 1)$$

The coefficients of (3.4, 1) are either constants or functions of $|y^{(2n)}|, \ldots, |y|$, and as before, φ is the phase angle between response and excitation. Equation (3.4, 1) does not have to be of an even order, i.e. f_{2n} may be equal to zero. The even order is used here to facilitate formal expression. The stationary solution at resonance is approximated by

$$y = A \cos \omega t .$$ (3.4, 2)

Using the harmonic balance method, the following equations are obtained for determining A, φ:

$$A[(-1)^n \omega^{2n} F_{2n} + (-1)^{n-1} \omega^{2(n-1)} + \ldots - \omega^2 F_2 + F_0] = \begin{cases} P \cos \varphi \\ \varepsilon m_0 \omega^2 \cos \varphi \end{cases},$$ (3.4, 3)

$$A[(-1)^n \omega^{2n-1} F_{2n-1} + (-1)^{n-1} \omega^{2n-3} F_{2n-3} + \ldots + \omega F_1] = \begin{cases} P \sin \varphi \\ \varepsilon m_0 \omega^2 \sin \varphi \end{cases}$$ (3.4, 4)

The coefficients $F_{2n}, \ldots, F_0$ are generally functions of $A, \omega A, \ldots, \omega^{2n} A^{2n}$ which are established from functions $f_{2n}, \ldots, f_0$ by the averaging process.

For $\cos \varphi = 0$, the equation

$$(-1)^n \omega^{2n} F_{2n} + \ldots + F_0 = 0$$ (3.4, 5)

defines the backbone curve which, as will be shown later, may consist of several branches. The equation

$$A[(-1)^n \omega^{2n-1} F_{2n-1} + \ldots + \omega F_1] = \begin{cases} \pm P \\ \pm \varepsilon m_0 \omega^2 \end{cases}$$ (3.4, 6)

defines the limit envelope which is also apt to consist of several branches. If the left-hand side of (3.4, 6) is positive for all A, only the plus sign on the right-hand side makes sense. Hence the resonance curve $A(\omega)$ must lie only in the region between the ω axis and the limit envelope and touch the latter at its points of intersection with the backbone curve. The method of solution is outlined below.

Discrete mechanical systems are usually described by a set of second-order differential equations. In some cases it is expedient — if it can be done at all — to convert the set to a single differential equation of higher order. The following example will show how to proceed in such cases.

A mechanical system containing two masses, m_1, m_2 (Fig. 3.4, 1) with linear viscous damping of the motion of mass m_2 is excited by a harmonic force $\omega^2 \cos \omega t$. The spring

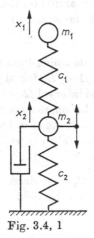

Fig. 3.4, 1

connecting the two masses is linear, with a stiffness coefficient c_1. The other spring is progressive non-linear, and its restoring force is expressed by a linear term with coefficient c_2 and a cubic term. After rearrangement to the dimensionless form, the equations describing the system are:

$$x_1' + x_1 - x_2 = 0 \,,$$

$$x_2' - M(x_1 - x_2) + \varkappa x_2' + q^2 x_2(1 + \varepsilon x_2^2) = \omega^2 \cos \omega t \qquad (3.4, 7)$$

where $M = m_1/m_2$ is the ratio of the masses, $q = \left(\dfrac{c_2/m_2}{c_1/m_1}\right)^{1/2}$ the tuning coefficient of the restoring force linear component, $\varkappa$ — the damping coefficient, and ε — the coefficient of the non-linear term of the spring characteristic.

The first of (3.4, 7) gives

$$x_2 = x_1' + x_1 \,. \qquad (3.4, 8)$$

Substituting (3.4, 8) in the second of (3.4, 7) and using, as before, the shift of the time origin results in the following single fourth-order differential equation

$$x_1^{IV} + \varkappa x_1''' + \{1 + M + q^2[1 + \varepsilon(x_1'' + x_1)^2]\}\, x_1'' + \varkappa_1 x_1'$$
$$+ q^2[1 + \varepsilon(x_1'' + x_1)^2]\, x_1 = \omega^2 \cos(\omega t + \varphi) \,. \qquad (3.4, 9)$$

If the stationary solution is sought in the form of (3.4, 2) (i.e. $x_1 = X_1 \cos \omega t$ in the case being analyzed), equations (3.4, 3) and (3.4, 4) take the form

$$X_1[\omega^4 - (1 + M + q^2)\,\omega^2 + q^2 + \tfrac{3}{4}\varepsilon q^2(1 - \omega^2)^3\, X_1^2] = \omega^2 \cos \varphi \,, \quad (3.4, 10)$$

$$\varkappa \omega X_1(1 - \omega^2) = \omega^2 \sin \varphi \,. \qquad (3.4, 11)$$

X_1 and φ are obtained from the equations

$$X_1^2\{[\omega^4 - (1 + M + q^2)\,\omega^2 + q^2 + \tfrac{3}{4}\varepsilon q^2(1 - \omega^2)^3\, X_1^2]^2 + [\varkappa \omega(1 - \omega^2)]^2\} = \omega^4 \,, \qquad (3.4, 12)$$

$$\tan \varphi = \varkappa \omega(1 - \omega^2)\,[\omega^4 - (1 + M + q^2)\,\omega^2 + q^2 + \tfrac{3}{4}\varepsilon q^2(1 - \omega^2)^3\, X_1^2]^{-1} \,. \qquad (3.4, 13)$$

The backbone curve is defined by the equation

$$X_{1S}^2 = \tfrac{4}{3}\,[\varepsilon q^2(1 - \omega^2)^3]^{-1}\,[\omega^4 - (1 + M + q^2)\,\omega^2 + q^2] \,. \qquad (3.4, 14)$$

In the case being considered, the curve has two branches, both starting on the ω axis. The coordinate ω of the starting points is given by the roots of the equation

$$\omega^4 - (1 + M + q^2)\,\omega^2 + q^2 = 0 \,, \qquad (3.4, 15)$$

i.e. by the natural frequencies of the abbreviated (linearized, damping-less) system. Denoting these frequencies by Ω_1, Ω_2 and considering that $\Omega_1 < 1 < \Omega_2$ for any M, q, it is clear that for $\varepsilon > 0$, the real values of X_{1S} are obtained in the interval $\Omega_1 < \omega < 1$, $\Omega_2 < \omega$. As ω is increased from Ω_1 to 1, X_{1S} (the lower branch) becomes a continually increasing function of ω and grows beyond all bounds for $\omega \to 1$. As ω is increased from Ω_2 (the upper branch), X_{1S} first increases in a certain interval of ω, reaches its maximum and then becomes a decreasing function because $\lim\limits_{\omega \to \infty} X_{1S} = 0$.

For $\varepsilon < 0$, X_{1S} is defined in the intervals $\omega < \Omega_1$, $1 < \omega < \Omega_2$. For decreasing ω, $\lim\limits_{\omega \to 1} X_{1S} = \infty$ for the lower branch, and $\lim\limits_{\omega \to 0} X_{1S} = \infty$ for the upper branch. X_{1S} is an increasing function of decreasing ω.

Recalling that only the positive values of X_1 have a meaning, the equation of the limit envelope can be written as follows:

$$X_{1L} = \omega(\varkappa \, |\omega^2 - 1|)^{-1} \,. \tag{3.4, 16}$$

Clearly, the course of the limit envelope depends only on the damping coefficient $\varkappa$ and the type of excitation, and is independent of other parameters of the system. It should be stressed that it is independent of the non-linearity coefficient ε. It is found that

$$\lim_{\omega \to 0} X_{1L} = 0 \,, \qquad \lim_{\omega \to 1} X_{1L} = \infty \,, \qquad \lim_{\omega \to \infty} X_{1L} = 0 \,. \tag{3.4, 17}$$

Using (3.4, 8) one can write

$$x_2 = (1 - \omega^2) \, x_1 \tag{3.4, 18}$$

and

$$X_2 = (1 - \omega^2) \, X_1 \,. \tag{3.4, 19}$$

Let us now examine the backbone curve and the limit envelope corresponding to vibration in the x_2 coordinate. The backbone curve is defined by the equation

$$X_{2S}^2 = \tfrac{4}{3} \, [\varepsilon q^2 (1 - \omega^2)]^{-1} \, [\omega^4 - (1 + M + q^2) \, \omega^2 + q^2] \,. \tag{3.4, 20}$$

The definition intervals are the same as for X_{1S}; for $\varepsilon > 0$, however, X_{2S} increases with increasing ω for both branches. For $\varepsilon < 0$, $\lim\limits_{\omega \to 0} X_{2S} = 0$ and $\lim\limits_{\omega \to 1} X_{2S} = \infty$ as ω decreases.

The equation of the limit envelope is

$$X_{2L} = \omega/\varkappa \,, \tag{3.4, 21}$$

i.e. as in the case of one-mass systems with linear viscous damping, (Section 3.1), the limit envelope is a straight line.

The stability for small disturbances can be established by analyzing the variational equation. The equation used for the purpose is the linearized equation obtained after substituting $\bar{x}_1 + u$ for x_1 in (3.4, 9) and neglecting the non-linear terms. $\bar{x}_1$ is the solution of (3.4, 9), i.e. $\bar{x}_1 = X_1 \cos \omega t$ in the case being considered. The differential equation thus obtained is

$$u^{IV} + \varkappa u''' + (1 + M + q^2) \, u'' + \varkappa u' + q^2 u + 3\varepsilon q^2 (\bar{x}_1'' + \bar{x}_1)^2 \, (u'' + u) = 0; \tag{3.4, 22}$$

after substituting $\bar{x}_1 = X_1 \cos \omega t$, it becomes

$$u^{IV} + \varkappa u''' + (1 + M + q^2) \, u'' + \varkappa u' + q^2 u + \tfrac{3}{2} \, \varepsilon q^2 X_1 (1 - \omega^2)^2$$
$$(1 + \cos 2\omega t) \, (u'' + u) = 0 \,. \tag{3.4, 23}$$

This is a linear differential equation with periodically variable coefficients. Its approximate analysis is effected in two steps: The first involves a check of the stability for the average constant values of the coefficients. Applying the Routh-Hurwitz criterion to the characteristic equation

$$\lambda^4 + \varkappa \lambda^3 + (1 + M + q^2) \, \lambda^2 + \varkappa \lambda + \tfrac{3}{2} \, \varepsilon q^2 X_1^2 (1 - \omega^2)^2 \, (\lambda^2 + 1) = 0$$

and assuming that $\varkappa > 0$, it is found that the following inequalities must be satisfied:

$$q^2[1 + \tfrac{3}{2} \varepsilon X_1^2 (1 - \omega^2)^2] > 0 , \tag{3.4, 24}$$

$$\varkappa^2\{1 + M + q^2[1 + \tfrac{3}{2} \varepsilon (1 - \omega^2)^2 X_1^2] - 1\} - \varkappa^2 q^2 [1 + \tfrac{3}{2} \varepsilon (1 - \omega^2)^2 X_1^2]$$
$$= \varkappa^2 M > 0 . \tag{3.4, 25}$$

For $\varepsilon > 0$, both inequalities are satisfied in all cases. For $\varepsilon < 0$, (3.4, 24) implies the necessity of satisfying the condition

$$X_1^2 < 2[3(-\varepsilon) (1 - \omega^2)^2]^{-1} . \tag{3.4, 24a}$$

The second step of the stability investigation concerns the effect of the variability of the coefficients. An approximate solution on the boundary of the main instability intervals can take the form

$$u = a \cos \omega t + b \sin \omega t . \tag{3.4, 26}$$

Substituting this solution in (3.4, 23) and comparing the coefficients of $\cos \omega t$ and $\sin \omega t$ results in two homogeneous equations for a and b. The condition that the solution should be non-trivial leads to the equation

$$\begin{vmatrix} H(\omega) + \tfrac{9}{4} \varepsilon q^2 (1 - \omega^2)^3 X_1^2, & \varkappa\omega(1 - \omega^2) \\ -\varkappa\omega(1 - \omega^2) , & H(\omega) + \tfrac{3}{4} \varepsilon q^2 (1 - \omega^2)^3 X_1^2 \end{vmatrix} = 0 \tag{3.4, 27}$$

where $H(\omega) = \omega^4 - (1 + M + q^2) \omega^2 + q^2$. X_1 is obtained from (3.4, 27), viz.

$$X_1 = \left\{ -\frac{8}{9} \frac{H(\omega)}{\varepsilon q^2 (1 - \omega^2)} \left[1 \mp \frac{1}{2} \left(1 - 3 \frac{\varkappa^2 \omega^2 (1 - \omega^2)^2}{[H(\omega)]^2} \right)^{1/2} \right] \right\}^{1/2} . \tag{3.4, 28}$$

By calculating X_1 as a function of ω, one obtains the boundaries of the regions in the (X_1, ω) plane in which the amplitudes X_1 determined from (3.4, 12) belong to unstable solutions. For $\varepsilon < 0$, inequality (3.4, 24a) must naturally be also taken into account. Applying (3.4, 19) to (3.4, 28) one obtains the regions in the (X_2, ω) plane in which the amplitudes X_2 correspond to unstable solutions.

The system described by the differential equation (3.4, 7) was also solved for several sets of parameters by help of an analogue computer. The extreme values of vibration in the x_1 and x_2 coordinate were plotted as functions of ω by means of a graph-plotter. The extreme values corresponding to X_1 and X_2 are denoted by $[x_1]$ and $[x_2]$.

The alternative presented below as an example has the following parameters: $M = 0.2$, $q = 0.8$, $\varepsilon = 0.05$, $\varkappa = 0.25$. For information about other cases, the reader is referred to a paper by TONDL (1977).

Fig. 3.4, 2 shows the diagrams of X_1 (heavy solid lines — stable solutions, heavy dashed lines — unstable solutions), X_{1L} (light solid lines) and X_{1S} (dot-and dash lines) as functions of frequency ω. The boundaries of the regions in which the solutions (obtained by using (3.4, 28)) are unstable are drawn in light solid lines, and the regions themselves are hatched. The upper branch of the backbone curve is seen to differ from that obtained by means of the approximate method based on the normal modes of the abbreviated system (the latter has the character of a parabola, i.e. is an increasing function of increasing ω). The qualitative difference between the results of the two approaches is particularly noticeable at higher resonances for which the proposed method yields a function which is in part increasing, in part decreasing. As analogue computing proves, the proposed method leads to more correct results than does the

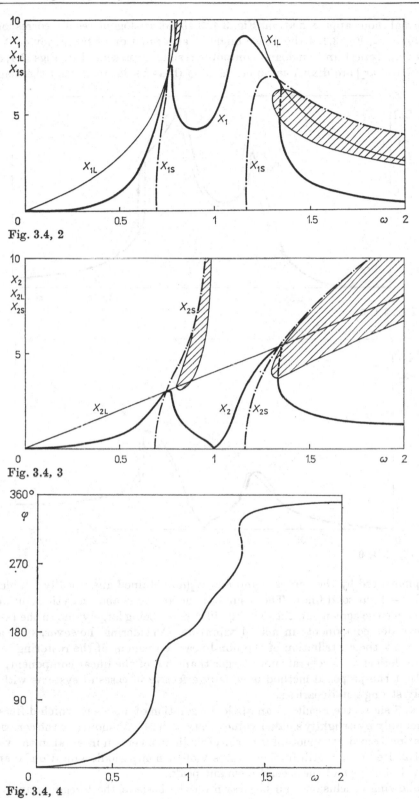

Fig. 3.4, 2

Fig. 3.4, 3

Fig. 3.4, 4

quasi-normal mode approximation. Fig. 3.4, 3 shows analogous results corresponding to coordinate x_2, Fig. 3.4, 4 the phase angle φ as a function of frequency ω. A comparison of analytical and analogue computer results is presented in Figs. 3.4, 5 and 3.4, 6. $[x_1]$ and $[x_2]$ are drawn as functions of ω (heavy solid lines) and the diagrams

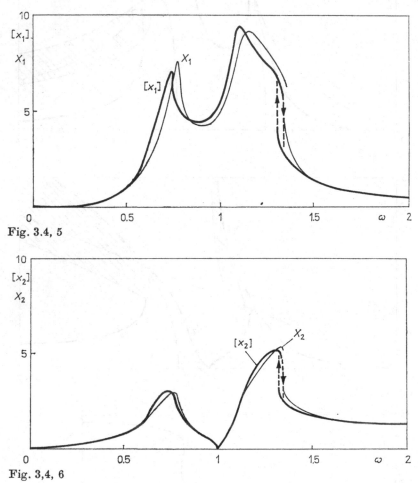

Fig. 3.4, 5

Fig. 3,4, 6

are complemented by the curves $X_1(\omega)$ and $X_2(\omega)$ obtained analytically (stable solutions only — light solid lines). The results of the analogue and analytic solutions are seen to be in close agreement, the existing differences being largely due to the presence of higher-order components in actual vibration. Considering, however, that at the resonant peak the contribution of the non-linear component of the restoring force at maximum deflection is several times larger than that of the linear component, it can be said that the proposed method is equally effective in cases of systems with comparatively strong non-linearities.

Fig. 3.4, 7 shows the results of an analogue solution of a system which differs from the former only by a slightly smaller value of $\varkappa$ ($\varkappa = 0.24$). As shown by the corresponding vibration records, the motion is quasi-periodic in a certain interval in the vicinity of $\omega = 1.2$. Reducing $\varkappa$ still further causes violent non-periodic vibration to arise in an interval of ω beyond the second resonant peak.

The following conclusions can be drawn on the basis of the foregoing analysis:

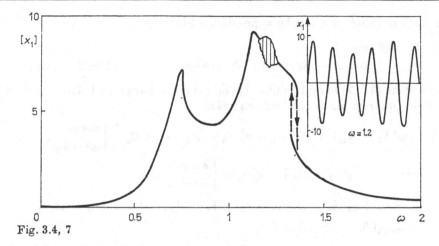

Fig. 3.4, 7

For some systems with two or more degrees of freedom, particularly for those which are no longer quasi-linear, the proposed method is apt to yield more realistic results than do the methods using quasi-normal mode approximation. As shown by way of an example, there exists a limit to the system's parameters for which the method ensures good agreement between the analytic solution and actual vibration. Outside this limit, the response is no longer periodic. A shortcoming of the proposed method is that it can be applied to a particular class of systems only.

3.5. Use of auxiliary curves in analysis of systems with several degrees of freedom

The backbone curve and the limit envelope find useful application in analysis of systems with several degrees of freedom whose equations of motion cannot be converted to a single higher-order differential equation. The class of systems for which an analysis using these curves is feasible, is, however, very small. The backbone curves can be obtained without difficulty for systems with masses interconnected chain-wise by dampers and springs, only one of which is non-linear. In the simplest case the non-linear spring is connected to the first or the last mass of the chain. If it is only this terminal mass which is acted on by harmonic excitation, one can (with some restriction concerning the damping non-linearity) also obtain the limit envelope and on the basis of it estimate the position of the main resonances and the amplitudes at the resonance peak. If the system has symmetric spring and damping characteristics, is not loaded with constant forces or tuned into internal resonance, the possibility of occurrence of other than the main resonances is small. A system of this kind is described by differential equations (after rearrangement to the dimensionless form; to simplify, linear viscous damping is assumed throughout) of the type

$$x_1'' + \varkappa_1 x_1' + \varkappa_2(x_1' - x_2') + x_1[1 + f(x_1)] + q_2^2(x_1 - x_2) = \begin{cases} \cos(\omega t + \varphi) \\ \omega^2 \cos(\omega t + \varphi) \end{cases},$$

$$\vdots$$

$$x_k'' + \varkappa_k(x_k' - x_{k-1}') + \varkappa_{k+1}(x_k' - x_{k+1}') + q_k^2(x_k - x_{k-1}) + q_{k+1}^2(x_k - x_{k+1}) = 0,$$

$$\vdots$$

$$x_n'' + \varkappa_n(x_n' - x_{n-1}') + q_n^2(x_n - x_{n-1}) = 0 \qquad (3.5, 1)$$

where $f(-x_1) = f(x_1)$.

The stationary solution can be approximated by

$$x_1 = X_1 \cos \omega t \,,$$
$$x_k = A_k \cos \omega t + B_k \sin \omega t = X_k \cos (\omega t - \varphi_k) \qquad (k = 2, \dots, n) \,.$$

(3.5, 2)

Substituting (3.5, 2) in system (3.5, 1) and using the harmonic balance method leads to the following non-linear algebraic equations:

$$-\omega^2 X_1 + X_1[1 + F(X_1)] + q_2^2(X_1 - A_2) - \varkappa_2 \omega B_2 = \begin{cases} \cos \varphi \\ \omega^2 \cos \varphi \end{cases} ,$$

$$\omega[\varkappa_1 X_1 + \varkappa_2(X_1 - A_2)] + q_2^2 B_2 = \begin{cases} \sin \varphi \\ \omega^2 \sin \varphi \end{cases} ,$$

$$-\omega^2 A_k + q_k^2(A_k - A_{k-1}) + q_{k+1}^2(A_k - A_{k+1}) + \omega[\varkappa_k(B_k - B_{k-1})$$
$$+ \varkappa_{k+1}(B_k - B_{k+1})] = 0 \,,$$

$$-\omega^2 B_k + q_k^2(B_k - B_{k-1}) + q_{k+1}^2(B_k - B_{k+1}) - \omega[\varkappa_k(A_k - A_{k-1})$$
$$+ \varkappa_{k+1}(A_k - A_{k+1})] = 0$$

(3.5, 3)

where $F(X_1)$ is a function resulting from function $f(x_1)$, $k = 2, \dots, n$, $q_{n+1} = 0$, $\varkappa_{n+1} = 0$.

Writing, as before, $\cos \varphi = 0$ and $\sin \varphi = \pm 1$, the last $2k$ equations of (3.5, 3) together with the equation

$$-\omega^2 X_1 + X_1[1 + F(X_1)] + q_2^2(X_1 - A_2) - \varkappa_2 \omega B = 0$$

(3.5, 4)

define the backbone curve (consisting of n branches), and the last $2(n-1)$ equations of (3.5, 3) together with the equation

$$\varkappa_1 \omega X_1 + \varkappa_2 \omega(X_1 - A_2) + q_2^2 B_2 = \pm \begin{cases} 1 \\ \omega^2 \end{cases}$$

(3.5, 5)

define the limit envelope.

In digital computations the values of X_1 are increased in steps starting from zero. The system of $2(n-1)$ algebraic equations is linear with respect to coefficients A_k, B_k. For a given X_1 these coefficients are obtained as functions of ω; their substitution in (3.5, 4) or (3.5, 5) results in a single equation in ω. This equation is used for determining the values of ω which correspond to the specified X_1 of the backbone curve and the limit envelope. The data thus obtained are then plotted in the form of diagrams representing $X_{1S}(\omega)$ and $X_{1L}(\omega)$.

In systems of the kind discussed, the backbone curve is generally no longer independent of damping; neither is the limit envelope dependent solely on damping and the excitation amplitude of the system.

As an example, consider a two-mass system which differs from that shown in Fig. 3.4, 1 only by a linear damper introduced between the two masses. The governing differential equations of this system are

$$x_1'' + \varkappa_1(x_1' - x_2') + x_1 - x_2 = 0 \,,$$

$$x_2'' - M[x_1 - x_2 + \varkappa_1(x_1' - x_2')] + \varkappa_2 x_2' + q^2(1 + \varepsilon x_2^2) x_2 = \omega^2 \cos (\omega t + \varphi)$$

(3.5, 6)

where, as before, $M = m_1/m_2$, $q^2 = \dfrac{c_2/m_2}{c_1/m_1}$.

If the stationary solution is sought in the form

$$x_1 = A \cos \omega t + B \sin \omega t = X_1 \cos (\omega t - \alpha) \, ,$$
$$x_2 = X_2 \cos \omega t \qquad\qquad\qquad\qquad\qquad \Bigg\} \qquad (3.5, 7)$$

the proposed method leads to

$$(1 - \omega^2) A + \varkappa_1 \omega B = X_2 \, ,$$
$$-\varkappa_1 \omega A + (1 - \omega^2) B = -\varkappa_1 \omega X_2 \, ,$$
$$[q^2(1 + \tfrac{3}{4} \varepsilon X_2^2) - \omega^2] X_2 - M\omega^2 A = \omega^2 \cos \varphi \, , \qquad (3.5, 8)$$
$$\varkappa_2 \omega X_2 + M\omega^2 B = \omega^2 \sin \varphi \, .$$

The first two equations give

$$A = \frac{X_2}{\varDelta} (1 - \omega^2 + \varkappa_1^2 \omega^2) \, ,$$
$$B = \frac{X_2}{\varDelta} \varkappa_1 \omega \qquad\qquad\qquad \Bigg\} \qquad (3.5, 9)$$

where

$$\varDelta = (1 - \omega^2)^2 + \varkappa_1^2 \omega^2 \, .$$

Substituting for A, B in the third of (3.5, 8) and setting $\cos \varphi = 0$ yields the following equations for X_{2S} and X_{1S}:

$$X_{2S} = \frac{2}{\sqrt{3\varepsilon}} \left\{ \frac{\omega^2}{q^2} \left[1 + M \frac{1 - \omega^2 + \varkappa_1^2 \omega^2}{\varDelta} \right] - 1 \right\}^{1/2} , \qquad (3.5, 10)$$

$$X_{1S} = \frac{X_{2S}}{\varDelta} [(1 - \omega^2 + \varkappa_1^2 \omega^2)^2 + \varkappa_1^2 \omega^6]^{1/2} . \qquad (3.5, 11)$$

The equations for the limit envelope are

$$X_{2L} = \omega \left[\varkappa_1 + \frac{M\varkappa_1 \omega^4}{\varDelta} \right]^{-1} , \qquad (3.5, 12)$$

$$X_{1L} = \frac{X_{2L}}{\varDelta} [(1 - \omega^2 + \varkappa_1^2 \omega^2)^2 + \varkappa_1^2 \omega^6]^{1/2} . \qquad (3.5, 13)$$

Fig. 3.5, 1 shows the curves $X_{1L}(\omega)$ (light solid lines) and $X_{1S}(\omega)$ (dot-and-dash lines) complemented by the curve of the extreme deflections $[x_1]$ as function of ω, obtained by analogue modelling. Fig. 3.5, 2 shows analogous curves for the x_2 coordinate. The parameters of the corresponding system are: $q = 0.5$, $M = 0.5$, $\varepsilon = 0.1$, $\varkappa_1 = \varkappa_2 = 0.1$. The backbone curve of the higher resonance in the x_1 coordinate resembles that of the example discussed in the preceding section. In this case also, the share of the non-linear term of the non-linear spring characteristic is several times larger than that of the linear term for the extreme deflection at resonance. This fact notwithstanding, the error of the resonant peak estimate is comparatively small.

Additional examples may be found in a paper by TONDL (1978a).

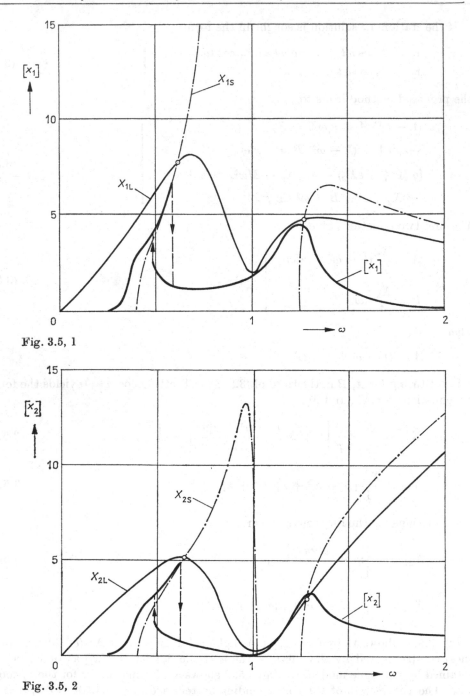

Fig. 3.5, 1

Fig. 3.5, 2

3.6.   Identification of damping

It was established on the basis of several equations derived in Section 3.1 that the form of the limit envelope is affected by damping and the type of excitation. This finding can be put to good use whenever damping identification is required in order to provide a more realistic expression of the damping forces in the mathematical models of real systems. The theoretical considerations presented in the foregoing can find most useful application in the field of simple elements, for example, rubber elastic elements which represent a spring and a damper as well. With many such elements one is at a loss for an expression of damping capable of describing reality at least approximately. If an elastic element of this kind is to be replaced simply by a damper and a spring, the frequency interval of excitation must be limited to a range (lower frequencies) in which the element need not be regarded as a continuum.

The test facilities should be designed so as to satisfy the assumptions made in the theoretical considerations (one degree of freedom, harmonic excitation, etc.). Since the stiffness of most elastic elements cannot be changed without altering the damping, it is best to obtain the limit envelope as an envelope of the set of resonance curves corresponding to different masses (which can be changed more readily). The test device is best designed as a torsion system in which, by varying the distance between two masses on its arm, one can change the moment of inertia quadratically (and hence the natural frequency linearly). Since the effect of prestress is not examined, the element is not subjected to a constant load in order not to complicate the identification. The test device should preferably incorporate means of providing more types of excitation (with a constant amplitude, with an amplitude proportional to the square of the excitation frequency). The possibility of varying the amplitude of the harmonic excitation over a sufficient range should be ensured in all cases.

Because of the limited scope of the book, only the salient points of the identification procedure are presented. The first point to be established is whether or not the damping characteristic is symmetric. The equations put forward at the end of Section 3.1 should be of help in this respect. If a symmetric characteristic or the symmetric part of a characteristic is involved, it is recommended to ascertain whether the damping of the element being tested belongs to a fundamental class of damping, is linear or non-linear.

If, for any excitation frequency, the amplitudes of the limit envelopes corresponding to two arbitrary excitation amplitudes P_1 and P_2 satisfy the relation

$$\frac{A_{L1}}{A_{L2}} = \frac{P_1}{P_2} \qquad\qquad (3.6, 1)$$

the damping is linear (but not necessarily linear viscous).

The characteristic features of the limit envelopes for the fundamental types of damping are as follows:

(a) If damping is described by the function $\varkappa(|y|)$ sgn $\dot{y}$, the limit envelope is a straight line parallel to the ω axis whenever the excitation force has a constant amplitude P. If (3.6, 1) is not satisfied, the values of P represent directly the values of function $K(A_L)$ (cf. (3.1, 11 a) on the consideration that, in this case, function K depends only on amplitude A). Function $K(A_L)$ is obtained by means of regression and from there function $\varkappa(|y|)$ sgn $\dot{y}$ taking into account the relations (resulting from the averaging process) between the corresponding coefficients of, for example, the polynomial of the two functions.

(b) If damping is described by function $\varkappa(|\dot{y}|)\,\dot{y}$, the limit envelopes corresponding to different values of the constant amplitude of the excitation force are regular hyperbolae. If (3.6, 1) does not apply to all P, function $K(\omega A_L)\,\omega A_L$ is non-linear. As in the former case, it is obtained by means of regression using different values of P which again represent the values of $K(\omega A_L)\,\omega A_L$ for a particular ω.

(c) If damping is described by function $\varkappa(|\dot{y}|)\,\dot{y}$, the limit envelope is defined by the equation

$$K(A_\mathrm{L})\,\omega A_\mathrm{L} = \begin{cases} P \\ \varepsilon m_0 \omega^2 \end{cases}. \tag{3.6, 2}$$

Taking the inverse to function $A_\mathrm{L}(\omega)$ for both types of excitation one obtains the equations

$$\omega = \begin{cases} \dfrac{P}{A_\mathrm{L} K(A_\mathrm{L})} = \omega_P(A_\mathrm{L}) \\[2mm] \dfrac{A_\mathrm{L}}{\varepsilon m_0}\, K(A_\mathrm{L}) = \omega_s(A_\mathrm{L}) \end{cases} \tag{3.6, 3}$$

from which it follows that the product of the two functions is constant for any A_L, viz.

$$\omega_P(A_\mathrm{L})\,\omega_s(A_\mathrm{L}) = \mathrm{const}. \tag{3.6, 4}$$

The curve representing the function $K(A_\mathrm{L})$ can be obtained as follows. The limit envelopes corresponding to the first type of excitation are determined for various values of P. For a chosen ω, the values of A_L are read from the diagram and the values of $P/A_\mathrm{L}\omega$ corresponding to them are calculated. The latter determine the values of function $K(A_\mathrm{L})$ whose course is obtained by means of regression.

The more complicated types of damping and their combinations will not be treated here.

A brief mention, at least, should be made of the following important fact. In systems with visco-elastic elements, the actual resonance is shifted against that obtained on the basis of the static characteristic of the restoring force. Hence springs of visco-elastic materials exhibit a difference between the static and the dynamic characteristic of the restoring force. The dynamic characteristic of the restoring force is determined by the experimental backbone curve.

4. Analysis in the phase plane

4.1. Fundamental considerations

In the present context, the term "phase plane" denotes a plane whose coordinates are the dependent variables of a system of two coupled first-order differential equations of the type

$$\left.\begin{array}{l} \dot{x} = X(x, y) \\ \dot{y} = Y(x, y) \end{array}\right\} \qquad (4.1, 1)$$

where the functions $X(x, y)$ and $Y(x, y)$ are to satisfy the conditions of existence and uniqueness of the solution at any point of the (x, y, t) space.

In interpretation of the results of an analysis made in the phase plane, the analyst should know the way in which $(4.1, 1)$ were obtained. Knowing this, he can divide the various systems into the following three groups:

(a) Equations $(4.1, 1)$ describe directly the mathematical model of the system being studied.

(b) Equations $(4.1, 1)$ were obtained by rearrangement, e.g. introduction of the variables $\dot{y} = v, y$, of the homogeneous second-order differential equation of the type

$$\ddot{y} + f(\dot{y}, y) = 0 \qquad (4.1, 2)$$

which describes the model of a self-excited system with one degree of freedom.

(c) Equations $(4.1, 1)$ were obtained by transformation of a second-order differential equation, either non-homogeneous with the right-hand side representing a periodic function of time (the independent variable), or homogeneous with coefficients which are periodic functions of time (the independent variable). The corresponding systems are those with external or parametric excitation, and equations $(4.1, 1)$ are obtained when these systems are solved by means, for example, of the method of slowly varying amplitudes, or of amplitude and phase (the van der Pol or Krylov and Bogoljubov methods).

On the above assumptions it holds for the integral curves — the phase plane trajectories — of system $(4.4, 1)$ which are obtained by solving the equation

$$\frac{\mathrm{d}x}{\mathrm{d}y} = \frac{X(x, y)}{Y(x, y)} \qquad (4.1, 3)$$

that, except for the singular points defined by the equations

$$X(x, y) = 0, \qquad Y(x, y) = 0 \qquad (4.1, 4)$$

only a single trajectory passes through every point of the phase plane.

Points which lie on the curve defined by the equation

$$X(x, y) = 0$$

but fail to satisfy the second of equations (4.1, 1) are characterized by the fact that the tangent to the trajectory is parallel to the y-axis. Similarly, the tangent to the trajectory at points lying on the curve

$$Y(x, y) = 0$$

is parallel to the x-axis.

The singular points define the equilibrium state and represent the non-oscillatory solution of (4.1, 1), i.e. the real equilibrium state in the case of systems of the groups (a) and (b), and steady periodic solutions of a certain type (e.g. in the main or subharmonic resonance) for systems of the group (c).

In analyses in the phase plane it is important to know the position of the singular points as well as their type and stability. The last two properties can be determined from the roots of the characteristic equation

$$\begin{vmatrix} \dfrac{\partial X}{\partial x} - \lambda & \dfrac{\partial X}{\partial y} \\[2mm] \dfrac{\partial Y}{\partial x} & \dfrac{\partial Y}{\partial y} - \lambda \end{vmatrix} = 0 \tag{4.1, 5}$$

where the coordinates of the singular points are substituted for x and y in the expressions $\partial X/\partial x$, $\partial X/\partial y$, $\partial Y/\partial x$, $\partial Y/\partial y$. The characteristic equation (4.1, 5) can also be given the form

$$\lambda^2 + \beta\lambda + \gamma = 0 \tag{4.1, 5a}$$

where

$$\beta = -\left(\frac{\partial X}{\partial x} + \frac{\partial Y}{\partial y}\right), \quad \gamma = \frac{\partial X}{\partial x}\frac{\partial Y}{\partial y} - \frac{\partial X}{\partial y}\frac{\partial Y}{\partial x}.$$

Whenever the conditions

$$\beta > 0, \quad \gamma > 0 \tag{4.1, 6}$$

are satisfied, the singular point in question is also asymptotically stable.

The singular point is a *node* if the roots of the characteristic equation are real and of equal signs, viz.

$$\beta^2 \geqq 4\gamma, \quad \gamma > 0. \tag{4.1, 7}$$

The singular point is a *saddle* if the roots of the characteristic equation are real but of opposite signs, viz.

$$\beta^2 > 4\gamma, \quad \gamma < 0. \tag{4.1, 8}$$

The singular point is a *focus* if the roots of the characteristic equation are complex imaginaries with non-zero real part, viz.

$$\beta^2 < 4\gamma, \quad \beta \neq 0. \tag{4.1, 9}$$

A complete discussion of the singular points should also include the so-called *centre*, i.e. a singular point for which the roots of the characteristic equation are pure imaginaries, viz.

$$\gamma > 0 , \qquad \beta = 0 . \tag{4.1, 10}$$

The corresponding solution is stable but not asymptotically.

Knowledge of the type of the singular point enables the analyst to estimate the character of the trajectories in the point's immediate neighbourhood. In the neighbourhood of a node, the trajectories form a bundle (Fig. 4.1, 1); the node is stable if they move towards it (Fig. 4.1, 1a), and unstable if they move away from it (Fig. 4.1, 1b). A saddle point (Fig. 4.1, 2) features two pairs of trajectories, one entering, the other leaving it. The trajectories of each pair have common tangents at the saddle point. In the neighbourhood of a focus, the trajectories are spiral shaped and move either towards (stable focus — Fig. 4.1, 3a) or away from (unstable focus — Fig. 4.1, 3b) the point. In the neighbourhood of a centre, the trajectories form closed curves (Fig. 4.1, 4); this case has a more or less theoretical meaning (systems without damping).

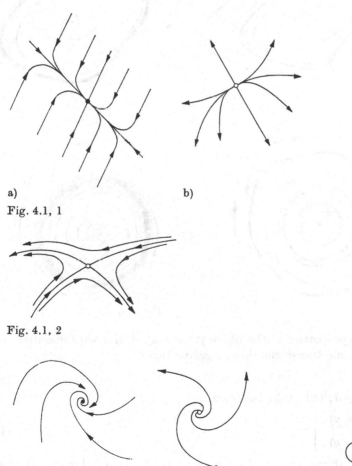

a)   b)

Fig. 4.1, 1

Fig. 4.1, 2

(a)   (b)

Fig. 4.1, 3   Fig. 4.1, 4

If there exist — apart from the singular points — additional steady but oscillating (periodical) solutions, their trajectories in the phase plane are closed curves — the *limit cycles*. A limit cycle is stable (Fig. 4.1, 5a) if the trajectories in its neighbourhood converge to the cycle, and unstable (Fig. 4.1, 5b) if they diverge from it. The case of systems without damping whose set of closed trajectories surrounds the singular point — the centre — is not considered. If a single singular point is surrounded by several limit cycles, their stable and unstable forms alternate regularly — see Fig. 4.1, 6 where the stable limit cycles marked L_S are drawn in heavy lines, the unstable limit cycles, L_N, in heavy dashed lines. If a particular parameter of a system is varied so as to cause the stable limit cycle to approach the unstable, than at a definite value of the parameter, the two cycles become one. The resulting *limit cycle is called semi-stable* and in practice its presence is revealed by close spacing of the trajectories in its neighbourhood (Fig. 4.1, 7).

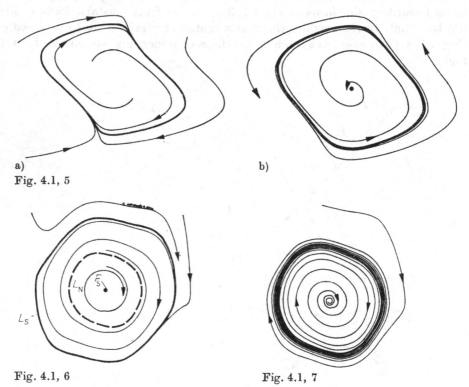

a)

Fig. 4.1, 5

b)

Fig. 4.1, 6   Fig. 4.1, 7

Other important properties of the phase plane trajectories are presented below. Introducing the time transformation (negative time)

$$t = -\tau \tag{4.1, 11}$$

causes the system (4.1, 1) to take the form

$$\left.\begin{aligned} \dot{x} &= -X(x, y) , \\ \dot{y} &= -Y(x, y) . \end{aligned}\right\} \tag{4.1, 12}$$

It can readily be shown that a phase portrait in the phase plane (x, y) of system (4.1, 1) is identical with that of system (4.1, 12) except for the opposite sense of the reference point motion. This finding has the following important consequences:

If the point (x_k, y_k) is a stable (unstable) focus in the (x, y) plane for system (4.1, 1), it is an unstable (stable) focus for system (4.1, 12).

If the point (x_k, y_k) is a stable (unstable) node in the (x, y) plane for system (4.1, 1), it is an unstable (stable) node for system (4.1, 12).

If the point (x_k, y_k) is a saddle for system (4.1, 1), it is also a saddle for system (4.1, 12); the motion of the reference point in the two systems is, however, of opposite sense.

A stable (unstable) limit cycle for system (4.1, 1) is an unstable (stable) cycle for system (4.1, 12).

These properties can be used to advantage when drawing the phase portraits by means of a digital or an analogue computer by alternating the solution of system (4.1, 1) with that of system (4.1, 12) to speed up the process of plotting the most important phase plane trajectories.

An analysis using the phase plane portraits is advantageous in that it provides comprehensive knowledge of both the stationary and the non-stationary solutions and enables the stability of steady solutions to be investigated not only for small but for any disturbances in the initial conditions; this is the way in which the *stability in the large* is investigated and the *domains of attraction* for various steady solutions are determined.

A few notes concerning systems more complicated then (4.1, 1) seem to be necessary at this point. In some cases the two first-order differential equations are supplemented by additional algebraic relations (as, for example, in systems of the (c) group for which the steady solution must be approximated by yet other terms, such as that of a constant deflection, etc.). As an example consider the system of the type

$$
\left.
\begin{aligned}
\dot{x} &= X(x, y, z) , \\
\dot{y} &= Y(x, y, z) , \\
Z(x, y, z) &= 0
\end{aligned}
\right\}
\tag{4.1, 13}
$$

in which even other algebraic relations are apt to exist. The singular points in the (x, y) plane which exist in this case, also, are similar to those of system (4.1, 1) and obtained from the roots of the characteristic equation

$$
\alpha \lambda^2 + \beta \lambda + \gamma = 0
\tag{4.1, 14}
$$

where

$$
\alpha = \frac{\partial Z}{\partial z}, \qquad
\beta = -
\begin{vmatrix}
\dfrac{\partial X}{\partial x} & \dfrac{\partial X}{\partial y} \\[2mm]
\dfrac{\partial Z}{\partial x} & \dfrac{\partial Z}{\partial y}
\end{vmatrix}
-
\begin{vmatrix}
\dfrac{\partial Y}{\partial y} & \dfrac{\partial Y}{\partial z} \\[2mm]
\dfrac{\partial Z}{\partial y} & \dfrac{\partial Z}{\partial z}
\end{vmatrix},
\qquad
\gamma =
\begin{vmatrix}
\dfrac{\partial X}{\partial x} & \dfrac{\partial X}{\partial y} & \dfrac{\partial X}{\partial z} \\[2mm]
\dfrac{\partial Y}{\partial x} & \dfrac{\partial Y}{\partial y} & \dfrac{\partial Y}{\partial z} \\[2mm]
\dfrac{\partial Z}{\partial x} & \dfrac{\partial Z}{\partial y} & \dfrac{\partial Z}{\partial z}
\end{vmatrix}.
$$

The type and stability of the various singular points (assuming $\alpha > 0$) are shown in table 4.1, 1.

Table 4.1, 1

Type of singular point	Satisfying the conditions	Stable for	Unstable for
The node	$\beta^2 \geqq 4\alpha\gamma,\ \gamma/\alpha > 0$	$\beta > 0$	$\beta < 0$
The saddle	$\beta^2 > 4\alpha\gamma,\ \gamma/\alpha < 0$	—	always
The focus	$\beta^2 < 4\alpha\gamma,\ \beta \neq 0$	$\beta > 0$	$\beta < 0$
The centre	$\gamma > 0,\ \beta = 0$	always	—

Let us turn to systems governed by a set of three differential equations of the first order

$$\left.\begin{array}{l} x = X(x, y, z)\,, \\ \dot{y} = Y(x, y, z)\,, \\ \dot{z} = Z(x, y, z)\,. \end{array}\right\} \tag{4.1, 15}$$

Denoting by x_s, y_s, z_s the coordinates of a singular point and introducing $x = x_s + \xi$, $y = y_s + \eta$, $z = z_s + \zeta$ the perturbation equations are obtained. As critical cases will be excluded from our considerations the stability of singular points can be determined according to the roots of the characteristic equation

$$\lambda^3 + a_1\lambda^2 + a_2\lambda + a_3 = 0 \tag{4.1, 16}$$

where

$$a_1 = -\left(\frac{\partial X}{\partial x} + \frac{\partial Y}{\partial y} + \frac{\partial Z}{\partial z}\right), \tag{4.1, 17}$$

$$a_2 = \begin{vmatrix} \dfrac{\partial X}{\partial x} & \dfrac{\partial X}{\partial y} \\[2mm] \dfrac{\partial Y}{\partial x} & \dfrac{\partial Y}{\partial y} \end{vmatrix} + \begin{vmatrix} \dfrac{\partial X}{\partial x} & \dfrac{\partial X}{\partial z} \\[2mm] \dfrac{\partial Z}{\partial x} & \dfrac{\partial Z}{\partial z} \end{vmatrix} + \begin{vmatrix} \dfrac{\partial Y}{\partial y} & \dfrac{\partial Y}{\partial z} \\[2mm] \dfrac{\partial Z}{\partial y} & \dfrac{\partial Z}{\partial z} \end{vmatrix}, \tag{4.1, 18}$$

$$a_3 = -\begin{vmatrix} \dfrac{\partial X}{\partial x} & \dfrac{\partial X}{\partial y} & \dfrac{\partial X}{\partial z} \\[2mm] \dfrac{\partial Y}{\partial x} & \dfrac{\partial Y}{\partial y} & \dfrac{\partial Y}{\partial z} \\[2mm] \dfrac{\partial Z}{\partial x} & \dfrac{\partial Z}{\partial y} & \dfrac{\partial Z}{\partial z} \end{vmatrix}, \tag{4.1, 19}$$

$$x = x_s\,, \qquad y = y_s\,, \qquad z = z_s\,.$$

Let us denote by $\lambda_1, \lambda_2, \lambda_3$ the sought-for roots of the characteristic equation. Then it holds that

$$\left.\begin{array}{l} a_1 = -(\lambda_1 + \lambda_2 + \lambda_3)\,, \\ a_2 = \lambda_1\lambda_2 + \lambda_1\lambda_3 + \lambda_2\lambda_3\,, \\ a_3 = -\lambda_1\lambda_2\lambda_3\,. \end{array}\right\} \tag{4.1, 20}$$

In our discussion of the quality and signs of the roots we shall make use of the discriminant of the equation

$$D_3 = -4(a_3 - \tfrac{1}{3}a_1^2)^3 - 27(a_3 - \tfrac{1}{3}a_1a_2 + \tfrac{2}{27}a_1^3)^2 \tag{4.1, 21}$$

as well as of Hurwitz's condition

$$a_1 > 0, \qquad a_3 > 0, \qquad a_1a_2 - a_0 > 0 \tag{4.1, 22}$$

and finally, of Descartes' theorem:

The number of positive roots of an algebraic equation is equal to, or smaller by an even number than the number of sign changes in the sequence

$$1, a_1, a_2, a_3.$$

Let us note that this theorem may be reversed to apply to the number of negative roots and sign changes in the sequence

$$-1, a_1, -a_2, a_3$$

i.e. to those of the equation

$$-\lambda^3 + a_1\lambda^2 - a_2\lambda + a_3 = 0 \tag{4.1, 16a}$$

Let us now classify the singular points. The ones most likely to occur are:

Stable node. Hurwitz's conditions are satisfied for (4.1, 16); $D_3 \geqq 0$. All the roots are real negative. All the trajectories enter the singular point (see Fig. 4.1, 8).

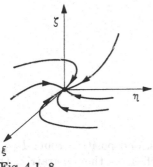

Fig. 4.1, 8

Unstable node. Hurwitz's conditions are satisfied for (4.1, 16a); $D_3 \geqq 0$. All the roots are real, positive; $a_1 < 0$, $a_2 > 0$, $a_3 < 0$. All the trajectories start from the singular point (see Fig. 4.1, 9).

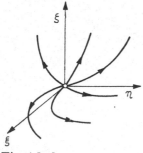

Fig. 4.1, 9

Stable focus. Hurwitz's conditions are satisfied for (4.1, 16); $D_3 < 0$; a pair of complex roots with negative real parts, one negative real root. All the trajectories enter the singular point, but in contrast to the node are spiral-shaped (see Fig. 4.1, 10).

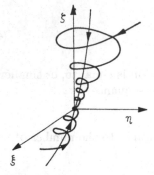

Fig. 4.1, 10

Unstable focus. Hurwitz's conditions are satisfied for (4.1, 16a); $D_3 < 0$; a pair of · complex roots with positive real parts. All the trajectories — spiral shaped — start from the singular point ($a_1 < 0$, $a_2 < 0$, $a_3 < 0$) (see Fig. 4.1, 11).

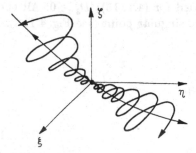

Fig. 4.1, 11

Saddle of the first kind. Two real negative roots, one positive root; $D_3 \geqq 0$. The trajectories entering the singular point form a surface — the separatrix. Only two trajectories leave the singular point, and they have a common tangent there (see Fig. 4.1, 12).

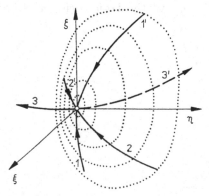

Fig. 4.1, 12

Saddle of the second kind. Two real positive roots, one negative root; $D_3 \geqq 0$. Only two trajectories enter the singular point. The trajectories leaving that point form a surface (see Fig. 4.1, 13).

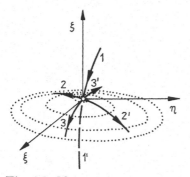

Fig. 4.1, 13

Saddle-focus of the first kind. A pair of complex roots with negative real parts, one positive real root. All the trajectories entering the singular point form a surface — the separatrix, and are spiral-shaped. Two trajectories leave the singular point (Fig. 4.1, 14); $D_3 < 0$, $a_3 < 0$.

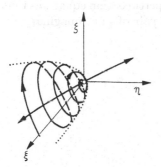

Fig. 4.1, 14

Saddle-focus of the second kind. A pair of complex roots with positive real parts, one negative real root. The sense of the course of the trajectories is opposite to that of the preceding case (see Fig. 4.1, 15), $D_3 < 0$, $a_3 > 0$.

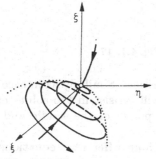

Fig. 4.1, 15

Unstable centre. A pair of pure imaginary roots, one positive real root (in this as well as in the next case, a stable centre); this is of course only a necessary condition for the existence of a root of this type. Only two trajectories leave the singular point; the remaining ones lie on parallel cylindrical surfaces (see Fig. 4.1, 16).

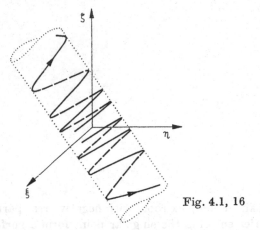

Fig. 4.1, 16

Stable centre. The sense of the course of the trajectories is opposite to that of the proceding case (see Fig. 4.1, 17). Here as well as in the critical case it is the terms of higher orders with respect to ξ, η, ζ in the system of perturbation equations that decide on the stability. The characteristic equation has a pair of pure imaginary roots and one negative real root.

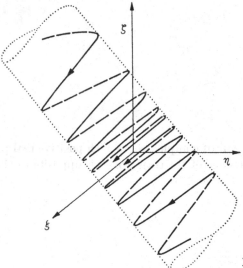

Fig. 4.1, 17

All cases with a single or multiple zero root are critical. Similarly as in case of two differential equations of the first order, our task will consist in the solution of the separatrix and in turn in the examination of the points of the saddle and saddle focus type, lying on those separating surfaces.

Let us turn now to systems governed by a set of four differential equations of the first order:

$$\left.\begin{aligned}
\dot{x} &= X(x, y, z, w) , \\
\dot{y} &= Y(x, y, z, w), \\
\dot{z} &= Z(x, y, z, w) , \\
\dot{w} &= W(x, y, z, w) .
\end{aligned}\right\} \tag{4.1, 23}$$

In a similar way as in the previous case the following characteristic equation can be obtained:

$$\lambda^4 + a_1\lambda^3 + a_2\lambda^2 + a_3\lambda + a_4 = 0 \tag{4.1, 24}$$

whose coefficients are defined by the relations

$$a_1 = -\left(\frac{\partial X}{\partial x} + \frac{\partial Y}{\partial y} + \frac{\partial Z}{\partial z} + \frac{\partial W}{\partial w}\right), \tag{4.1, 25}$$

$$a_2 = \begin{vmatrix} \dfrac{\partial X}{\partial x} & \dfrac{\partial X}{\partial w} \\[2mm] \dfrac{\partial W}{\partial x} & \dfrac{\partial W}{\partial w} \end{vmatrix} + \begin{vmatrix} \dfrac{\partial Y}{\partial y} & \dfrac{\partial Y}{\partial w} \\[2mm] \dfrac{\partial W}{\partial y} & \dfrac{\partial W}{\partial w} \end{vmatrix} + \begin{vmatrix} \dfrac{\partial Z}{\partial z} & \dfrac{\partial Z}{\partial w} \\[2mm] \dfrac{\partial W}{\partial z} & \dfrac{\partial W}{\partial w} \end{vmatrix}$$

$$+ \begin{vmatrix} \dfrac{\partial X}{\partial x} & \dfrac{\partial X}{\partial y} \\[2mm] \dfrac{\partial Y}{\partial x} & \dfrac{\partial Y}{\partial y} \end{vmatrix} + \begin{vmatrix} \dfrac{\partial X}{\partial x} & \dfrac{\partial X}{\partial z} \\[2mm] \dfrac{\partial Z}{\partial x} & \dfrac{\partial Z}{\partial z} \end{vmatrix} + \begin{vmatrix} \dfrac{\partial Y}{\partial y} & \dfrac{\partial Y}{\partial z} \\[2mm] \dfrac{\partial Z}{\partial y} & \dfrac{\partial Z}{\partial z} \end{vmatrix}, \tag{4.1, 26}$$

$$a_3 = -\begin{vmatrix} \dfrac{\partial X}{\partial x} & \dfrac{\partial X}{\partial y} & \dfrac{\partial X}{\partial z} \\[2mm] \dfrac{\partial Y}{\partial x} & \dfrac{\partial Y}{\partial y} & \dfrac{\partial Y}{\partial z} \\[2mm] \dfrac{\partial Z}{\partial x} & \dfrac{\partial Z}{\partial y} & \dfrac{\partial Z}{\partial z} \end{vmatrix} - \begin{vmatrix} \dfrac{\partial X}{\partial x} & \dfrac{\partial X}{\partial y} & \dfrac{\partial X}{\partial w} \\[2mm] \dfrac{\partial Y}{\partial x} & \dfrac{\partial Y}{\partial y} & \dfrac{\partial Y}{\partial w} \\[2mm] \dfrac{\partial W}{\partial x} & \dfrac{\partial W}{\partial y} & \dfrac{\partial W}{\partial w} \end{vmatrix}$$

$$- \begin{vmatrix} \dfrac{\partial X}{\partial x} & \dfrac{\partial X}{\partial z} & \dfrac{\partial X}{\partial w} \\[2mm] \dfrac{\partial Z}{\partial x} & \dfrac{\partial Z}{\partial z} & \dfrac{\partial Z}{\partial w} \\[2mm] \dfrac{\partial W}{\partial x} & \dfrac{\partial W}{\partial z} & \dfrac{\partial W}{\partial w} \end{vmatrix} - \begin{vmatrix} \dfrac{\partial Y}{\partial y} & \dfrac{\partial Y}{\partial z} & \dfrac{\partial Y}{\partial w} \\[2mm] \dfrac{\partial Z}{\partial y} & \dfrac{\partial Z}{\partial z} & \dfrac{\partial Z}{\partial w} \\[2mm] \dfrac{\partial W}{\partial y} & \dfrac{\partial W}{\partial z} & \dfrac{\partial W}{\partial w} \end{vmatrix}, \tag{4.1, 27}$$

$$a_4 = \begin{vmatrix} \dfrac{\partial X}{\partial x} & \dfrac{\partial X}{\partial y} & \dfrac{\partial X}{\partial z} & \dfrac{\partial X}{\partial w} \\[2mm] \dfrac{\partial Y}{\partial x} & \dfrac{\partial Y}{\partial y} & \dfrac{\partial Y}{\partial z} & \dfrac{\partial Y}{\partial w} \\[2mm] \dfrac{\partial Z}{\partial x} & \dfrac{\partial Z}{\partial y} & \dfrac{\partial Z}{\partial z} & \dfrac{\partial Z}{\partial w} \\[2mm] \dfrac{\partial W}{\partial x} & \dfrac{\partial W}{\partial y} & \dfrac{\partial W}{\partial z} & \dfrac{\partial W}{\partial w} \end{vmatrix} . \tag{4.1, 28}$$

Denoting by $\lambda_1, \lambda_2, \lambda_3, \lambda_4$ the roots of (4.1, 24), it holds that

$$\left.\begin{aligned}
a_1 &= -(\lambda_1 + \lambda_2 + \lambda_3 + \lambda_4)\,, \\
a_2 &= \lambda_1\lambda_2 + \lambda_1\lambda_3 + \lambda_1\lambda_4 + \lambda_2\lambda_3 + \lambda_2\lambda_4 + \lambda_3\lambda_4\,, \\
a_3 &= -\lambda_1\lambda_2(\lambda_3 + \lambda_4) - \lambda_3\lambda_4(\lambda_1 + \lambda_2)\,, \\
a_4 &= \lambda_1\lambda_2\lambda_3\lambda_4\,.
\end{aligned}\right\} \qquad (4.1,\,29)$$

In the discussion of the roots we shall again make use of the discriminant of (4.1, 24)

$$D_4 = \tfrac{4}{27}\,(12a_4 + a_2^2 - 3a_1a_3)$$
$$- \tfrac{1}{27}\,(27a_3^2 - 72a_2a_4 + 2a_2^3 - 9a_1a_2a_3 + 27a_1^2a_4)^2\,, \qquad (4.1,\,30)$$

of Hurwitz's conditions for (4.1, 24)

$$a_1 > 0\,, \qquad a_2 > 0\,, \qquad a_3 > 0\,, \qquad a_4 > 0\,,$$

$$\begin{vmatrix} a_1 & 1 \\ a_3 & a_2 \end{vmatrix} > 0\,, \qquad \begin{vmatrix} a_1 & 1 & 0 \\ a_3 & a_2 & a_1 \\ 0 & a_4 & a_3 \end{vmatrix} > 0\,, \qquad (4.1,\,31)$$

of Hurwitz's conditions for equation

$$-\lambda^4 + a_1\lambda^3 - a_2\lambda^2 + a_3\lambda - a_4 = 0 \qquad (4.1,\,24\text{a})$$

and finally, of Descartes' theorem concerning the changes of sign of the sequences

$$1,\, a_1,\, a_2,\, a_3,\, a_4$$

appertaining to (4.1, 24), and

$$-1,\, a_1,\, -a_2,\, a_3,\, -a_4 \qquad (4.1,\,32\text{a})$$

appertaining to (4.1, 24a).

Compared with the previous case a higher number of unstable singular points of the saddle or saddle-focus type exist, mainly because a distinction must be made between

$$\operatorname{sgn} \operatorname{Re}(\lambda_1) = \operatorname{sgn} \operatorname{Re}(\lambda_2) = \operatorname{sgn} \operatorname{Re}(\lambda_3) = -\operatorname{sgn} \operatorname{Re}(\lambda_4) \qquad (4.1,\,33)$$

and

$$\operatorname{sgn} \operatorname{Re}(\lambda_1) = \operatorname{sgn} \operatorname{Re}(\lambda_2) = -\operatorname{sgn} \operatorname{Re}(\lambda_3) = -\operatorname{sgn} \operatorname{Re}(\lambda_4)\,. \qquad (4.1,\,33\text{a})$$

The critical cases are again excluded. The singular points can be classified as follows:

Stable node. All the roots of the characteristic equation are negative, real; $D_4 > 0$; the stability conditions are satisfied for (4.1, 24). All the trajectories enter the singular point.

Unstable node. All the roots are real, positive; $D_4 > 0$; Hurwitz's conditions are satisfied for (4.1, 24a). All the trajectories leave the singular point.

Stable focus. Two pairs of complex roots with negative real parts; $D_4 > 0$; Hurwitz's conditions are satisfied for (4.1, 24). The trajectories are spiral-shaped and all enter the singular point.

Unstable focus. Two pairs of complex roots with positive real parts; $D_4 > 0$; Hurwitz's conditions are satisfied for (4.1, 24a). All the spiral-shaped trajectories leave the singular point.

Stable focus-node. A pair of complex roots with negative real parts, two real negative roots; Hurwitz's conditions are satisfied for (4.1, 24); $D_4 < 0$. All the trajectories enter the singular point and are again spiral-shaped.

Unstable focus-node. A pair of complex roots with positive real parts; two real positive roots; $D_4 < 0$; Hurwitz's conditions are satisfied for (4.1, 24a). All spiral-shaped trajectories leave the singular point.

Saddle of the first kind. Three real, negative roots; one positive root; $D_4 > 0$; sequence (4.1, 32a) has three changes of sign. Two trajectories leave the singular point; the trajectories that enter it form a three-dimensional separatrix surface.

Saddle of the second kind. Two real negative roots; two positive roots; $D_4 > 0$; sequence (4.1, 32) has an even number of sign changes. The trajectories entering and leaving the singular point form two two-dimensonal surfaces.

Saddle of the third kind. A real negative root, three real positive roots; $D_4 > 0$; sequence (4.1, 32) has three sign changes; $a_4 < 0$. Only two trajectories enter the singular point; those leaving it form a three-dimensional surface.

Saddle-focus of the first kind. Two real roots: one positive, one negative root; a pair of complex roots with negative real parts; sequence (4.1, 32) has an odd number of sign changes; $a_4 < 0$. Saddle foci of the first or fifth kind are by their nature analogous to saddles of the first or third kind, similarly as saddle-foci of the second, third and fourth kind are analogous to saddles of the second kind; they differ from saddles only by the spiral character of the trajectories at the singular point.

Saddle-focus of the second kind. Two real positive roots, a pair of complex roots with negative real parts; $D_4 < 0$; sequence (4.1, 32) has an even number of sign changes.

Saddle-focus of the third kind. A pair of complex roots with negative real parts, a pair with positive real parts; $D_4 > 0$; sequence (4.1, 32) has an even number of sign changes.

Saddle-focus of the fourth kind. Two real negative roots, a pair of complex roots with positive real parts; $D_4 < 0$; sequence (4.1, 32) has an even number of sign changes; $a_4 > 0$.

Saddle-focus of the fifth kind. Two real roots: one positive, one negative; a pair of complex roots with positive real parts; $D_4 < 0$; sequences (4.1, 32) has an odd number of sign changes. From our point of view the cases of a multiple complex root with non-zero real part that occur in addition to the points just enumerated, represent neither an exception nor new types.

Unstable centre-node. A pair of pure imaginary roots, two positive real roots; $D_4 > 0$; sequence (4.1, 32) has an even number of sign changes.

Unstable centre-focus. A pair of pure imaginary roots, a pair of complex roots with positive real parts; $D_4 < 0$; sequence (4.1, 32) has an even number of sign changes.

Centre-saddle. A pair of pure imaginary roots, one real positive, one negative root; $D_4 > 0$; sequence (4.1, 32) has an odd number of sign changes.

Additional information can be found in the monograph by TONDL (1970a).

4.2.   Practical solution of the phase portraits

In order to be sufficiently clear and comprehensive and to provide enough information about the behaviour of the system being investigated, a phase portrait must contain the various singular points and the most important trajectories in the phase plane. Whenever several singular points are present, it is particularly necessary to establish (approximately at least) the position of all saddle points which are important for a practical solution. The most important trajectories are the limit cycles (if they exist at all) — both stable and unstable — and the trajectories entering and leaving the saddle points. The trivial case of a single steady solution represented by a singular point in the phase plane is not considered.

The cases which are apt to come up can be divided into three groups:

(1) A single singular point surrounded by one or more limit cycles.
(2) Several singular points but no limit cycles.
(3) Several singular points, one or more limit cycles.

Stable singular points and stable limit cycles represent the simple attractors of the system. In practical solutions it is well to keep in mind the following important consequences of the fact that, on the assumptions made in connection with system (4.1, 1), no trajectory can intersect another at a regular point (all points in the phase plane except the singular ones). At least one singular point must lie inside each limit cycle. A single singular point can be surrounded by several limit cycles, the stable cycles alternating with the unstable ones. If a single singular point, but no other limit cycle lies inside an unstable (stable) limit cycle, the singular point is stable (unstable). The unstable limit cycle then represents the separatrix which bounds the domain of attraction of the stable singular point lying inside the unstable limit cycle. In other cases the unstable limit cycle always forms part of the separatrix which bounds the domain of attraction of a particular stable steady solution. An unstable limit cycle can be obtained in the same way as a stable one by solving (4.1, 1) in "the negative time", i.e. by solving (4.1, 12) instead of (4.1, 1).

A solution should proceed from the determination (approximate at least) of the position of the singular points and identification of the saddle points. As will be shown in Section 4.5, this can be done quite generally for some systems. In the majority of cases, a practical analysis is not essentially concerned with finding out whether a singular point is a focus or a node.

The simplest procedure of all is that applicable to systems of group (1) (a single singular point). It starts with the initial conditions corresponding, approximately, to the coordinates of the singular point; depending on whether the singular point is stable or unstable, the equations are solved in the positive or negative time. The limit cycle is drawn as soon as a steady solution (a closed trajectory) has been obtained. In the next step the solution starts from a point lying outside the limit cycle drawn for the opposite time (i.e. the negative time if the preceding solution was effected in the positive time). Notice is taken of the trajectory (the motion of the plotter pen) to see whether or not it converges to a new limit cycle. In this way the limit cycles are determined one after another.

In the case of systems of group (2) (several singular points but no limit cycles) it should be realized that — as the trajectory geometry in the phase plane implies — n saddle points must exist among $2n + 1$ singular points. Accordingly, the solution should always start from a saddle point and continue by drawing the trajectories issuing from it in both the positive andnegative times. The character of the two pairs of trajectories thus obtained frequently provides enough information about the behaviour of the system being investigated. If a divergent solution cannot exist, the trajectories issuing from the saddle point in the positive time must terminate in a stable singular point. Only the stable singular points are the attractors of the system.

In the case of systems of group (3), one or both trajectories issuing from the saddle point in the positive time are apt to terminate on a stable limit cycle. An unstable limit cycle or an unstable singular point must lie inside this cycle. If the two trajectories issuing from the saddle point in the negative time are continuously moving away from this point, they form a separatrix, i.e. a boundary between the different domains of attraction of a particular steady solution. The trajectories issuing from the saddle point in the negative time can, however, also terminate in an unstable singular point or on an unstable limit cycle. Section 4.5 will discuss these possibilities using actual examples. The procedure just outlined makes it possible to obtain comprehensive and informative phase portraits within a comparatively short time.

Early phase portraits were frequently obtained by application of the graphico-numerical method of Lienard (see, e.g. KLOTTER (1980)). It has been found, however that an analogue or a digital computer fitted with a graph plotter for direct drawing of the trajectories is much more effective for this task. When a digital computer is used, the systems of first-order differential equations can be solved by one of the specific methods, such as the Runge-Kutta method or a simple procedure based on the idea of the Lienard method; the latter, proposed by TONDL (1978c) has been used for solving the examples presented in this chapter and its principle is explained in the Appendix.

The next three sections will deal with examples of different systems belonging to the groups defined at the beginning of Section 4.1. The simpler cases of groups (b) and (c) will be discussed first. The following notation will be used throughout:

$F_S(F_N)$ stable (unstable) focus
$N_S(N_N)$ stable (unstable) node
SP saddle point
$L_S(L_N)$ stable (unstable) limit cycle (drawn in solid (dashed) heavy lines)
s trajectories converging to the saddle point, provided that they form a separatrix (drawn in dot-and-dash lines).

4.3. Examples of systems of group (b)

The first example refers to a self-excited system governed by the equation

$$\ddot{y} + (\varkappa - \beta y^2 + \varepsilon y^4)\,\dot{y} + y + \mu y^3 = 0 \tag{4.3, 1}$$

where $\varkappa$, β, ε and μ are positive constants. Introducing the new variables $\dot{y} = v$, y, equation (4.3.1) takes the form

$$\left.\begin{aligned}
\dot{y} &= v\,, \\
\dot{v} &= -(\varkappa - \beta y^2 + \varepsilon y^4)\,v + y + \mu y^3\,.
\end{aligned}\right\} \tag{4.3, 2}$$

Although an analytical solution of steady vibration is not the principal subject of this chapter, it will be recapitulated here in order to provide as complete a picture as possible. Approximating the steady solution of (4.3, 1) by

$$y = A \cos \Omega t$$

and applying the harmonic balance method for the determination of A and Ω, leads to the following equations

$$\Omega^2 = 1 + \tfrac{3}{4}\mu A^2 ,$$ (4.3, 3)

$$K(A) = \varkappa - \tfrac{1}{4}\beta A^2 + \tfrac{1}{8}\varepsilon A^4 = 0 .$$ (4.3, 4)

For the amplitude A, (4.3, 4) implies

$$A_{1,2} = \left\{\frac{\beta}{\varepsilon} \mp \left[\left(\frac{\beta}{\varepsilon}\right)^2 - 8\frac{\varkappa}{\varepsilon}\right]^{1/2}\right\}^{1/2} .$$ (4.3, 4a)

So long as

$$\beta^2 > 8\varkappa\varepsilon$$ (4.3, 5)

the equation yields two values of A, i.e. two limit cycles exist in the phase plane. If condition (4.3, 5) is not satisfied, no real values of A are obtained. Since function $K(A)$ represents the damping (positive for $K(A) > 0$, negative for $K(A) < 0$), it can be assumed that the equilibrium position $y = 0$ represents a stable solution, that the unstable limit cycle in the phase plane corresponds to the solution with the amplitude A_1 ($A_1 < A_2$) and the stable limit cycle to the solution with the amplitude A_2. The curve of function $K(A)$ is shown in Fig. 4.3, 1 for $\varkappa = 0.3$, $\varepsilon = 10^{-4}$ and two values of β, i.e. 0.0175 and $8\varkappa\varepsilon$. In the first case ($\beta = 0.0175$) there exist two limit cycles in the phase plane; in the second (a border-line case) the two limit cycles become a single semi-stable limit cycle. From the practical point of view, the latter case may be considered to belong to those for which condition (4.3, 5) is not satisfied and the trivial solution ($y = 0$) is absolutely stable. The phase portrait of the first case ($\beta = 0.0175$) is shown in Fig. 4.3, 2 while Fig. 4.3, 3 shows the records of the vibration $y(t)$ corresponding to the limit cycles. The system belongs to the class with the so-called hard self-

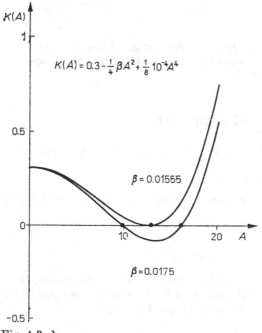

Fig. 4.3, 1

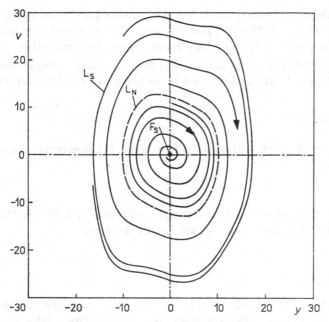

Fig. 4.3, 2

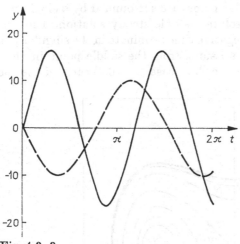

Fig. 4.3, 3

excitation because — so long as the initial conditions $(v(0),\ y(0))$ correspond to the points lying inside the unstable limit cycle — the solutions converge to the equilibrium position $y = 0$. The unstable limit cycle forms the boundary of the domain of attraction of the trivial solution, and outside it lies the domain of attraction of the steady self-excited vibration.

The second example refers to a self-excited system with three equilibrium positions, governed by the equation

$$\ddot{y} - (\beta - \delta\dot{y}^2)\,\dot{y} - \alpha y + \gamma y^3 = 0\,. \tag{4.3, 6}$$

When self-excitation is not in effect $(\beta = 0)$, two equilibrium positions, $y = \pm(\alpha/\gamma)^{1/2}$, are stable and one, $y = 0$, is unstable. The solution is carried out for constant $\beta = 0.2$

and $\alpha = \gamma = 1$ and alternatively varying δ. For $\beta > 0$, all equilibrium positions are unstable and the values of the coefficients β and δ decide (given α and γ) whether the steady vibration vibrates with a smaller amplitude about the equilibrium position $y = \pm \sqrt{\alpha/\gamma}$ ($y = \pm 1$ in the case being considered) or with a larger amplitude about the equilibrium position $y = 0$. If the first case takes place, the analytic solution can be approximated by

$$y = Y + A \cos \Omega t \tag{4.3, 7}$$

where Y is the constant deflection $\left(\text{for } A \to 0, \ Y \to \sqrt{\alpha/\gamma}\right)$. If the second, the solution can be approximated by

$$y = A \cos \Omega t \,. \tag{4.3, 8}$$

Applying the procedure used in the first example, equation (4.3, 6) (for the given values of the coefficients β, α and γ) can be written in the form

$$\dot{y} = v \,,$$
$$\dot{v} = (0.2 - \delta v^2)\, v + y(1 - y^2) \,. \tag{4.3, 9}$$

It can easily be shown that the singular points $y = \pm 1$ and $v = 0$ are unstable foci, and the point $v = 0$, $y = 0$ is a saddle. The phase portraits for three different values of δ are shown in Fig. 4.3, 4 ($\delta = 0.2$), Fig. 4.3, 5 ($\delta = 0.4$) and Fig. 4.3, 6 ($\delta = 0.7$). In the first two cases, all three singular points are surrounded by a single stable limit cycle which represents the only absolutely stable steady solution. The trajectories issuing from the saddle point in the negative time terminate in the singular points with coordinates $y = \pm 1$. The trajectories issuing from the saddle point in the positive time terminate on the limit cycle. In the third case ($\delta = 0.7$) each of the foci is sur-

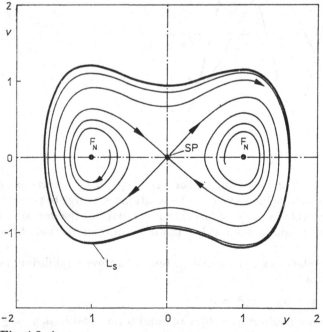

Fig. 4.3, 4

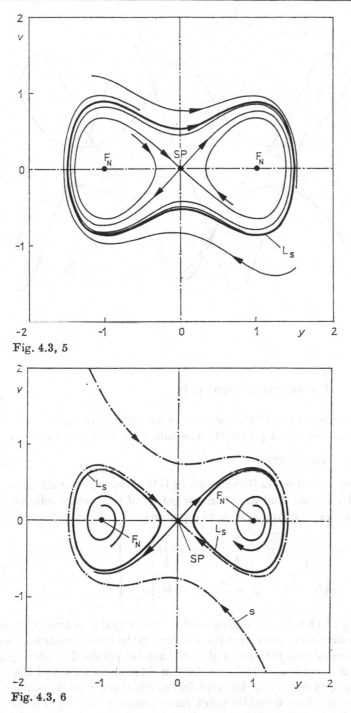

Fig. 4.3, 5

Fig. 4.3, 6

rounded by a stable limit cycle. The trajectories issuing from the saddle point in the negative time form the separatrix which separates the domains of attraction of the steady solutions represented in the phase plane by the above-mentioned limit cycle. Fig. 4.3, 7 shows the records of vibration corresponding to the limit cycles of the cases discussed.

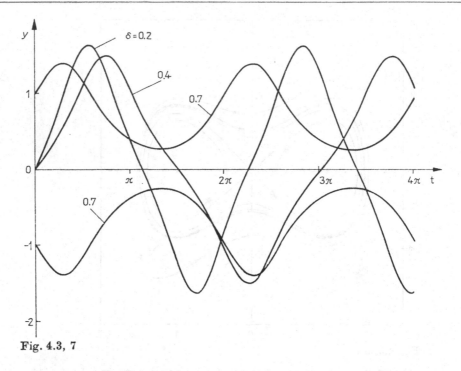

Fig. 4.3, 7

4.4.   Examples of systems of group (c)

As the first example consider the Duffing system (whose steady resonance solution was analyzed in Chapter 2) described (in the dimensionless form) by the equation

$$\ddot{y} + \varkappa\dot{y} + y + \alpha y^3 = \cos \eta\tau \qquad (4.4, 1)$$

where $\varkappa$ and γ are dimensionless coefficients, and η is the relative excitation frequency. Applying the van der Pol method to the investigation of the steady solution and its stability one obtains the following system of equations

$$\left.\begin{aligned}
\dot{a} &= \frac{1}{2\eta}\left\{-\varkappa\eta a + \left[1 - \eta^2 + \frac{3}{4}\gamma(a^2 + b^2)\right]b\right\}, \\
\dot{b} &= \frac{1}{2\eta}\left\{1 - \varkappa\eta b - \left[1 - \eta^2 + \frac{3}{4}\gamma(a^2 + b^2)\right]a\right\}.
\end{aligned}\right\} \qquad (4.4, 2)$$

The steady solutions of (4.4, 1) are represented by the singular points. The problem of stability of the steady solutions to any disturbances in the initial conditions is solved by determining the domains of attraction of these singular points. The phase portraits are drawn for the parameters $\varkappa = 0.05$, $\gamma = 10^{-2}$ and for three values of the relative excitation frequency η lying in the interval for which three steady solutions (two stable, one unstable to which a saddle point corresponds in the (a, b) phase plane) exist. The stable solutions differ from one another by the magnitude of the vibration amplitude $(A = (a^2 + b^2)^{1/2})$: small A — non-resonant solution, large A — resonant solution. The phase portraits are shown in Fig. 4.4, 1 $(\eta = 1.2)$, Fig. 4.4, 2 $(\eta = 1.3)$ and Fig. 4.4, 3 $(\eta = 1.5)$. As η increases, the amplitude of the resonant vibration grows larger, that of the non-resonant vibration, smaller. The reverse is true of the

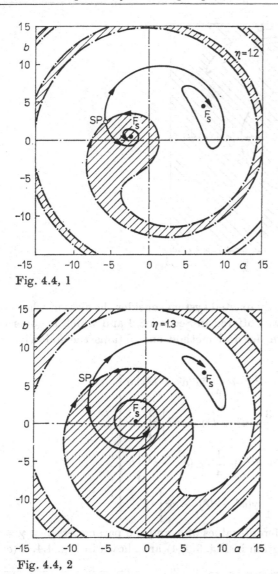

Fig. 4.4, 1

Fig. 4.4, 2

domains of attraction: as η increases, the domain of attraction of the resonant solution grows smaller, that of the non-resonant solution, larger (in Figs. 4.4, 1 to 4.4, 3, the latter domain is shown in hatching).

This finding has some important consequences: Consider a device whose motion is governed by the Duffing equation and which is desired to vibrate with a large amplitude (e.g. a vibratory compressor). Although this can be achieved in a broad interval of the excitation frequency in consequence of the nonlinearity of the restoring force with a hardening characteristic, it should be borne in mind that there exists an interval of this frequency in which two stable steady solutions (one resonant, the other non-resonant) are possible. The more the resonance peak is approached, the more readily can the effect of disturbances cause the device to "fall out" from the resonance regime; consequently, the device is incapable of being run in a regime corresponding to the vicinity of the resonance peak.

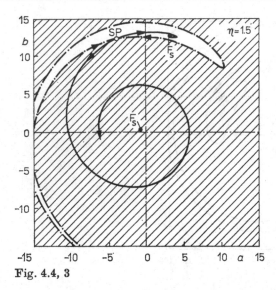

Fig. 4.4, 3

As an example of a parametrically excited system consider the system whose analytic solution of steady vibration was offered in Section 3.3 and which is governed by equation (3.3, 13). Using the van der Pol method one obtains the following set of differential equations:

$$\dot{a} = \frac{1}{2\omega} \left\{ - \left[\frac{4}{\pi} \vartheta (a^2 + b^2)^{-1/2} + \frac{1}{4} \delta \omega (a^2 + b^2) \right] a \right.$$
$$\left. + \left[1 - \frac{1}{2} \mu + \frac{3}{4} \gamma (a^2 + b^2) - \omega^2 \right] b \right\},$$

$$\dot{b} = \frac{1}{2\omega} \left\{ - \left[\frac{4}{\pi} \vartheta (a^2 + b^2)^{-1/2} + \frac{1}{4} \delta \omega (a^2 + b^2) \right] b \right.$$
$$\left. - \left[1 + \frac{1}{2} \mu + \frac{3}{4} \gamma (a^2 + b^2) - \omega^2 \right] a \right\}.$$

(4.4, 3),

The phase portraits for three different values of ω and the parameters $\delta = \gamma = 0.01$, $\vartheta = 0.1$ and $\mu = 0.2$ (cf. the diagram in Fig. 3.3, 4) are shown in Fig. 4.4, 4 ($\omega = 1$)

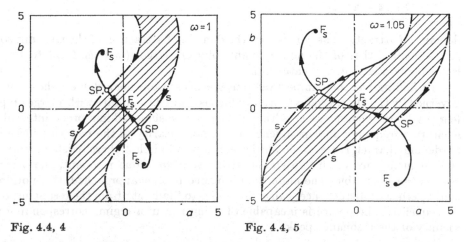

Fig. 4.4, 4   Fig. 4.4, 5

Fig. 4.4, 5 ($\omega = 1.05$) and Fig. 4.4, 6 ($\omega = 1.1$). Three domains of attraction are found to exist: that of the trivial solution, which grows larger with increasing ω (shown in hatching), and two domains of parametric resonance vibration. These vibrations have the same amplitude and differ from one another only by the phase shift. Their domains grow smaller with increasing ω.

A number of additional examples of systems of this group may be found in a book by HAYASHI (1964) and in a monograph by TONDL (1970a).

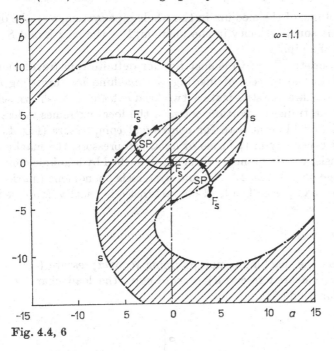

Fig. 4.4, 6

4.5. An example of a system of group (a)

The example does not represent a mechanical system but a model of a system with a flowing medium pumped into a pipe line by a centrifugal compressor, a pump or a blower. This simplified model is governed by a set of two first-order differential equations describing the time variation of the pressure difference P and the volumetric flow rate of the conveyed medium, Q. The example has been chosen not only because it represents an interesting case of (4.1, 1) when both the singular points and the limit cycles can exist in the phase plane but also because with its aid one can show quite generally some interesting properties of the system and the application of the phase portraits to its analysis. Only the salient points of the analysis will be discussed in this book. For a thorough treatment of the topical problems the reader is referred to a monograph by TONDL (1981a) discussing, among others, the effect of the system parameters, machine characteristics, etc.

The system used in the example is governed by the equations (whose derivation may be found in a book by KAZAKEVIČ (1974); McQUEEN (1976), SKALICKÝ (1979) and other authors have used similar equations)

$$C \frac{\mathrm{d}P}{\mathrm{d}t} = Q - \psi(P) , \qquad L \frac{\mathrm{d}Q}{\mathrm{d}t} = f(Q) - P \tag{4.5, 1}$$

where P is the inlet-outlet pressure difference, Q — the volumetric flow rate of the medium, t — the time, $\psi(P)$ or the inverse of it, $P = \varphi(Q)$ — the load characteristic (pipeline resistance), $P = f(Q)$ — the machine characteristic, and the parameters C (acoustic compliance) and L (acoustic mass) are defined by

$$C = \frac{V}{\varrho c^2}, \qquad L = \frac{\varrho l}{S} \tag{4.5, 2}$$

where V is the total volume of the medium in the whole pipe system, ϱ — the density of the medium, c — the sound velocity in the medium, l — the length and S — the cross-sectional area of the piping.

The machine characteristic $P = f(Q)$ is specified for definite rpm (revolutions per minute) and obtained in the process of loading the machine by throttling at the outlet. Typical machine characteristics feature two local extremes. As shown schematically in Fig. 4.5, 1a the transient portion between the local extremes is less steep for centrifugal pumps and blowers than for centrifugal compressors (Fig. 4.5, 1b); in the latter case, and especially in the case of axial compressors, the machine characteristic is apt to consist of two branches (it is not described by a unique continuous curve in the whole range of Q — Fig. 4.5, 1c) and hysteresis phenomena (marked with arrows in the diagram) take place. The load characteristic is usually expressed by a quadratic relation of the type

$$P = kQ^2 \operatorname{sgn} Q + d \tag{4.5, 3}$$

where $d = 0$ so long as the machine is not pumping to overpressure (e.g. a pump conveying water to an elevated reservoir); accordingly, the load characteristic is always an increasing function of Q.

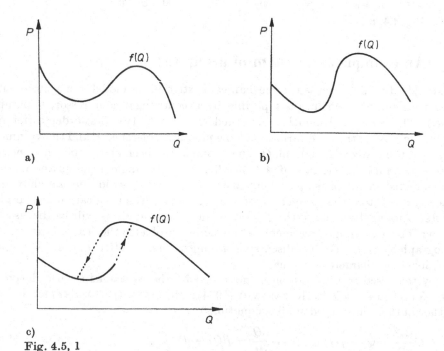

a)

b)

c)

Fig. 4.5, 1

Introducing the relative quantities

$$q = \frac{Q}{Q_0}, \qquad p = \frac{P}{P_0}.$$ (4.5, 4)

where P_0, Q_0 are chosen constant values, and the time transformation

$$\mathrm{d}t = \frac{LQ_0}{P_0}\, \mathrm{d}\tau$$ (4.5, 5)

equations (4.5, 1) can be converted to the dimensionless form

$$\frac{\mathrm{d}p}{\mathrm{d}\tau} = A[q - \Phi(p)], \qquad \frac{\mathrm{d}q}{\mathrm{d}\tau} = F(q) - p$$ (4.5, 6)

where

$$A = \frac{L}{C}\left(\frac{Q_0}{P_0}\right)^2.$$

is a dimensionless parameter, and the functions $\Phi(p)$, $F(p)$ are derived from the functions $\psi(P)$ and $f(q)$.

The points of intersection of the machine and the load characteristics are the singular points of (4.5, 6) in the (p, q) plane and represent the steady state — the non-oscillatory solutions. It is desirable that the steady state should occur for sufficiently large values of p and q, i.e. that it should be represented by a point on the right-hand portion of the characteristic. This means that the machine is supplying a large quantity of the medium at a high enough pressure. For a machine to operate reliably the steady state should be stable and that not only for very small disturbances (in the ideal case, the steady state would be absolutely stable). Such a case is apt to occur whenever the load characteristic intersects the machine characteristic at several points, i.e. whenever several steady states exist (if the limit case of the load characteristic only touching the machine characteristic is not considered, the usual number of such states is three).

Accordingly, the problem consists in establishing the initial conditions for which the different steady states occur, or the disturbances which are apt to lead to transition from the desirable steady state — represented by the singular point on the right-hand part of the machine characteristic — to the undesirable ones, i.e. the non-oscillatory steady states represented by the other singular points, or even the oscillatory ones (self-excited vibration) represented by the limit cycles in the (p, q) plane. This self-excited vibration — called the surge — manifests itself by violent fluctuations of the pressure and the flow rate and always constitute a serious hazard for safe operation.

In the analysis which follows the stability and the type of the singular points will be dealt with first. They can both be established by using the characteristic equation

$$\begin{vmatrix} -A\dfrac{\partial \Phi}{\partial p} - \lambda & A \\[2ex] -1 & \dfrac{\partial F}{\partial q} - \lambda \end{vmatrix} = 0$$ (4.5, 7)

where the coordinates of the singular point are substituted for p and q in expressions $\partial\Phi(p)/\partial p = \mathrm{d}\Phi/\mathrm{d}p$, $\partial F(q)/\partial q = \mathrm{d}F/\mathrm{d}q$. Denoting by $\varphi(q)$ the reciprocal to function

$\Phi(p)$, one obtains

$$\frac{\mathrm{d}\Phi}{\mathrm{d}p} = \frac{1}{\dfrac{\mathrm{d}\varphi}{\mathrm{d}q}} \tag{4.5, 8}$$

and the conditions for the singular point stability take the form

$$\frac{A}{\varphi'} - F' > 0 , \tag{4.5, 9}$$

$$1 - \frac{F'}{\varphi'} > 0 \tag{4.5, 10}$$

where

$$\varphi' = \frac{\mathrm{d}\varphi}{\mathrm{d}q} , \qquad F' = \frac{\mathrm{d}F}{\mathrm{d}q} .$$

Since always $A > 0$, $\varphi' > 0$ (as already mentioned, $\varphi(q)$ is an increasing function), it follows from conditions (4.5, 9) and (4.5, 10) that a sufficient condition for the stability of a singular point is a negative slope of the tangent to the machine characteristic at that point ($F' < 0$).

In the next step it will be generally shown which of the singular points are saddles (a proof concerning this matter was offerend by TONDL (1979 b)). Arrange the singular points according to the magnitude of the coordinate $q: q_1 < q_2 < q_3 < \ldots$ and mark them with the numerals 1, 2, ... (Fig. 4.5, 2). Since in the case being considered the

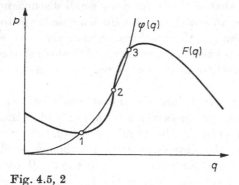

Fig. 4.5, 2

inequality $F(q) > \varphi(q)$ is satisfied for $q < q_1$, it always holds for the odd singular points that

$$\varphi' > F' \tag{4.5, 11}$$

and for the even singular points that

$$\varphi' < F' . \tag{4.5, 12}$$

As condition (4.5, 10) implies, all even singular points are always stable.

Theorem: *The even singular points are always saddles, and an odd singular point can never be a saddle.*

Proof: In this case, conditions (4.1, 8) for the existence of a saddle take the form

$$\left(\frac{A}{\varphi'} - F'\right)^2 > 4A\left(1 - \frac{F'}{\varphi'}\right), \tag{4.5, 13}$$

$$1 - \frac{F'}{\varphi'} < 0. \tag{4.5, 14}$$

The inequality $(1 - F'/\varphi' < 0)$ always applies to the even singular points; consequently, the above conditions are always satisfied. For the odd singular points, on the other hand, it always holds that $(1 - F'/\varphi' > 0)$ and the condition (4.5, 14) is, therefore, never satisfied.

The theorem is thus proved.

Mention (without proof) will also be made of another property of the system being studied. As Tondl (1980c) has shown the boundary — when the stability of the singular point is defined by the condition (4.5, 9) while the condition (4.5, 10) is always satisfied — is represented in the phase plane by a straight line passing through the point $q = 0$, $p = \varphi(0)$ and having a slope of $\frac{1}{2}\sqrt{A}$. If the singular point lies above this line, the condition (4.5, 9) prevails; this means that the stability of the singular point is affected by the magnitude of the parameter A (so long as a change of A does not cause the singular point to move below the limit straight line). As the analysis of a number of examples reveals, condition (4.5, 9) is the decisive one in most cases, i.e. the stability of the singular points is affected by the parameter A.

The more complicated case of a machine characteristic consisting of two branches will not be treated here. Tondl's (1979b) (see also (1981a)) approach to the solution of this problem is as follows: Divide the phase plane by the dividing curve into two definition domains; in each domain define the machine characteristic in terms of a function obtained by means of regression of the corresponding experimentally determined branch. As proved by the author, the point of intersection of the dividing curve and the machine characteristic can — for practical purposes — be regarded as a saddle point. The solution then proceeds in the way outlined in connection with the continuous machine characteristics.

As mentioned in Section 4.2, in the most effective procedure for obtaining the phase portraits, the solution starts from the saddle point and continues by drawing the trajectories in the positive as well as the negative time. If both characteristics are specified in the phase plane (p, q), the saddle points can readily be determined on the basis of the considerations put forward in the foregoing. The subsequent strategy is controlled by the course of the trajectories issuing from the saddle point in the positive and the negative time.

Below is a discussion of two sets of phase portraits obtained in the case of a continuous compressor characteristic expressed by the function

$$F(q) = 0.8 \tan^{-1}(100q - 20) + 1.8 - q + 25q^2 - 50q^3 \tag{4.5, 15}$$

and a load characteristic defined by the function

$$p = \varphi(q) = Kq^2 \operatorname{sgn} q \tag{4.5, 16a}$$

i.e. for the function $\Phi(p)$

$$\Phi(p) = \left(\frac{p}{K}\right)^{1/2} \operatorname{sgn} p. \tag{4.5, 16b}$$

Each set of phase portraits is introduced by the diagrams of the characteristics. The individual portraits correspond to different values of the parameter A (shown in the top right-hand corner). The first set corresponds to the load characteristic parameter $K = 40$ (Fig. 4.5, 3), the second to $K = 50$ (Fig. 4.5, 4). The notation is the same as that used in the preceding examples, but for better clarity the trajectories leaving the saddle point in the negative time (i.e. those approaching the saddle point in the positive time) and not forming the separatrix are drawn in light dashed lines. For $K = 40$ and the lowest value of $A = 50$, both trajectories issuing from the saddle point (in the positive time) terminate on the stable limit cycle. One of the trajectories issuing from the saddle in the negative time is moving towards the unstable focus, the other terminates on the unstable limit cycle which surrounds the stable focus. It will be seen that a sufficiently clear phase portrait is obtained by simply drawing the trajectories issuing from the saddle point in both the positive and the negative time. Two locally stable steady solutions exist in this case: one, non-oscillatory, represented by the stable focus lying on the right-hand part of the compressor characteristic, the other, oscillatory, represented by the stable limit cycle. A comparatively small unstable limit cycle bounds the domain of attraction of the desirable steady state. The whole remaining area of the phase plane constitutes the domain of attraction of the dangerous oscillatory solution. The oscillations are so violent that the minimum value of the instantaneous relative pressure p tends to zero and the value of the relative flow rate q is, in fact, negative in a certain interval of time, i.e. the medium flows in the opposite direction. An increase of the parameter A causes the area bounded by the stable limit cycle to grow smaller and the area bounded by the unstable limit cycle (which, for $A = 100$, reaches as far as the saddle point) to grow larger. For greater values of A (150 to 200) the stable limit cycle no longer exists and both foci are stable. The trajectories issuing from the saddle point in the negative time form the separatrix which divides the domain of attraction of the foci corresponding to the operating points on the right-hand and the left-hand part of the compressor characteristic. The latter point is unfavourable not only because the compressor delivers a small quantity of the medium at a low pressure but also because of the danger of its blades becoming overheated.

The phase portraits shown in Fig. 4.5, 4 for $K = 50$ are similar to those of the preceding set except that, for lower values of A ($A < 150$), the singular point on the right-hand part of the compressor characteristic is unstable ($F' > 0$). As A increases, the point on the left-hand part of the compressor characteristic becomes stabilized first (at an A as low as 75). The domain of attraction of the singular point on the right-hand part of the characteristic is comparatively small even for large values of A. Although the limit cycle no longer exists for $A = 100$, the singular point on the left-hand part of the characteristic is the only stable steady (i.e. absolutely stable) solution.

As a study of the different shapes of the compressor characteristics reveals, a rounded flat peak (the neighbourhood of the local maximum) when the slope of the tangent is positive ($F' > 0$) over a wide range of q, is not advantageous. In such a case the limit cycles are present for a wide interval of the values of the parameters K and A. This finding is confirmed by the results obtained for a compressor characteristic defined by the equation

$$F(q) = 0.7 \tan^{-1}(50q - 10) + 2 + q + 20q^2 - 70q^3 + 40q^4 .   (4.5, 17)$$

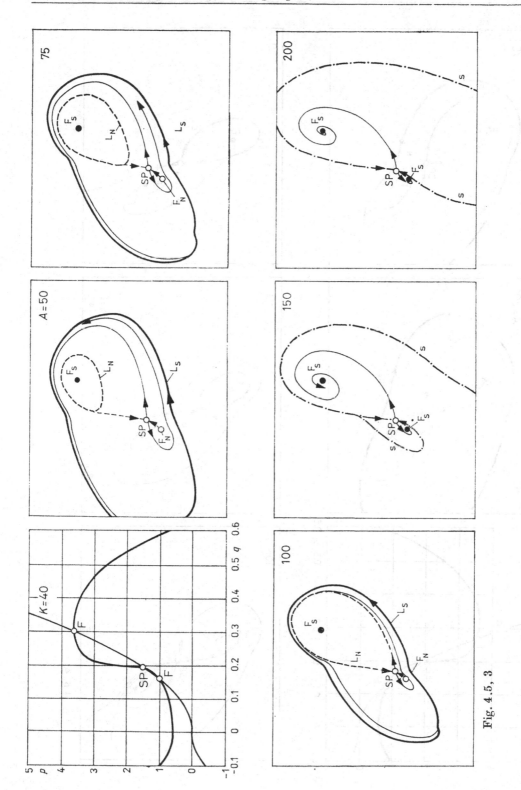

Fig. 4.5, 3

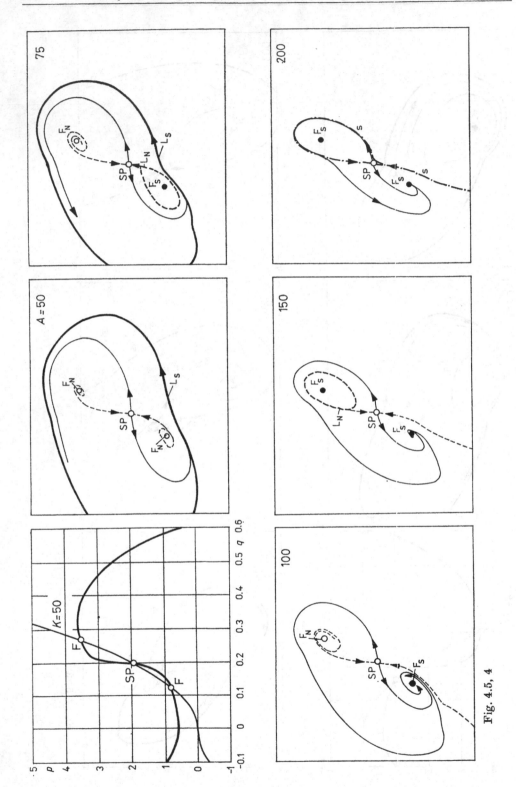

Fig. 4.5, 4

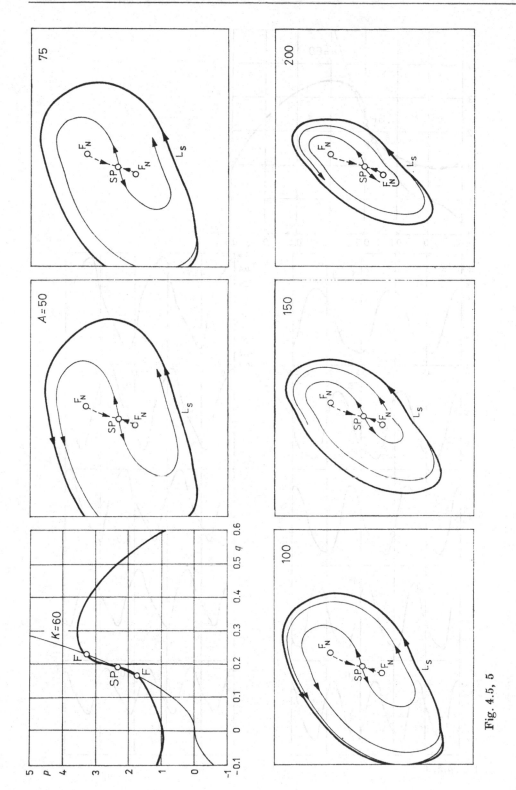

Fig. 4.5, 5

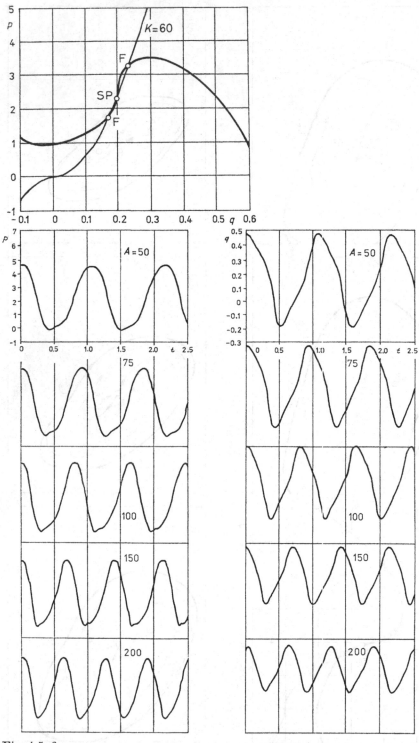

Fig. 4.5, 6

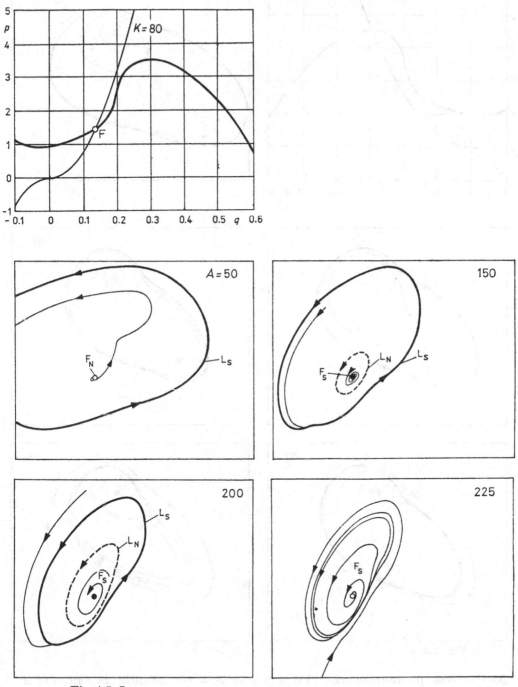

Fig. 4.5, 7

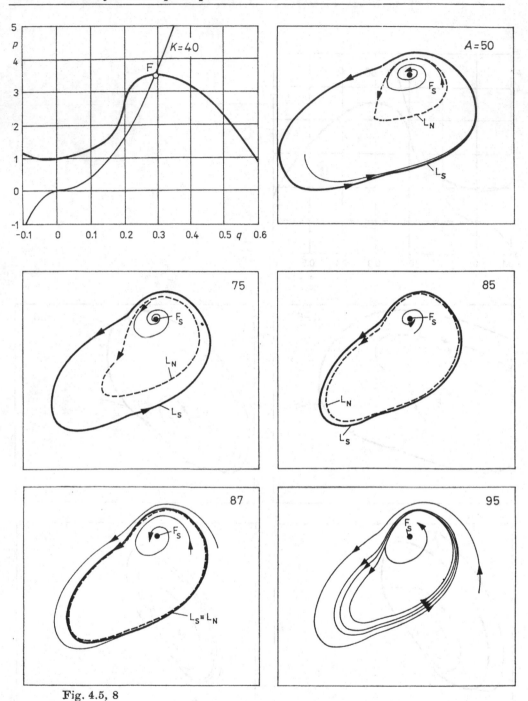

Fig. 4.5, 8

As the phase portraits shown in Fig. 4.5, 5 for $K = 60$ and different values of A suggest, limit cycles representing an absolutely stable steady solution exist for a wide range of A. Fig. 4.5, 6 shows the time relations $p(\tau)$, $q(\tau)$ for different values of A corresponding to the limit cycles. The phase portraits drawn in Fig. 4.5, 7 ($K = 80$) and Fig. 4.5, 8 ($K = 40$) (cases of a single singular point) disclose the presence of a limit cycle at low values of A in these cases, also. The effect of the parameter A is clearly seen in both figures. As this parameter is increased from a low value, one finds that at a definite A the singular point becomes stabilized and an unstable limit cycle lying inside the stable limit cycle forms around it. At a further increase of A the two limit cycles begin to approach one another until, at a definite value of A, they become a single semi-stable limit cycle (Fig. 4.5, 8 for $A = 87$). For A larger than this value, the limit cycles no longer exist.

The most important results of this section can be summed up as follows:

(a) Stability of a non-oscillatory steady state is no guarantee that self-excited oscillations (the surge) will not arise.

(b) Increasing the parameter A has a favourable effect on stability of the odd singular points lying on the right-hand and left-hand parts of the machine characteristic (provided that the rare case of condition (4.5, 4) being decisive is not considered).

(c) The domain of attraction of the singular point lying on the right-hand part of the machine characteristic does not necessarily increase monotonously with increasing A (Fig. 4.5, 3).

(d) The intensity of the self-excited oscillation decreases with increasing A. Although the self-excited oscillation vanishes at a certain value of A, its amplitudes do not tend to zero -- the change occurs by a jump.

(e) A rounded peak (at the local maximum of the machine characteristic) along which the slope of the tangent to the characteristic is positive in a wide interval of q, is not advantageous.

5. Forced, parametric and self-excited vibrations

5.1. Amplitude equations

The methods introduced in Chapter 2 make possible quite systematic investigations of vibrating systems under forced, parametric and self-excitation. In this chapter we investigate systems with one degree of freedom. The vibrations of a wide class of such systems can be modelled by a differential equation of the form

$$\ddot{x} + \Omega^2 x = \sum_{j=0}^{\infty} (P_j \cos j\omega t + Q_j \sin j\omega t) - B\dot{x} - Cx^2\dot{x} - Dx^4\dot{x}$$

$$- Ex^3 - Fx^5 - \sum_{j=1}^{\infty} (G_j \cos j\omega t + H_j \sin j\omega t) x$$

$$- \sum_{j=0}^{\infty} (K_j \cos j\omega t + L_j \sin j\omega t) x^2 \qquad (5.1, 1)$$

where dots denote derivatives with respect to time t, the coefficients P_j and Q_j a general periodic forced excitation (containing a factor ω^2 in case of an inertial excitation; P_0 describes a constant prestress), E and F a symmetric restoring force component B, a linear and C, D an amplitude-dependent non-linear damping if positive (for the equivalence of a nonlinear damping $-C'\dot{x}^3 - D'\dot{x}^5$ see KLOTTER (1980)). A soft self-excitation occurs for the van der Pol equation where

$$B < 0, \quad C > 0, \quad D = 0 \qquad (5.1, 2)$$

and for the generalized van der Pol equation where

$$B < 0, \quad C > 0, \quad D < 0, \qquad (5.1, 3)$$

a hard self-excitation for

$$B > 0, \quad C < 0, \quad D > 0. \qquad (5.1, 4)$$

The coefficients G_j and H_j denote a general periodic (linear) parametric excitation K_j and L_j $(j \geq 1)$ a periodic non-linear parametric excitation, K_0 especially a non-symmetric restoring force component.

We will introduce a dimensionless time $\tau = \omega t$, derivatives with respect to which are denoted by dashes, and a frequency variation

$$\alpha = \frac{\omega - \omega_0}{\omega_0}, \quad \text{i.e.} \quad \omega = \omega_0(1 + \alpha)$$

which characterizes the distance from a fixed (circular) frequency ω_0, and write

$$\lambda = \frac{\Omega^2}{\omega_0^2}, \qquad x(t) = x\left(\frac{\tau}{\omega}\right) = y(\tau).$$

Then the differential equation (5.1, 1) reads

$$y'' + \lambda y = \frac{1}{\omega_0^2} \sum_{j=0}^{\infty} (P_j \cos j\tau + Q_j \sin j\tau) - \alpha(2 + \alpha) y''$$

$$- \frac{(1+\alpha) B}{\omega_0} y' - \frac{(1+\alpha) C}{\omega_0} y^2 y' - \frac{(1+\alpha) D}{\omega_0} y^4 y' - \frac{E}{\omega_0^2} y^3$$

$$- \frac{F}{\omega_0^2} y^5 - \frac{1}{\omega_0^2} \sum_{j=1}^{\infty} (G_j \cos j\tau + H_j \sin j\tau) y$$

$$- \frac{1}{\omega_0^2} \sum_{j=0}^{\infty} (K_j \cos j\tau + L_j \sin j\tau) y^2 . \tag{5.1, 5}$$

We assume the coefficients $P_j, Q_j, \alpha, B, G_j, H_j$ on the right-hand side as sufficiently small. If λ is not a square of an integer (*non-resonance case*), the integral equation method introduced in Section 2.3 leads to the first approximative solution

$$y_{10} = \frac{1}{\omega_0^2} \sum_{j=0}^{\infty} \frac{P_j \cos j\tau + Q_j \sin j\tau}{\vartheta \lambda - j^2} \tag{5.1, 6}$$

where the factor $\vartheta = \vartheta_\lambda^{j^2}$, now equal to 1 for every j, is added because of the later use of the formula. Inserting this approximation in the right-hand side of (5.1, 5), we get the second approximation, and so on. The second and all higher approximations contain only terms of the order of magnitude P_j, Q_j multiplied by small terms. Therefore, the influence of non-linear restoring forces, damping, self-excitation and parametric excitation on the solution is of smaller order of magnitude than that of forced excitation, which mainly determines the vibrations.

For the rest of this chapter, we investigate the *resonance case* where λ is the square of an integer,

$$\lambda = n^2 .$$

Then the first approximative solution is

$$y_1 = r \cos n\tau + s \sin n\tau + y_{10} \tag{5.1, 7}$$

(the non-resonance case being included for $r = s = 0$). Omitting additional higher P_j, Q_j-terms mentioned above we get the second approximation

$$y_2 = y_1 - 2\alpha(r \cos n\tau + s \sin n\tau) + \frac{C}{4n\omega_0} A^2 (s \cos n\tau - r \sin n\tau)$$

$$+ \frac{C}{32n\omega_0} [s(3r^2 - s^2) \cos 3n\tau - r(r^2 - 3s^2) \sin 3n\tau] + \frac{D}{8n\omega_0} A^4(s \cos n\tau - r \sin n\tau)$$

$$+ \frac{3D}{128n\omega_0} A^2 [s(3r^2 - s^2) \cos 3n\tau - r(r^2 - 3s^2) \sin 3n\tau]$$

$$+ \frac{D}{240n\omega_0} [s(5r^4 - 10r^2 s^2 + s^4) \cos 5n\tau - r(r^4 - 10r^2 s^2 + 5s^4) \sin 5n\tau]$$

$$+ \frac{3E}{4n^2\omega_0^2} A^2(r \cos n\tau + s \sin n\tau)$$

$$+ \frac{E}{32n^2\omega_0^2} [r(r^2 - 3s^2) \cos 3n\tau + s(3r^2 - s^2) \sin 3n\tau]$$

$$+ \frac{5F}{8n^2\omega_0^2} A^4(r \cos n\tau + s \sin n\tau)$$

$$+ \frac{5F}{128n^2\omega_0^2} A^2[r(r^2 - 3s^2) \cos 3n\tau + s(3r^2 - s^2) \sin 3n\tau]$$

$$+ \frac{F}{240n^2\omega_0^2} [r(r^4 - 10r^2s^2 + 5s^4) \cos 5n\tau + s(5r^4 - 10r^2s^2 + s^4) \sin 5n\tau]$$

$$- \frac{1}{2\omega_0^2} \sum_{j=1}^{\infty} G_j \left[\frac{r \cos (n-j)\tau + s \sin (n-j)\tau}{\vartheta n^2 - (n-j)^2} + \frac{r \cos (n+j)\tau + s \sin (n+j)\tau}{n^2 - (n+j)^2} \right]$$

$$- \frac{1}{2\omega_0^2} \sum_{j=1}^{\infty} H_j \left[\frac{s \cos (n-j)\tau - r \sin (n-j)\tau}{\vartheta n^2 - (n-j)^2} + \frac{-s \cos (n+j)\tau + r \sin (n+j)\tau}{n^2 - (n+j)^2} \right]$$

$$- \frac{1}{2\omega_0^2} A^2 \sum_{j=0}^{\infty} \frac{K_j \cos j\tau + L_j \sin j\tau}{\vartheta n^2 - j^2}$$

$$- \frac{1}{4\omega_0^2} \sum_{j=0}^{\infty} \frac{[(r^2 - s^2) K_j + 2rsL_j] \cos (2n-j)\tau + [2rsK_j - (r^2 - s^2) L_j] \sin (2n-j)\tau}{\vartheta n^2 - (2n-j)^2}$$

$$- \frac{1}{4\omega_0^2} \sum_{j=0}^{\infty} \frac{[(r^2 - s^2) K_j - 2rsL_j] \cos (2n+j)\tau + [2rsK_j + (r^2 - s^2) L_j] \sin (2n+j)\tau}{n^2 - (2n+j)^2}$$

$$(5.1, 8)$$

where we denote the amplitude of the resonance part of the first approximation by A,

$$A = \sqrt{r^2 + s^2},$$

which we simply call "amplitude", and also omit terms containing α^2 because they are small in comparison with the term containing α and the third approximation would lead to additional α^2 terms.

The bifurcation parameters r and s have to be determined from the periodicity equations (2.3, 3). If they are known, the different approximative solutions describe the behaviour of the resulting vibrations. But as they are rather complicated, we have to confine ourselves mainly to the most important quantity characterizing the vibrations, the amplitude A which we derive from the periodicity equations.

Using the second approximation (5.1, 8), the periodicity equations contain the linear parametric excitation in form of the coefficients P_{2n}, Q_{2n} (main parametric resonance) only. In order to describe other parametric resonances, we have to use the lengthy parametric excitation terms of the third approximations. Thus we come to the two periodicity equations written in vector form

$$\binom{p}{0} + a \binom{r}{s} - \beta \binom{s}{-r} - g \binom{r}{-s} - h \binom{s}{r}$$

$$- k \binom{3r^2 + s^2}{2rs} - l \binom{2rs}{r^2 + 3s^2} - K \binom{r^2 - s^2}{-2rs} - L \binom{2rs}{r^2 - s^2} = 0 \qquad (5.1, 9)$$

where for abbreviation

$$a = \alpha - eA^2 - fA^4 + \Gamma, \qquad \beta = b + cA^2 + dA^4 \qquad (5.1, 10)$$

and

$$p = \frac{P_n}{2n^2\omega_0^2}, \qquad e = \frac{3E}{8n^2\omega_0^2}, \qquad f = \frac{5F}{16n^2\omega_0^2},$$

$$\Gamma = \frac{1}{8n^2\omega_0^2} \sum_{j=1}^{\infty} \left[\frac{1}{\vartheta n^2 - (n-j)^2} + \frac{1}{n^2(n+j)^2} \right] (G_j^2 + H_j^2), \qquad b = \frac{B}{2n\omega_0},$$

$$c = \frac{C}{8n\omega_0}, \qquad d = \frac{D}{16n\omega_0}, \qquad g = \frac{G_{2n}}{4n^2\omega_0^2} + G, \qquad h = \frac{H_{2n}}{4n^2\omega_0^2} + H,$$

$$G = \frac{1}{8n^2\omega_0^2} \sum_{j=1}^{\infty} \left[\frac{G_j G_{2n-j} + G_j G_{j-2n} - H_j H_{2n-j} + H_j H_{j-2n}}{\vartheta n^2 - (n-j)^2} \right.$$
$$\left. + \frac{G_j G_{j+2n} + H_j H_{j+2n}}{n^2 - (n+j)^2} \right],$$

$$H = \frac{1}{8n^2\omega_0^2} \sum_{j=1}^{\infty} \left[\frac{G_j H_{2n-j} - G_j H_{j-2n} + H_j G_{2n-j} + H_j G_{j-2n}}{\vartheta n^2 - (n-j)^2} \right.$$
$$\left. + \frac{G_j H_{j+2n} - H_j G_{j+2n}}{n^2 - (n+j)^2} \right],$$

$$k = \frac{K_n}{8n^2\omega_0^2}, \qquad l = \frac{L_n}{8n^2\omega_0^2}, \qquad K = \frac{K_{3n}}{8n^2\omega_0^2}, \qquad L = \frac{L_{3n}}{8n^2\omega_0^2}.$$

$$(5.1, 11)$$

The coefficient Q_n may be assumed to be zero, P_n to be positive without loss of generality by a suitable shifting of the time origin in the basic differential equation (5.1, 5):

$$P_n \cos n\tau + Q_n \sin n\tau = \tilde{P}_n \cos (n\tau - T) \tag{5.1, 12}$$

with

$$\tilde{P}_n = \sqrt{P_n^2 + Q_n^2}, \qquad \tan T = \frac{Q_n}{P_n}.$$

Up to Section 5.7, we neglect the influence of the non-linear parametric excitation in the periodicity equations (5.1, 9), $k = l = K = L = 0$, which holds in any case when no non-linear parametric excitation exists at all. Then the periodicity equations are, besides the expression $A^2 = r^2 + s^2$, linear in r and s, their solution is

$$(a^2 + \beta^2 - g^2 - h^2) \, r = -(a + g) \, p \, ,$$
$$(a^2 + \beta^2 - g^2 - h^2) \, s = (\beta - h) \, p \, . \tag{5.1, 13}$$

Squaring them and adding leads to the frequency amplitude equation or briefly *amplitude equation*

$$(a^2 + \beta^2 - g^2 - h^2)^2 \, A^2 = [(a + g)^2 + (\beta - h)^2] \, p^2 \tag{5.1, 14}$$

which contains the bifurcation parameters r, s in form of A^2 only. It is of power four in a, that is in the frequency variation α, and of power nine (five, if $d = 0$, resp. linear, if additionally $c = 0$) in A^2.

In what follows, different vibration phenomena will be explained by help of this amplitude equation. Some general conclusions are:

8*

1. The relation

$$A \approx \frac{|p|}{|a|}, \quad \text{if} \quad a^2 \gg \beta^2, g^2, h^2 \tag{5.1, 15}$$

holds, that is, A tends to zero (absolute minimum) with increasing $|a|$.

2. In case of vanishing forced excitation, $p = 0$, and linear damping, $c = d = 0$, the amplitude A vanishes for $a^2 \neq g^2 + h^2 - \beta^2$ (which always holds for $\beta^2 > g^2 + h^2$) whereas from the opposite relation $a^2 = g^2 + h^2 - \beta^2$ follows

$$eA^2 + fA^4 = \alpha + \Gamma \pm \sqrt{g^2 + h^2 - \beta^2},$$

an unlimited increasing of the amplitude.

3. If $p = 0$, no self-excitation occurs and damping is non-linear, $d > 0$ or $c > 0$, $d = 0$, the maximum amplitude is determined by

$$\beta = b + cA^2 + dA^4 = \sqrt{g^2 + h^2}. \tag{5.1, 16}$$

4. If $p \neq 0$ and damping is linear, the amplitude is limited for $\beta^2 > g^2 + h^2$.

5. If $p \neq 0$, no self-excitation occurs and damping is non-linear, the amplitude is limited.

The number of coefficients can be diminished by introducing transformed quantities. We introduce, if forced excitation p and linear damping b do not vanish,

$$\left.\begin{array}{ll} \bar{A} = \dfrac{|b|}{p} A, \quad \bar{g} = \dfrac{g}{|b|}, \quad \bar{h} = \dfrac{h}{|b|}, \\[2ex] \bar{a} = \bar{\alpha} - \bar{e}\bar{A}^2 - \bar{f}\bar{A}^4 + \bar{\Gamma}, \quad \bar{\beta} = \bar{b} + \bar{c}\bar{A}^2 + \bar{d}\bar{A}^4 \end{array}\right\} \tag{5.1, 17}$$

with

$$\left.\begin{array}{llll} \bar{\alpha} = \dfrac{\alpha}{|b|}, \quad \bar{\Gamma} = \dfrac{\Gamma}{|b|}, \quad \bar{b} = \dfrac{b}{|b|}, \quad \bar{e} = \dfrac{ep^2}{|b^3|}, \\[2ex] \bar{f} = \dfrac{fp^4}{|b^5|}, \quad \bar{c} = \dfrac{cp^2}{|b^3|}, \quad \bar{d} = \dfrac{dp^4}{|b^5|}. \end{array}\right\} \tag{5.1, 18}$$

The transformed quantities are also called amplitude, frequency variation, and so on. Thus (5.1, 14) reads

$$(\bar{a}^2 + \bar{\beta}^2 - \bar{g}^2 - \bar{h}^2)^2 \bar{A}^2 = (\bar{a} + \bar{g})^2 + (\bar{\beta} - \bar{h})^2, \tag{5.1, 19}$$

where $\bar{b} = \text{sign } b = \pm 1$ and especially for linear damping (vanishing self-excitation)

$$\bar{c} = \bar{d} = 0, \quad \text{that is} \quad \bar{\beta} = \bar{b} = 1 \tag{5.1, 20}$$

because then $b > 0$ is valid.

The amplitude A is, as (5.1, 14) shows, generally of the order of magnitude of p divided by the small parameters a, β, g, h. A comparison with the non-resonance solution (5.1, 6) reveals that A is of higher order of magnitude than the amplitude of y_{10} so that we are justified in denoting only the amplitude of the resonance part of the solution as "amplitude".

5.2.   Resonance curves, extremal amplitudes, and stability

The resonance curves show the dependence of the amplitude on the exciting frequency (variation).

If the system is linear, $\bar{a}$ and $\bar{\beta}(=1)$ are independent of the amplitude $\bar{A}$, and the resonance curves can be evaluated immediately from (5.1, 19), the only parameters being the parametric excitation coefficients $\bar{g}$ and $\bar{h}$. As an example, we choose in Fig. 5.2, 1 with $\bar{g} = 0$ a special phase relation between forced and parametric excitation. The resonance curves give the (transformed) amplitude as a function of $\bar{a} = \bar{\alpha} + \bar{\Gamma}$, that is of the frequency, for different values of the parametric excitation coefficient $\bar{h}$.

If damping is linear, but non-linear restoring forces with the coefficients $\bar{e}, \bar{f}$ appear, the dependence of $\bar{A}$ on $\bar{a}$ is the same, only the ordinate axis $\bar{a} = \bar{\alpha} + \bar{\Gamma} = 0$ is replaced by the *backbone curve*

$$\bar{a} = \bar{\alpha} - \bar{e}\bar{A}^2 - \bar{f}\bar{A}^4 + \bar{\Gamma} = 0 \qquad (5.2, 1)$$

from which the points of the resonance curve have equal horizontal distance. Fig. 5.2, 2 gives an example corresponding to Fig. 5.2, 1 but now $\bar{e} = -0.5, \bar{f} = 0$ is valid. If $\bar{e}$ and $\bar{f}$ have different sign, the backbone curve and in general the resonance curves tilt over, as Fig. 5.2, 3 shows for $\bar{e} = -0.5, \bar{f} = 0.2$.

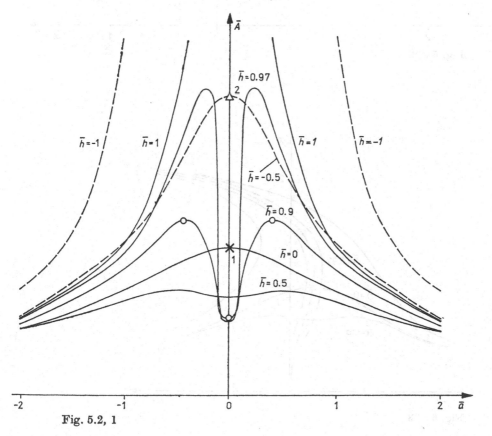

Fig. 5.2, 1

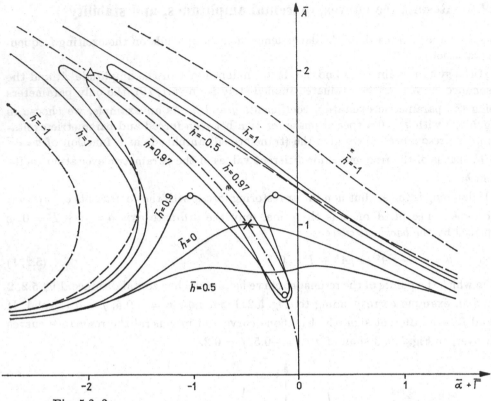

Fig. 5.2, 2

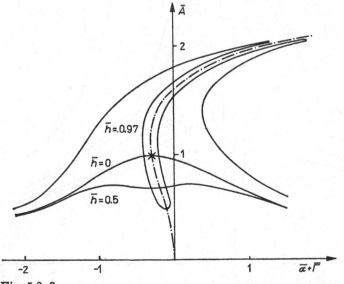

Fig. 5.2, 3

It is apparent that already for linear, the more for non-linear systems the resonance curves give information on the influence of the system parameters that is too complex and too complicated, at least as far as a combined forced and parametric excitation is involved. Very often, it is less important to know how the amplitudes depend on the frequency (variation) than how the maximum amplitude depends on the other system parameters.

In order to find this dependence, we differentiate (5.1, 19) by $\bar{a}$,

$$[(\bar{a}^2 + \bar{\beta}^2 - \bar{g}^2 - \bar{h}^2)^2 + 4(\bar{a}^2 + \bar{\beta}^2 - \bar{g}^2 - \bar{h}^2)\,\bar{\beta}(\bar{c} + 2\bar{d}\bar{A}^2)\,\bar{A}^2$$

$$+ 2(\bar{h} - \bar{\beta})\,(\bar{c} + 2\bar{d}\bar{A}^2)]\,\bar{A}\,\frac{\mathrm{d}\bar{A}}{\mathrm{d}\bar{a}} = -2\bar{a}(\bar{a}^2 + \bar{\beta}^2 - \bar{g}^2 - \bar{h}^2)\,\bar{A}^2 + \bar{a} + \bar{g}\,.$$

$$(5.2, 2)$$

Therefore the (necessary) condition $\mathrm{d}\bar{A}/\mathrm{d}\bar{a} = 0$ for extreme values $\bar{A} = \bar{A}_\nu$ of the resonance curve yields (if we exclude the amplitude values for which $\mathrm{d}\bar{a}/\mathrm{d}\bar{\alpha} = 0$) the equation

$$2\bar{A}^2\bar{a}^3 + [2(\bar{\beta}^2 - \bar{g}^2 - \bar{h}^2)\,\bar{A}^2 - 1]\,\bar{a} - \bar{g} = 0\,. \qquad (5.2, 3)$$

The extreme values can be found by solving the system (4.1, 19), (5.2, 3) for $\bar{A}$ and, what is of minor interest, $\bar{a}$.

For the rest of this section, we assume a *linear damping*, $\bar{c} = \bar{d} = 0$, that is, $\bar{\beta} = \bar{b}$ independent of $\bar{A}$. Then $\bar{A}$ can be eliminated in (5.2, 3) by help of (5.1, 19),

$$\bar{a}^3 + 3\bar{g}\bar{a}^2 + (\bar{\beta}^2 - 4\bar{\beta}\bar{h} + 3\bar{g}^2 + 3\bar{h}^2)\,\bar{a} + (\bar{g}^2 + \bar{h}^2 - \bar{\beta}^2)\,\bar{g} = 0 \qquad (5.2, 4)$$

or

$$\bar{a}^3 + 3\,|\gamma|\,\varphi\bar{a}^2 + (1 - 4\gamma\sqrt{1 - \varphi^2} + 3\gamma^2)\,\bar{a} + (\gamma^2 - 1)\,|\gamma|\,\varphi = 0 \qquad (5.2, 5)$$

with (5.1, 20) and the abbreviations

$$\gamma = \operatorname{sign} \bar{h} \cdot \sqrt{\bar{g}^2 + \bar{h}^2}\,, \quad \text{that is } |\gamma| = \sqrt{\bar{g}^2 + \bar{h}^2}\,, \quad \text{and} \quad \varphi = \frac{\bar{g}}{|\gamma|}\,.$$

The real solutions (three at most) $\bar{a}_0$, $\bar{a}_1$ and $\bar{a}_2$ of (5.2, 5) depend only on the two parameters γ, φ. The corresponding amplitude extreme values $\bar{A}_\nu$ follow from equation (5.1, 19), which now reads

$$(\bar{a}_\nu^2 + 1 - \gamma^2)^2\,\bar{A}_\nu^2 = (\bar{a}_\nu + |\gamma|\,\varphi)^2 + (1 - \gamma\sqrt{1 - \varphi^2})^2\,, \quad \nu = 0, 1, 2\,.$$

$$(5.2, 6)$$

The phase relation between the parametric excitation, with amplitude $|\gamma|$, and the forced excitation is characterized by $\varphi(-1 \leqq \varphi \leqq 1)$. Changing the sign of $\bar{g}$ and therefore of $\bar{\varphi}$ is, as follows from (5.2, 5), (5.2, 6), equivalent to changing the sign of $\bar{a}$, the amplitudes $\bar{A}$ being independent of this change. Therefore we can assume $0 \leqq \varphi \leqq 1$ without loss of generality. Equation (5.2, 6) shows that $\bar{A}_\nu$ is finite for $|\gamma| < 1$ whereas $|\gamma| = 1$ leads to $\bar{a}_0 = 0$ and $\bar{A}_0$ tending to infinity at least for $\varphi \neq 0$. The maximum amplitude in dependence on γ and φ can be found by numerically eliminating $\bar{a}$ from (5.2, 5) and (5.2, 6).

Fig. 5.2, 4 gives the resulting maximum amplitude $\bar{A}$ as dependent on γ for different values of φ. It reveals the *phenomenon* that the maximum amplitude is smaller for

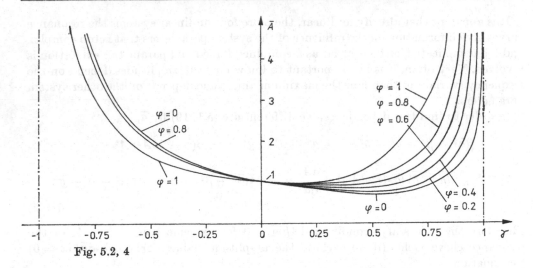

Fig. 5.2, 4

a region of positive values of γ (then the sine component of parametric excitation and the forced excitation operating in (5.1, 9) have the same sign) than for $\gamma = 0$, that is *an additional parametric excitation diminishes the amplitudes caused by forced excitation.* This effect occurs for all values of φ and shows *more distinctly the smaller φ is.* When γ is negative, no such effect arises and the maximum amplitude increases monotonically with $|\gamma|$.

Formulae (5.2, 5), (5.2, 6) respectively Fig. 5.2, 4 give the maximum amplitude, the most important vibration characteristic, for forced and linear parametric excitation *in full generality*, by means of (5.1, 17), (5.1, 18) and (5.1, 11) depending on all system parameters, with the one exception that we omitted non-linear damping. Fig. 5.2, 4 shows that the maximum amplitude depends for $|\gamma| < 0.15$ slightly, but for $\gamma < -0.5$ and $\gamma > 0.25$ essentially on the phase relation between forced and parametric excitation expressed by φ.

The *special case* $\varphi = 0$, that is $\bar{g} = 0$, $\gamma = \bar{h}$, which we assume in what follows, can be evaluated much more easily. In this case the amplitude frequency relation (5.1, 19) is biquadratic in $\bar{a}$. The extreme value equation (5.2, 5) simplifies to

$$\bar{a}^3 + (1 - 4\gamma + 3\gamma^2)\,\bar{a} = 0$$

with the solutions

$$\bar{a}_0 = 0\,, \qquad \bar{a}_{1,2} = \pm\sqrt{4\gamma - 1 - 3\gamma^2} \tag{5.2, 7}$$

the latter values being real for

$$\tfrac{1}{3} \leqq \gamma \leqq 1\,. \tag{5.2, 8}$$

Equation (5.2, 6) gives the corresponding amplitude extremes

$$(1 + \gamma)^2\,\bar{A}_0^2 = 1 \tag{5.2, 9}$$

and

$$8\gamma(1 - \gamma)\,\bar{A}_{1,2}^2 = 1\,. \tag{5.2, 10}$$

The amplitude extreme $\bar{A}_{1,2}$, drawn as a dashed line in Fig. 5.2, 5, represents, in the interval (5.2, 8) of its existence, the maximum amplitude. The amplitude extreme $\bar{A}_0$, drawn as a full line, represents the maximum amplitude in the interval $-1 \leqq \gamma \leqq \frac{1}{3}$ and a (relative, compare (5.1, 15)) minimum in the interval $\frac{1}{3} < \gamma < 1$. The maximum curve in Fig. 5.2, 5 is identical with the curve $\varphi = 0$ in Fig. 5.2, 4. Using (5.2, 9) and (5.2, 10), we can confirm the *phenomenon* explained above, *for* $\varphi = 0$: The maxi-

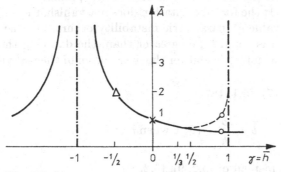

Fig. 5.2, 5

mum amplitude decreases when γ changes from 0 to $+\frac{1}{2}$ and increases with increasing $|\gamma|$ only outside of this interval, especially for negative γ. The minimum value of the maximum amplitude is $\overline{A}_{1,2} = 1/\sqrt{2} = 0.7071$ whereas $A_0 = 1$ for $\gamma = 0$, that is an amplitude decrease of about 30% by additional parametric excitation. Figures 5.2, 1 to 5.2, 3 show examples for this phenomenon. The extension to non-linear damping is given in the next Section.

In the interval $\frac{1}{3} < \gamma < 1$, the minimum amplitude $\overline{A}_0$ lies on the backbone curve $\overline{a} = 0$, the horizontal frequency distance of the maximum amplitude from this curve being given by (5.2, 7), sketched in Fig. 5.2, 6. The greatest possible distance $\overline{a}_1 = 1/\sqrt{3}$ occurs for $\gamma = \frac{2}{3}$.

Each of the resonance curves in Figures 5.2, 1 to 5.2, 3 corresponds to curve points (marked in the same way) in Figures 5.2, 5 and 5.2, 6 which give the maximum and minimum amplitudes and corresponding frequency values. Figures 5.2, 4 to 5.2, 6 do not show the special shape of the resonance curves, but reveal the dependence of the maximum and minimum amplitudes and the corresponding frequency values on the system parameters.

The forced excitation coefficient p influences, as (5.1, 17), (5.1, 18) show, besides the non-linear restoring force and damping terms, only the amplitude scale. If the system is linear, the (original) amplitude A is proportional to p because of (5.1, 14).

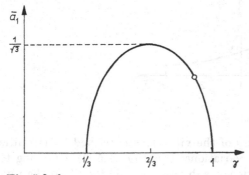

Fig. 5.2, 6

The extremes A_0, $A_{1,2}$ of the original amplitude, being independent of the non-linear restoring forces, are proportional to p when only damping is linear, as follows from (5.2, 9), (5.2, 10) and (5.1, 17). If the forced excitation vanishes, A also vanishes for $\gamma < 1$ (lower curve in Fig. 5.3, 5), but remains indefinite for the value $\gamma = 1$, that is $\sqrt{g^2 + h^2} = b$, which we denote as the *threshold value* because the parametric excitation reaches the damping threshold. If the forced excitation does not vanish, the amplitude is unlimited for the threshold value. A parametric instability occurs, in other words, when the parametric excitation is equal to, or greater than, the damping threshold, then the linear damping model is not sufficient for the description of the real vibration behaviour.

Introducing instead of (5.1, 17), (5.1, 18)

$$\tilde{A} = \frac{|h|}{p} A \ (= |\bar{h}| \, \overline{A}) \, , \qquad \tilde{b} = \frac{b}{h} \quad (\gtreqless 0 \text{ when } h \gtreqless 0) \tag{5.2, 11}$$

gives, because of $\gamma = \bar{h} = h/b$, instead of (5.2, 9), (5.2, 10)

$$(\tilde{b} + 1)^2 \, \tilde{A}_0^2 = 1 \tag{5.2, 12}$$

and

$$8(\tilde{b} - 1) \, \tilde{A}_{1,2}^2 = 1 \, . \tag{5.2, 13}$$

The amplitude extremum $\tilde{A}_0$ is the maximum for $\tilde{b} \leq -1$ and $\tilde{b} \geq 3$ and a relative minimum for $1 \leq \tilde{b} < 3$ whereas $\tilde{A}_{1,2}$ is the maximum (drawn by dashed line in Fig. 5.2, 7) in the latter interval. In the interval $-1 \leq \tilde{b} \leq 1$ the linear damping model fails to describe the vibrations. The intervals $(-1, 0)$ and $(0, 1)$ in Fig. 5.2, 5 now correspond with the intervals $(-\infty, -1)$ resp. $(1, \infty)$.

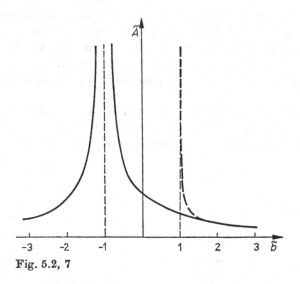

Fig. 5.2, 7

In order to investigate the *stability* of the vibrations described by the differential equation (5.1, 5), we use the linear variational equation (1.5, 3). Omitting terms of smaller order of magnitude, in particular substituting $-\lambda z$ for z'' on the right-hand side, results in the linear variational equation of the form

$$z'' + \lambda z = 2\lambda\alpha z - \frac{B}{\omega_0} z' - \frac{C}{\omega_0} y^2 z' - \frac{2C}{\omega_0} yy'z - \frac{D}{\omega_0} y^4 z' - \frac{4D}{\omega_0} y^3 y'z$$

$$- \frac{3E}{\omega_0^2} y^2 z - \frac{5F}{\omega_0^2} y^4 z - \frac{1}{\omega_0^2} \sum_{j=1}^{\infty} (G_j \cos j\tau + H_j \sin j\tau) z$$

$$- \frac{2}{\omega_0^2} \sum_{j=0}^{\infty} (K_j \cos j\tau + L_j \sin j\tau) yz . \qquad (5.2, 14)$$

After inserting the resonance part, as in evaluating (5.1, 8), of the first approximative solution (5.1, 7), the linear variational equation (5.2, 14) takes the form (2.4, 1). The coefficients appearing in the stability conditions (2.4, 4), (2.4, 5) now follow as

$$u_0 = 2\lambda\alpha - 2n^2(2eA^2 + 3fA^4 + 4kr + 4ls) ,$$

$$v_0 = -2n(b + 2cA^2 + 3dA^4) ,$$

$$u_{2n} = -2n^2[g + e(r^2 - s^2) + 2fA^2(r^2 - s^2) + 4crs + 8dA^2rs$$
$$+ 2(kr - ls + Kr + Ls)] ,$$

$$U_{2n} = -2n^2[h + 2crs + 4fA^2rs + 2c(s^2 - r^2) + 4dA^2(s^2 - r^2)$$
$$+ 2(lr + ks + Lr - Ks] ,$$

$$v_{2n} = -2n[c(r^2 - s^2) + 2dA^2(r^2 - s^2)] ,$$

$$V_{2n} = -4n(crs + 2dA^2rs)$$

where the transformed quantities (5.1, 11) have been introduced and non-linear parametric excitation has been included.

The first stability condition (2.4, 4) now reads

$$b + 2cA^2 + 3dA^4 > 0 . \qquad (5.2, 15)$$

It holds for every amplitude if no self-excitation exists (b, c, d non-negative and $b > 0$ or $c > 0$, $A > 0$ or $d > 0$, $A > 0$). The second stability condition (2.4, 5) can be written

$$(b + 2cA^2 + 3dA^4)^2 > [g + (e + 2fA^2)(r^2 - s^2) + 2(c + 2dA^2) rs$$
$$+ 2(kr - ls + kr + Ls)]^2$$

$$+ [h - (c + 2dA^2)(r^2 - s^2) + 2(e + 2fA^2) rs + 2(lr + ks + Lr - Ks)]^2$$

$$- (\alpha - 2eA^2 - 3fA^4 - 4kr - 4ls)^2 . \qquad (5.2, 16)$$

These conditions will be the starting point for different stability results.

Differentiating the amplitude equation (5.1, 14) by A and using

$$\frac{da}{dA} = \frac{d\alpha}{dA} - 2A(e + 2fA^2) , \qquad \frac{d\beta}{dA} = 2A(c + 2dA^2) ,$$

we get

$$[2(g^2 + h^2 - a^2 - \beta^2) aA^2 + (a + g) p^2] \frac{d\alpha}{dA} = -2[2(a^2 + \beta^2 - g^2 - h^2) aA^2$$

$$- (a + g) p^2] (e + 2fA^2) A + 2[2(a^2 + \beta^2 - g^2 - h^2) \beta A^2 + (h - \beta) p^2]$$

$$\times (c + 2dA^2) A + (a^2 + \beta^2 - g^2 - h^2)^2 A . \qquad (5.2, 17)$$

At present we restrict ourselves to the linear case $c = d = e = f = k = l = K = L = 0$. Then (5.2, 16) reduces to

$$\alpha^2 > g^2 + h^2 - b^2$$

from which follows an instability region that vanishes for $b^2 \geqq g^2 + h^2$. After (5.2, 16), a vertical tangent, $d\alpha/dA = 0$, implies

$$\alpha^2 = g^2 + h^2 - b^2$$

because otherwise $A \neq 0$ holds if $p \neq 0$. Consequently, the points on the resonance curve with vertical tangents correspond to the boundary points of the instability region.

5.3.   Non-linear damping

If damping is non-linear the amplitude equation (5.1, 19) is of ninth (for $d = 0$ of fifth) degree in $\bar{A}^2$ and of fourth degree in $\bar{a}$, a numerical evaluation of the resonance curves is possible for numerically given parameters.

In this section, we exclude a self-excitation by $b > 0$ (that is $\bar{b} = 1$), $c > 0$, $d \geqq 0$ and assume the special phase relation $\bar{g} = 0$ (that is $g = 0$, $\gamma = \bar{h}$) between forced and parametric excitation. Then the amplitude equation (5.1, 19) is biquadratic in $\bar{a}$, with the solution

$$\bar{a}^2 = \frac{1}{2\bar{A}^2} - \bar{\beta}^2 + \gamma^2 \pm \sqrt{\frac{1}{4\bar{A}^4} - \frac{2\bar{\beta}\gamma}{\bar{A}^2} + \frac{2\gamma^2}{\bar{A}^2}} \tag{5.3, 1}$$

which gives, together with (5.1, 17), vice versa the frequency variation as a function of the amplitude.

The condition (5.2, 3) for amplitude extremes yields $\bar{a}_0 = 0$, as for linear damping, and

$$\bar{a}_{1,2} = \pm \sqrt{\gamma^2 - \bar{\beta}^2 + \frac{1}{2\bar{A}_{1,2}^2}} \; . \tag{5.3, 2}$$

Substitution in (5.1, 19) leads to the corresponding amplitude extremes

$$(\bar{\beta} + \gamma)^2 \, \bar{A}_0^2 = 1 \tag{5.3, 3}$$

and

$$8\gamma(\bar{\beta} - \gamma) \, \bar{A}_{1,2}^2 = 1 \tag{5.3, 4}$$

which generalize (5.2, 9) and (5.2, 10). Using (5.3, 4), formula (5.3, 2) can be written

$$\bar{a}_{1,2} = \pm \sqrt{4\bar{\beta}\gamma - \bar{\beta}^2 - 3\gamma^2}$$

which is real for

$$\frac{\bar{\beta}}{3} \leqq \gamma \leqq \bar{\beta} \quad \text{(that is } \gamma < \bar{\beta} \leqq 3\gamma) \, . \tag{5.3, 5}$$

In this domain, $\bar{A}_1 = \bar{A}_2$ is also real and represents the maximum amplitude whereas $\bar{A}_0$ is a (relative) minimum amplitude. For $\gamma < \bar{\beta}/3$, the amplitude $\bar{A}_0$ represents the maximum amplitude.

The form

$$\gamma = -1 - \bar{c}\bar{A}_0^2 - \bar{d}\bar{A}_0^4 \pm \frac{1}{\bar{A}_0} \qquad (5.3, 6)$$

of equation (5.3, 3) reveals the composition of the extremal curve (full lines which asymptotically approach the γ axis and the dotted "damping curve"

$$\gamma_0 = -1 - \bar{c}\bar{A}_0^2 - \bar{d}\bar{A}_0^4 \qquad (5.3, 7)$$

in Fig. 5.3, 1 where only non-linear damping is specialized by $\bar{c} = 0.2$, $\bar{d} = 0$). For sufficiently great negative values of γ, a second extremal curve appears which corresponds to a second detached resonance curve for smaller amplitudes with a medium-sized minimum amplitude and a second small maximum amplitude.

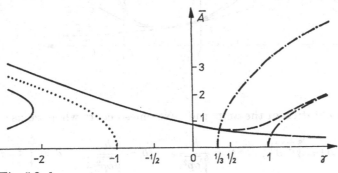

Fig. 5.3, 1

Analogously, (5.3, 4) can be written

$$\gamma = 1 + \bar{c}\bar{A}_{1,2}^2 + \bar{d}\bar{A}_{1,2}^4 - \frac{1}{8\gamma\bar{A}_{1,2}^2} \qquad (5.3, 8)$$

or, solving for γ,

$$2\gamma = 1 + \bar{c}\bar{A}^2 + \bar{d}\bar{A}^4 \pm \sqrt{(1 + \bar{c}\bar{A}^2 + \bar{d}\bar{A}^4)^2 - \frac{1}{2\bar{A}^2}} \;.$$

The boundaries of the domain of existence (5.3, 5) for the maximum amplitude given by (5.3, 8) are drawn in Fig. 5.3, 1 as dashed-dotted lines, the maximum curve (5.3, 8) in this domain, which asymptotically approaches the upper boundary, is drawn as a dashed line.

The influence of different values $\bar{c}$ of cubic damping on the maximum amplitudes is given in Fig. 5.3, 2 in greatest possible generality. Only the fifth order damping coefficient $\bar{d}$ (which would further diminish the maximum amplitudes) is put equal to zero.

Fig. 5.3, 2 shows:

1. The parametric diminishing effect (Section 5.2) holds for non-linear as well as for linear damping with the smallest maximum amplitude not appearing for vanishing parametric excitation but for $\gamma = 0.5$.
2. Non-linear damping has little influence on the maximum amplitudes for $0 < \gamma < 0.5$ but a decisive influence for $\gamma > 0.8$ and $\gamma < -0.5$.

3. If parametric excitation is equal to, or greater than, the damping threshold, $|\gamma| \geqq 1$, the maximum amplitude can only be determined by aid of non-linear damping.

4. Restoring non-linearities do not influence the maximum amplitude.

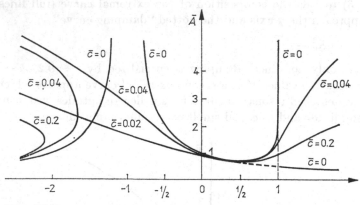

Fig. 5.3, 2

A different representation of the amplitude extremes occurs when we use

$$\check{A} = \sqrt[3]{\frac{c}{p}}\, A \left(= \sqrt[3]{\check{c}}\, \bar{A}\right), \qquad \check{\gamma} = \frac{h}{\sqrt[3]{cp^2}}, \qquad \check{b} = \frac{b}{\sqrt[3]{cp^2}}, \qquad \check{d} = \sqrt[3]{\frac{p^2}{c^5}}\, d$$

$$(5.3, 9)$$

instead of (5.1, 17), (5.1, 18). Then (5.3, 6) and (5.3, 8) read

$$\check{\gamma} = -\check{b} - \check{A}_0^2 - \check{d}\check{A}_0^4 \pm \frac{1}{\check{A}_0}, \qquad \check{\gamma} = \check{b} + \check{A}_{1,2}^2 + \check{d}\check{A}_{1,2}^4 - \frac{1}{8\check{\gamma}\check{A}_{1,2}^2}.$$

In case of a cubic damping non-linearity, $\check{d} = 0$, the corresponding amplitude extremal curves can be composed by parabolae and hyperbolae where only the starting point $(\mp)\,\check{b}$ of the parabolae and the second hyperbola depend on damping, as Figures 5.3, 3 and 5.3, 4 show. If $\check{d}$ is positive, the maximum amplitudes are smaller.

In order to determine the influence of forced excitation, we introduce

$$\hat{A} = \sqrt{\frac{5c}{b}}\, A \ \left(= \sqrt{5c}\bar{A}\right), \qquad \hat{d} = \frac{bd}{25c^2}\left(= \frac{\bar{d}}{25c^2}\right), \qquad \hat{p} = \sqrt{\frac{5c}{b^3}}\, p$$

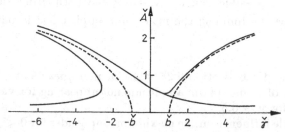

Fig. 5.3, 3

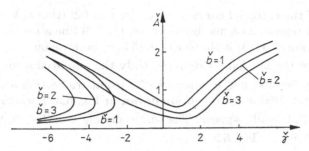

Fig. 5.3, 4

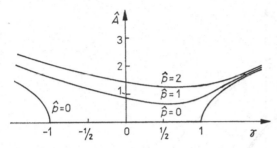

Fig. 5.3, 5

(the factors 5 and 0.2 enable the correspondence with the example of Fig. 5.3, 1) and get

$$\gamma = -1 - 0.2\hat{A}_0^2 - \hat{d}\hat{A}_0^4 \pm \frac{\hat{p}}{\hat{A}_0}, \qquad \gamma = 1 + 0.2\hat{A}_{1,2}^2 + \hat{d}\hat{A}_{1,2}^4 - \frac{\hat{p}^2}{8\gamma\hat{A}_{1,2}^2}$$

instead of (5.3, 7), (5.3, 8). Fig. 5.3, 5 gives the corresponding maximum amplitudes for different values of $\hat{p}$ under the only restriction $\hat{d} = 0$. In case $\hat{d} > 0$, the maximum amplitudes are smaller. The figure shows that the influence of forced excitation is comparatively great if the parametric excitation is below the threshold value $|\gamma| = 1$.

Using, in correspondence with (5.2, 11),

$$\tilde{A} = \frac{|h|}{p} A, \qquad \tilde{\beta} = \tilde{b} + \tilde{c}\tilde{A}^2 + \tilde{d}\tilde{A}^4, \qquad \tilde{b} = \frac{b}{h}, \qquad \tilde{c} = \frac{cp^2}{h^3}, \qquad \tilde{d} = \frac{dp^4}{h^5},$$

$$(5.3, 10)$$

(5.3, 3) and (5.3, 4) read

$$(\tilde{\beta} + 1)^2 \tilde{A}_0^2 = 1, \tag{5.3, 11}$$

$$8(\tilde{\beta} - 1) \tilde{A}_{1,2}^2 = 1 \tag{5.3, 12}$$

in generalization of (5.2, 22), (5.2, 13). Now $\tilde{A}_0$ is the maximum amplitude in the domain $\tilde{\beta} \leqq -1$ and $\tilde{\beta} \geqq 3$ and a relative minimum for $1 \leqq \tilde{\beta} < 3$ whereas $\tilde{A}_{1,2}$ is the maximum in the latter interval. When we write (5.3, 11), (5.3, 12) in the form

$$\tilde{b} = -1 - \tilde{c}\tilde{A}_0^2 - \tilde{d}\tilde{A}_0^4 \pm \frac{1}{\tilde{A}_0}, \tag{5.3, 13}$$

$$\tilde{b} = 1 - \tilde{c}\tilde{A}_{1,2}^2 - \tilde{d}\tilde{A}_{1,2}^4 + \frac{1}{8\tilde{A}_{1,2}^2}, \tag{5.3, 14}$$

we find the composition of the extremal curve as given for $\tilde{c} = 0.2$ (that is $h > 0$) in Fig. 5.3, 6. The dashed line represents a maximum curve, the full line a maximum or minimum curve. The equation (5.3, 14) is simpler than (5.3, 8) because now the right-hand side is independent of the left-hand side term. Only the amplitudes for $\tilde{b} \geqq 0$ in Fig. 5.3, 6 really appear, because $\tilde{b} < 0$ implies $h < 0$ (compare Section 5.6). The extremum curves for different values of $\tilde{c} > 0$ are drawn in Fig. 5.3, 7 where again only the amplitudes for $\tilde{b} \geqq 0$ really appear. The corresponding case $h < 0$, that is $\tilde{b} < 0$, $\tilde{c} \leqq 0$, $\tilde{d} = 0$ is sketched in Fig. 5.3, 8.

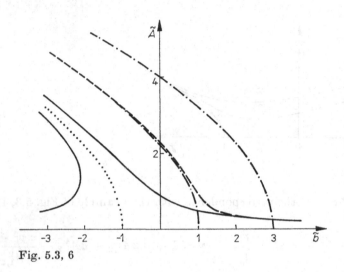

Fig. 5.3, 6

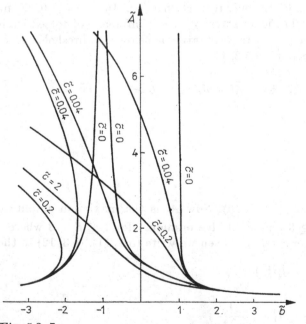

Fig. 5.3, 7

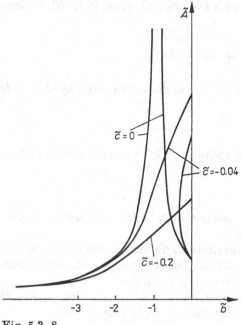

Fig. 5.3, 8

5.4.   Forced and self-excited vibrations

The amplitude equation (5.1, 14) specializes for vanishing parametric excitation influence, $g = h = 0$, when we exclude the trivial case $a = \beta = 0$, to

$$(a^2 + \beta^2)\, A^2 = p^2 \,. \tag{5.4, 1}$$

This basic equation is, contrary to (5.1, 14), only quadratic in a (that is in the frequency variation α) and of power five (cubic, if $d = 0$, resp. linear, if additionally $c = 0$) in A^2.

The assumption $Q_n = 0$, that is, the time shifting (5.1, 12), now brings no simplification because otherwise (5.1, 9), (5.1, 13) read

$$\begin{pmatrix} P \\ Q \end{pmatrix} + a \begin{pmatrix} r \\ s \end{pmatrix} - \beta \begin{pmatrix} s \\ -r \end{pmatrix} = 0 \quad \text{with} \quad P = \frac{P_n}{2n^2\omega_0^2}, \qquad Q = \frac{Q_n}{2n^2\omega_0^2},$$

$$(a^2 + \beta^2)\, r = -aP - \beta Q\,, \qquad (a^2 + \beta^2)\, s = \beta P - aQ\,, \tag{5.4, 2}$$

from what follows (5.4, 1) with

$$p = \sqrt{P^2 + Q^2}\,.$$

The influence of self-excitation will be discussed in this section independently of the somewhat more complicated analysis in Sections 5.2 and 5.3; compare HORTEL and SCHMIDT (1983) and TONDL (1970b, 1976a, 1982).

To begin with, (5.4, 1) shows that the amplitude A has the same order of magnitude as p divided by the small terms a, β, that is, a greater order of magnitude than y_{10} provided that all P_j, Q_j are of the same order of magnitude. This justifies denoting only the amplitude of the resonance part of the solution as amplitude.

The amplitude equation (5.4, 1) is with the original terms (5.1, 10), because $p \neq 0$ implies $A \neq 0$,

$$\alpha = eA^2 + fA^4 \pm \sqrt{\frac{p^2}{A^2} - (b + cA^2 + dA^4)^2} \,. \tag{5.4, 3}$$

The α distance of the two branches of the resonance curve given by (5.4, 3) from the backbone curve

$$\alpha = eA^2 + fA^4$$

is the same, it tends to p/A for small amplitudes. The resonance curves coalesce in the zeros

$$(b + cA^2 + dA^4)^2 A^2 = p^2 \tag{5.4, 4}$$

of the radicand which determine the maximum and minimum amplitudes. Equation (5.4, 4) is the extreme value equation.

Non-linear damping (vanishing self-excitation) diminishes the (unique real) solution of (5.4, 4), that is, the maximum amplitude. On the other hand, self-excitation can cause one or two additional regions of greater amplitudes.

Particularly for $d = 0$, equation (5.4, 4) yields — in the irreducible case $-b^3/c \geqq 27p^2/4$ of not too small self-excitation — three real solutions

$$A_1^2 = -\frac{2b}{3c}\left(1 + \cos\frac{\varphi}{3}\right), \qquad A_{2,3}^2 = -\frac{2b}{3c}\left[1 - \cos\left(60° \mp \frac{\varphi}{3}\right)\right]$$

$$\text{for} \quad -\frac{b^3}{c} \leqq \frac{27}{2}p^2 \,,$$

$$A_1^2 = -\frac{2b}{3c}\left(1 - \cos\frac{\varphi}{3}\right), \qquad A_{2,3}^2 = -\frac{2b}{3c}\left[1 + \cos\left(60° \mp \frac{\varphi}{3}\right)\right]$$

$$\text{for} \quad -\frac{b^3}{c} > \frac{27}{2}p^2$$

with the auxiliary variable φ introduced by

$$\cos\varphi = \left|1 + \frac{27c}{2b^3}p^2\right|.$$

For example is

$$A_1 = 3\frac{p}{|b|}, \qquad A_2 = A_3 = 1.5\frac{p}{|b|} \quad \text{for} \quad -\frac{b^3}{c} = \frac{27}{4}p^2 \tag{5.4, 5}$$

and

$$A_1 = 4.0981\frac{p}{|b|}, \qquad A_2 = 3\frac{p}{|b|}, \qquad A_3 = 1.0981\frac{p}{|b|} \quad \text{for} \quad -\frac{b^3}{c} = \frac{27}{2}p^2 \,. \tag{5.4, 6}$$

The transitition from (5.4, 5) to (5.4, 6) leads to the separation of an upper part of the resonance curve. When realized by diminishing $|c|$ to half its value, A_1 and A_2 increase and A_3 diminishes corresponding to the numerical factors. When realized by duplication of the value of b^3, the amplitudes A_1 and A_2 increase by 8.42% and 58.74% respectively and A_3 diminishes by 41.89% (transition from the points $\bigcirc$ to the points

$\triangle$ on the same curves in Fig. 5.4, 1, corresponding resonance curves in Fig. 5.4, 2). When at last realized by diminishing p^2 to half its value, A_1 and A_3 diminish by 3.41% and 48.24% respectively and A_2 increases by 41.42% (transition from $\bigcirc$ to $\square$ in Fig. 5.4, 1).

The extreme value equation (5.4, 4) can be written

$$b = -cA^2 - dA^4 \pm \frac{p}{A} \, . \tag{5.4, 7}$$

For $d = 0$, Fig. 5.4, 1 shows the composition of the values b from a (dashed) parabola and the hyperbolae $\pm p/A$ expressed by (5.4, 7) with $c = 1$ and $p = 0.002$, 0.001, and $0.001/\sqrt{2}$ corresponding to the transition from (5.4, 5) to (5.4, 6). Without self-excita-

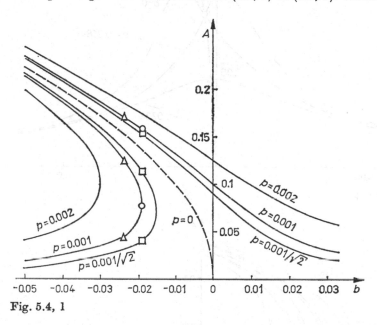

Fig. 5.4, 1

tion, $b \geqq 0$, only one maximal amplitude exists, which increases with increasing forced excitation p and with diminishing linear damping b. In particular,

$$A = \sqrt[3]{\frac{p}{|c|}}$$

holds for $b = d = 0$. On the other hand, sufficiently great (van der Pol) self-excitation $b < 0$ effects, in addition to the maximal amplitude, also a medium-sized minimal amplitude and a second small maximal amplitude.

The resonance curves belonging to the points $\bigcirc$ and $\triangle$ in Fig. 5.4, 1 by (5.4, 3) are drawn in Fig. 5.4, 2 where the (dashed) backbone curve is chosen by $e = 0.5$, $f = -2$.

If we choose a generalized van der Pol self-excitation (for which $b < 0$) by $d = -9$ and the above values for c and p, Fig. 5.4, 3 shows the composition of the extreme value curves corresponding to Fig. 5.4, 1. A hard self-excitation leads to an analogous figure where only the right and the left side are permutated because then the sign of b, c and d is changed.

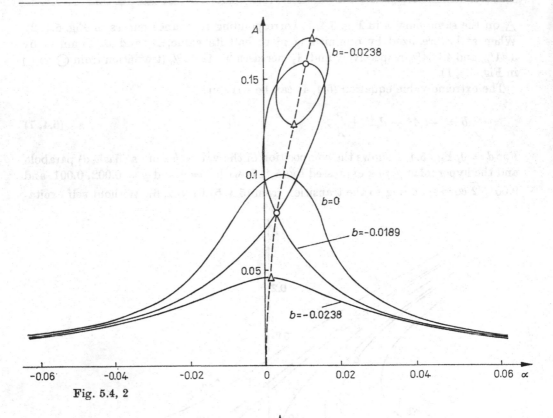

Fig. 5.4, 2

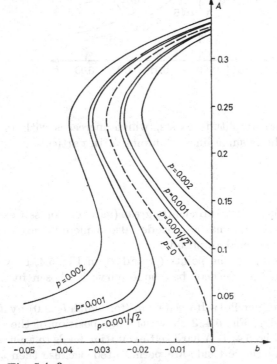

Fig. 5.4, 3

The extreme value curves given by (5.4, 7) reveal how A depends on b for given parameters c, d, p. By a suitable transformation of A and b, two of the three parameters can be eliminated so that a one-parametric family of curves describes the *general* behaviour. When we put

$$\bar{A} = \sqrt{\frac{|d|}{|c|}}\,\frac{A}{3}\,, \qquad \bar{b} = \frac{|d|}{9c^2}\,b\,, \qquad \bar{p} = \sqrt{\frac{|d|^3}{|c|^5}}\,\frac{p}{27}$$

for $c \neq 0$, $d \neq 0$, (5.4, 7) reads

$$\bar{b} = -\mathrm{sign}\,c \cdot \bar{A}^2 - \mathrm{sign}\,d \cdot 9\bar{A}^4 \pm \frac{\bar{p}}{\bar{A}}\,. \tag{5.4, 8}$$

herefore Fig. 5.4, 3 represents the *general* case (not only an example) of a generalized van der Pol self-excitation, the corresponding figure with permutated right and left side represents the *general* case of a hard self-excitation (5.1, 4) for the transformed parameters.

If by contrast we put, corresponding to (5.3, 9), for $c \neq 0$, $p \neq 0$

$$\check{A} = \sqrt[3]{\frac{|c|}{p}}\,A\,, \qquad \check{b} = \frac{b}{\sqrt[3]{|c|\,p^2}}\,, \qquad \check{d} = \sqrt[3]{\frac{p^2}{|c|^5}}\,d\,,$$

(5.4, 7) reads

$$\check{b} = -\mathrm{sign}\,c \cdot \check{A}^2 - \check{d}\check{A}^4 \pm \frac{1}{\check{A}}\,. \tag{5.4, 9}$$

Fig. 5.4, 4 gives the corresponding extreme value curves for the general case of a (generalized for $\check{d} < 0$) van der Pol self-excitation in the region $\check{b} < 0$ respectively of vanishing self-excitation for $\check{b} \geq 0$, $\check{d} = 0$. The general case of a hard self-excitation (5.1, 4) is given by the left half, permutated to the right-hand side, of the figure (because then $\check{b} > 0$ holds) with $\check{d} < 0$.

The extreme value equations (5.4, 7), (5.4, 8), (5.4, 9) and Figures 5.4, 1, 5.4, 2 and 5.4, 4 lead to the following general results:

1. *With augmenting forced excitation, the maximum amplitude augments, and an upper part of the resonance curve separates only for greater self-excitation.*

2. *For van der Pol self-excitation, an upper part of the resonance curve separates if $|b|$ exceeds a positive threshold value, and the maximum amplitude increases with $|b|$.*

3. *For generalized van der Pol self-excitation as well as for hard self-excitation, the corresponding threshold value is zero, comparatively great amplitudes can arise even for small values of $|b|$, and the maximum amplitude descreases as $|b|$ increases.*

The *stability* of the solutions found can be evaluated by means of the two stability conditions (5.2, 15) and (5.2, 16) where now $g = h = k = l = K = L = 0$.

Using (5.4, 2), (5.4, 1), (5.4, 3) and (5.1, 10), the condition (5.2, 16) can be given the form

$$\pm 2(e + 2fA^2)\sqrt{\frac{p^2}{A^2} - (b + cA^2 + dA^4)^2} - \frac{p^2}{A^4}$$

$$- 2(b + cA^2 + dA^4)\,(c + 2dA^2) < 0\,.$$

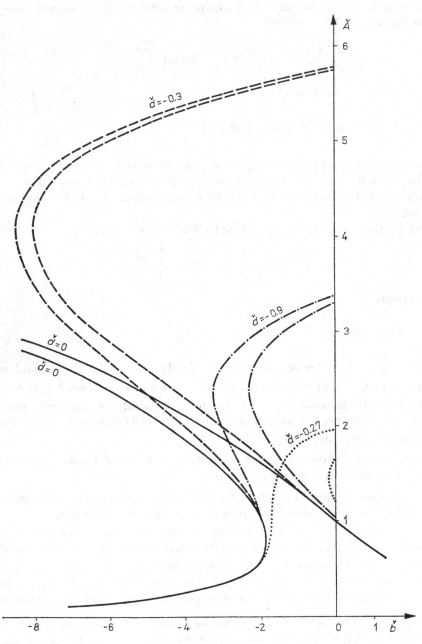

Fig. 5.4, 4

A comparison with (5.2, 17) reveals: The second stability condition (5.2, 16) holds for the upper sign, that is, for the parts of the response curve on the right-hand side of the backbone curve if and only if $d\alpha/dA < 0$, in other words, till respectively from the point of the resonance curve with vertical tangent. For the parts of the resonance curve on the left-hand side of the backbone curve, it holds correspondingly for $d\alpha/dA > 0$.

The first stability condition (5.2, 15) holds in case of van der Pol self-excitation for

$$A > \sqrt{\frac{|b|}{2c}}, \tag{5.4, 10}$$

in case of generalized van der Pol self-excitation (5.1, 3) if $c^2 > 3bd$ and A lies in the interval I built by the positive values

$$\sqrt{\frac{1}{3|d|}\left(c \pm \sqrt{c^2 - 3bd}\right)},$$

and in case of a hard self-excitation (5.1, 4) if $c^2 < 3bd$ or A lies outside of the interval I. Consequently, a self-excitation of the forms considered (with the exception of too great a hard self-excitation) always causes unstable amplitude intervals.

In investigating the vibrations in the amplitude intervals unstable because of self-excitation, we should first bear in mind the character of the solutions found so far.

The (circular) frequency n ($n\omega$ regarding t) of the resonance part of the first approximation coincides with the (circular) frequency of a component of forced excitation and thus of the non-resonance solution ("one-frequency solution"). It does not differ much from the eigenfrequency $\Omega = n\omega_0$ of the system. In the neighbourhood of the resonance, the resonance part of the solution and through this its harmonic character prevails.

If no stable solution exists, for instance if for van der Pol self-excitation the stability condition (5.4, 10) does not hold, other solutions have to be sought. We keep the notation n for the circular frequency of the resonance part of the first approximative solution, but we presume in what follows that the forced excitation includes no resonance component, $P_n = Q_n = 0$.

If there exists no forced excitation at all, (5.4, 1) yields the two equations

$$\beta = 0 \quad \text{and} \quad a = 0. \tag{5.4, 11}$$

The first one is only valid in case of self-excitation (otherwise $A = 0$ follows), for

$$A^2 = -\frac{b}{c} \tag{5.4, 12}$$

in case of van der Pol self-excitation. The second one determines the frequency (variation) at which the self-excited vibration appears with the amplitude given by (5.4, 12).

We now investigate a system (5.1, 1) where forced excitation exists, but without a resonance component, $P_n = Q_n = 0$. In order to determine the influence of forced excitation on the periodicity equations and through this on the amplitude equation, we have to take into consideration also the forced excitation terms of the second approximation. For simplicity we assume forced excitation as harmonic of the form $P_k \cos k\omega t + Q_k \sin k\omega t$ ($k \neq n$) and the non-linearities of fifth degree as zero, $D = F = 0$, that is, we investigate van der Pol self-excitation. In the non-resonance case, the first approximations are

$$y_{10} = \frac{P_k \cos k\tau + Q_k \sin k\tau}{\omega_0^2(\lambda - k^2)},$$
$$(5.4, 13)$$

$$y_{20} = y_{10} + \frac{2k^2\alpha}{\omega_0^2(\lambda - k^2)^2}(P_k \cos k\tau + Q_k \sin k\tau)$$

$$- \frac{kB}{\omega_0^3(\lambda - k^2)^2}(Q_k \cos k\tau - P_k \sin k\tau)$$

$$- \frac{kC}{4\omega_0^7(\lambda - k^2)^3}\left[\frac{P_k^2 + Q_k^2}{\lambda - k^2}(Q_k \cos k\tau - P_k \sin k\tau)\right.$$

$$\left. + \frac{Q_k(3P_k^2 - Q_k^2)\cos 3k\tau + P_k(3Q_k^2 - P_k^2)\sin 3k\tau}{\lambda - 9k^2}\right]$$

$$- \frac{E}{4\omega_0^8(\lambda - k^2)^3}\left[3\frac{P_k^2 + Q_k^2}{\lambda - k^2}(P_k \cos k\tau + Q_k \sin k\tau)\right.$$

$$\left. + \frac{P_k(P_k^2 - 3Q_k^2)\cos 3k\tau + Q_k(3P_k^2 - Q_k^2)\sin 3k\tau}{\lambda - 9k^2}\right]$$

while in the resonance case $\lambda = n^2$, if $k \neq 3n$, $3k \neq n$, additional to (5.1, 8) the terms containing $\frac{\cos}{\sin} n\tau$ arise:

$$y_{2\,\mathrm{add}} = \frac{C}{2n\omega_0^5}\frac{P_k^2 + Q_k^2}{(n^2 - k^2)^2}(s \cos n\tau - r \sin n\tau)$$

$$+ \frac{3E}{2n^2\omega_0^6}\frac{P_k^2 + Q_k^2}{(n^2 - k^2)^2}(r \cos n\tau + s \sin n\tau).$$
$$(5.4, 14)$$

The periodicity equations now also yield the two equations (5.4, 11), which can be written, with the additional terms (5.4, 14) and the notations

$$p_k = \frac{\sqrt{P_k^2 + Q_k^2}}{2n^2\omega_0^2}, \qquad N = \frac{2}{\left|1 - \dfrac{k^2}{n^2}\right|}, \qquad A_f = Np_k,$$
$$(5.4, 15)$$

in the form

$$b + c(A^2 + 2A_f^2) = 0, \qquad \alpha - e(A^2 + 2A_f^2) = 0.$$
$$(5.4, 16)$$

The quantity A_f is, as (5.4, 13) shows, the amplitude of the forced excitation part of the first approximation which we denote by *forced amplitude*, in contrast to the *self-excitation amplitude* $A = A_s$. Writing the first equation (5.4, 16)

$$A_s^2 = -\frac{b}{c} - 2A_f^2$$

and comparing with (5.4, 12), we realize the influence of forced excitation. A self-excited vibration of the form considered here only appears if the forced amplitude yields

$$A_f < \sqrt{\frac{|b|}{2c}}.$$

This condition is identical with the condition that the resonance solution, which has the same period as the forced excitation and the amplitude of which is given by (5.4, 1), is unstable.

By help of the transformed quantities

$$\eta = \frac{k}{n}, \qquad \overline{A}_f = \frac{A_f}{p_k}, \qquad \overline{A}_s = \frac{A_s}{p_k}, \qquad \overline{b} = \sqrt{\frac{|b_k|}{2cp_k^2}},$$

the equations (5.4, 15) and (5.4, 16) for the forced and the self-excitation amplitude simplify to

$$\overline{A}_f = \frac{2}{|1 - \eta^2|}, \qquad \overline{A}_s = \sqrt{2(\overline{b}^2 - \overline{A}_f^2)}.$$

The behaviour of these amplitudes is given in general form in Fig. 5.4, 5. The self-excitation amplitudes $\overline{A}_s$ are greater the greater $\overline{b}$ is. They diminish for $\eta \to 1$ and become zero for values $\eta = \eta_{b\pm}$, which are closer to the resonance value $\eta = 1$ the greater $\overline{b}$ is. The corresponding forced amplitudes $\overline{A}_f$ increase for $\eta \to 1$ hyperbolically.

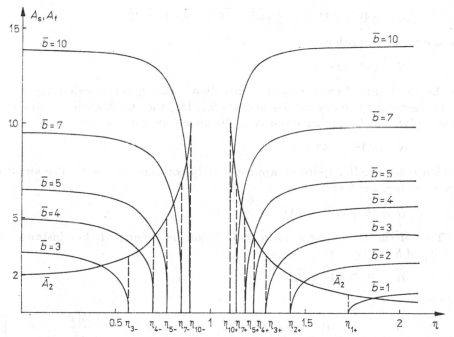

Fig. 5.4, 5

The value $\overline{A}_s = 0$ corresponds to the limit $\overline{A}_f = \overline{b}$ given by (5.4, 17). The combination of forced and self-excited vibrations (two-frequency solution) leads to beating which is stronger the greater the self-excitation $\overline{b}$ is and the less η differs from one. For $\eta \to \eta_{b\pm}$, the beating disappears because $\overline{A}_s \to 0$; in the interval (η_{b-}, η_{b+}) only the former one-frequency solution exists and it is stable in just this interval. The amplitudes $\overline{A}_f$ of the two-frequency solutions and those of the one-frequency solution are equal at the limit points $\eta_{b\pm}$. The suppression of the self-excited vibration in the interval (η_{b-}, η_{b+}) is called *synchronization* or *frequency pulling* (pulling of the frequency of the self-excited vibration by the forced excitation frequency), the interval is called the pull-in range.

If we take into consideration an unharmonic forced excitation, vibration components with other frequencies which are not in the neighbourhood of the resonance $\eta = 1$ are added the amplitudes of which are therefore smaller and which cause no beating. Higher approximations can lead to additional beating parts of the solution, but they are small in comparison with the parts evaluated.

5.5.   Parametric and self-excited vibrations

When parametric excitation and eventually self-excitation, but no forced excitation terms appear in the periodicity conditions, the amplitude equation (5.1, 14) yields, because of $p = 0$, besides the trivial solution $A = 0$

$$\alpha^2 + \beta^2 - G^2 = 0$$

or explicitely, using (5.1, 10),

$$\alpha = eA^2 + fA^4 - \Gamma \pm \sqrt{G^2 - (b + cA^2 + dA^4)^2} \qquad (5.5, 1)$$

where for abbreviation

$$G = \sqrt{g^2 + h^2}$$

is the amplitude of the parametric excitation occuring in the periodicity equations.[1]

Corresponding to Section 5.4 and to (5.1, 16), the two branches of the resonance curve given by (5.5, 1) have the same distance from the backbone curve

$$\alpha = aA^2 + fA^4 - \Gamma$$

which is for small amplitudes approximately constant, $\sqrt{G^2 - b^2}$. The amplitude extreme values are found when

$$G = |b + cA^2 + dA^4| . \qquad (5.5, 2)$$

Two of the five parameters in (5.5, 2) can be eliminated. For instance, dividing by $|b|$ (if $b \neq 0$) gives

$$\overline{G} = |\overline{b} + \overline{c}\overline{A}^2 + \overline{d}\overline{A}^4|$$

where

$$\overline{G} = \frac{G}{|b|} \ (\geqq 0) , \quad \overline{b} = \operatorname{sign} b , \quad \overline{c} = \operatorname{sign} c , \quad \overline{d} = \frac{|b|\, d}{c^2} , \quad \overline{A} = \sqrt{\left|\frac{c}{b}\right|}\, A .$$

$$(5.5, 3)$$

When no self-excitation exists, $\overline{b} = \overline{c} = 1$, $d \geqq 0$, Fig. 5.5, 1 shows the *general* behaviour of the amplitude extremes: Only for $\overline{G} > 1$ do positive amplitudes occur, in particular a positive maximum amplitude which decreases with increasing nonlinear damping d.

---

[1] As the time shifting (5.1, 11) for getting a vanishing sin-component of the forced excitation now does not take place, a corresponding time shifting would lead to $h = 0$ (at least in first approximation, that is $H_{2n} = 0$).

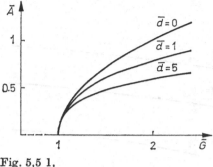

Fig. 5,5 1,

In case of van der Pol self-excitation, $\bar{b} = -1, \bar{c} = 1, \bar{d} = 0$, the general dependence of the amplitude extremes is given in Fig. 5.5, 2 where, as in the figures to follow, the dashed line shows only the emergence of the upper curve. A positive maximum amplitude exists for every $\bar{G}$ and an additional minimum amplitude for $\bar{G} < 1$.

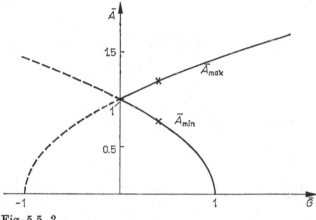

Fig. 5.5, 2

For generalized van der Pol self-excitation, $\bar{b} = -1, \bar{c} = 1, \bar{d} < 0$ (as well as for hard self-excitation (5.1, 4), $\bar{b} = 1, \bar{c} = -1, \bar{d} > 0$), we get, as drawn in Fig. 5.5, 3, two maximum and two minimum amplitudes for $|\bar{d}| < 1/4$ and every $\bar{G}$, but only one maximum and one minimum amplitude if $|\bar{d}| \geq 1/4$ and $\bar{G}$ is sufficiently great, $\bar{G} > 1 - 1/(4|\bar{d}|)$, and no positive amplitudes at all in the remaining case (when non-linear damping is too great respectively parametric excitation is too small). The cases in Fig. 5.5, 1 and Fig. 5.5, 2 correspond to Fig. 5.5, 3 as limiting cases.

The result is that without self-excitation, vibrations arise only when the parametric excitation exceeds a threshold value depending on the linear damping ($\bar{G} = 1$), whereas self-excitation renders vibrations possible for parametric excitations below this threshold value, but with a positive minimum amplitude. The group of curves reveals the form of the vibrations especially in dependence on $\bar{d}$ and $\bar{G}$.

In order to construct the resonance curves for at least one example, we choose $\bar{G} = 0.4, e = f = 0, \Gamma = |b|$ and write $\bar{\alpha} = \alpha/|b|$ analogous to (5.5, 3). The resonance curves coming from (5.5, 1) for different values of $\bar{d}$ are given in Fig. 5.5, 4, the extreme

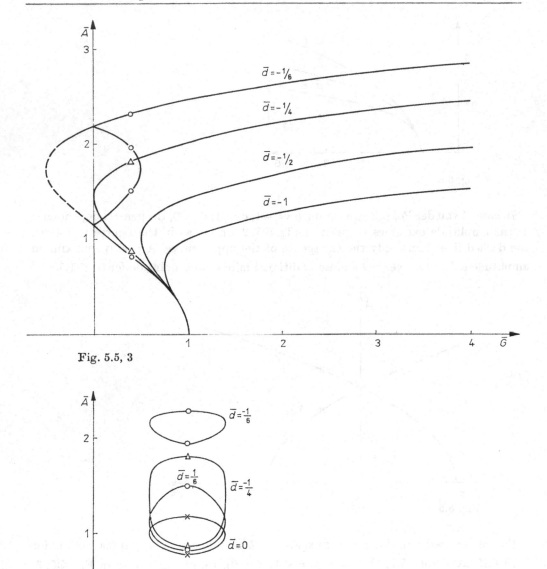

Fig. 5.5, 3

Fig. 5.5, 4

values are marked by the same signs as in Figures 5.5, 2 and 5.5, 3. It shows again that the resonance curves give only an isolated result whereas the extreme value curves of Figures 5.5, 2 and 5.5, 3 reveal the tendency of building one or several resonance curves.

An even simpler description is possible if we multiply (5.5, 2) by d^2/c^4 (if $cd \neq 0$). We get

$$\tilde{G} = |\tilde{b} + \tilde{c}\tilde{A}^2 + \tilde{A}^4|$$

with

$$\tilde{G} = \frac{|d|\,G}{c^2}\,(\geqq 0)\,, \qquad \tilde{b} = \frac{db}{c^2}\,, \qquad \tilde{c} = \operatorname{sign} cd\,, \qquad \tilde{A} = \sqrt{\left|\frac{d}{c}\right|}\,A\,.$$

For generalized van der Pol self-excitation as well as for hard self-excitation (5.1, 4), $\tilde{b} > 0$ and $\tilde{c} = -1$ hold. Therefore the different amplitude extreme curves are identical, only shifted to the starting point $\tilde{A} = 0$, $\tilde{G} = \tilde{b}$ (Fig. 5.5, 5).

Many results on self-excited and parametric vibrations can be found in Tondl (1978b).

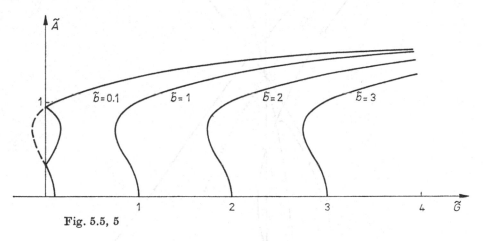

Fig. 5.5, 5

5.6.   Forced, parametric and self-excited vibrations

As in Section 5.3, we assume non-linear damping and restoring forces and restrict ourselves to the special phase relation $g = 0$ between forced and parametric excitation, but now take into consideration an additional self-excitation (5.1, 2), (5.1, 3) or (5.1, 4). The relations (5.3, 1) to (5.3, 4) now are valid without change whereas for $\bar{\beta} < 0$, (5.3, 5) reads $\bar{\beta}/3 \geqq \gamma \geqq \bar{\beta}$ and the equations (5.3, 6), (5.3, 8) for the amplitude extremes are modified to read

$$\gamma = -\bar{b} - \bar{c}\bar{A}_0^2 - \bar{d}\bar{A}_0^4 \pm \frac{1}{A_0}\,, \tag{5.6, 1}$$

$$\gamma = \bar{b} + \bar{c}\bar{A}_{1,2}^2 + \bar{d}\bar{A}_{1,2}^4 - \frac{1}{8\gamma\bar{A}_{1,2}^2}\,, \tag{5.6, 2}$$

that is,

$$2\gamma = \bar{b} + \bar{c}\bar{A}_{1,2}^2 + \bar{d}\bar{A}_{1,2}^4 \pm \sqrt{(\bar{b} + \bar{c}\bar{A}_{1,2}^2 + \bar{d}\bar{A}_{1,2}^4)^2 - \frac{1}{2\bar{A}_{1,2}^2}} \tag{5.6, 3}$$

with $\bar{b} = \operatorname{sign} b$.

For van der Pol self-excitation, $\bar{b} = -1$, $\bar{c} > 0$, $\bar{d} = 0$, Fig. 5.6, 1 shows how the amplitude extreme curves given by (5.6, 1) (full line) and (5.6, 2) (dashed line, in the region (5.3, 5) limited by the dashed-dotted lines) compose by parabolae $\gamma = \pm(1 - \bar{c}\bar{A}^2)$ and hyperbolae for $\bar{c} = 1/9$. Fig. 5.6, 2 gives the general dependence of the amplitude maximum curves on the coefficient c of nonlinear damping, the influence of which

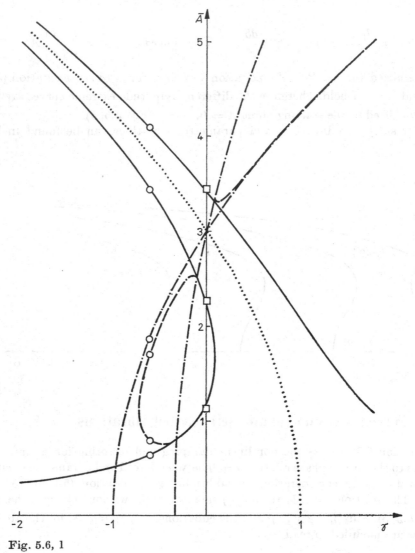

Fig. 5.6, 1

proves to be essential. The influence of parametric excitation (including the amplitude diminishing effect found in Section 5.2) is greater the smaller the non-linear damping.

Resonance curves corresponding to Fig. 5.6, 1 are given in Fig. 5.6, 3 for $e = f = 0$, writing $\bar{\alpha} = \alpha/|b|$ as in Fig. 5.5, 4. The extreme values are marked by the same signs as in Fig. 5.6, 1. The resonance curves yield examples of the result, given in Figures 5.6, 1 and 5.6, 2, that parametric excitation in general (disregarding the diminishing effect) augments the maximal amplitudes but gives rise, in connection with self-excitation, to the possibility of isolated smaller resonance curves, that is, of much smaller maximal amplitudes (relative maxima, under the condition that no suitable disturbances bring the system to the higher amplitude level).

In case of generalized van der Pol self-excitation as well as hard self-excitation, a tilting over of the amplitude extreme curves occurs. Fig. 5.6, 4 shows an example for this behaviour with $\bar{c} = 1$, $\bar{d} = 0.1$.

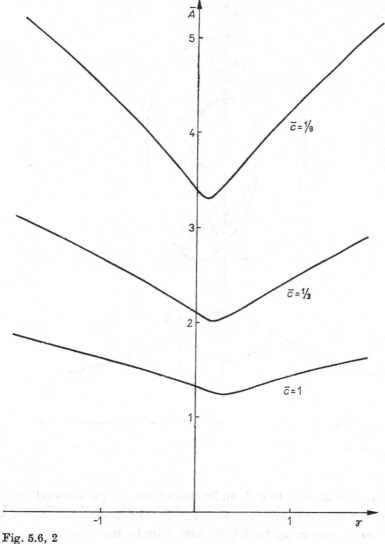

Fig. 5.6, 2

The dependence on $\bar{b}$ becomes apparent for $\bar{c} \neq 0$ when we divide (5.6, 1), (5.6, 2) by $|\bar{c}|^{1/3}$ and introduce, corresponding to (5.3, 9),

$$\check{A} = |\bar{c}|^{1/3}\,\bar{A}, \qquad \check{\gamma} = \frac{\gamma}{|\bar{c}|^{1/3}}, \qquad \check{b} = \frac{\bar{b}}{|\bar{c}|^{1/3}}, \qquad \check{c} = \text{sign}\ \bar{c}\,, \qquad \check{d} = \frac{\bar{d}}{|\bar{c}|^{5/3}}.$$

We get

$$\check{\gamma} = -\check{b} - \check{c}\check{A}_0^2 - \check{d}\check{A}_0^4 \pm \frac{1}{\check{A}_0}, \tag{5.6, 5}$$

$$\check{\gamma} = \check{b} + \check{c}\check{A}_{1,2}^2 + \check{d}\check{A}_{1,2}^4 - \frac{1}{8\check{\gamma}\check{A}_{1,2}^2} \tag{5.6, 6}$$

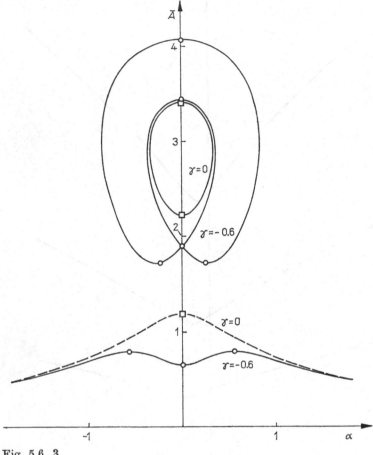

Fig. 5.6, 3

and a formula corresponding to (5.6, 3). In case of van der Pol self-excitation $(\check{b} < 0,$ $\check{c} = 1,$ $\check{d} = 0)$ for instance, we get the case in Fig. 5.6, 5 for $\check{b} = -1$ and $\check{b} = -2$. Now the curves corresponding to (5.6, 5) differ only by the starting point $-\check{b}$ of the backbone curve, whereas the curves answering to (5.6, 6) exist in the interval (5.6, 3) differing with $\check{b}$. It shows that *an augmenting self-excitation coefficient* $\check{b}$ *on the one hand augments the maximum amplitude, on the other hand gives rise to the possibility of a smaller relative maximum amplitude* even for smaller values of $\check{\gamma}$ (even for $\check{\gamma} = 0$ in case $\check{b} = -2$).

The notation (5.3, 10) leads to the equations (5.3, 13), (5.3, 14) where, in contrast to Section 5.3, for van der Pol self-excitation $\tilde{b} < 0,$ $\tilde{c} \geqq 0$ now holds. The left-hand sides of Figures 5.3, 6 and 5.3, 7 show the behaviour of the amplitude extreme values. Self-excitation enlarges the maximum amplitudes whereas non-linear damping diminishes these amplitudes.

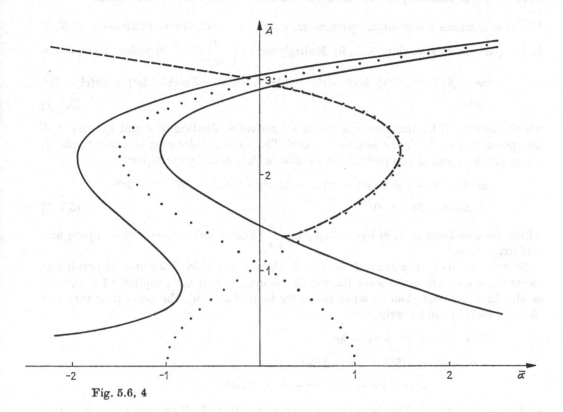

Fig. 5.6, 4

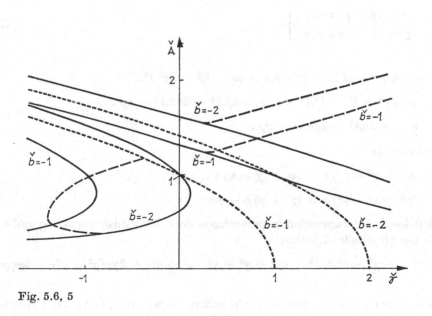

Fig. 5.6, 5

5.7.   Non-linear parametric excitation. Harmonic resonance

We now consider a non-linear parametric excitation with the coefficients k, l, K, L in the periodicity equations (5.1, 9). Multiplying by $\begin{pmatrix} s \\ -r \end{pmatrix}$, (5.1, 9) yields the equation

$$ps - \beta A^2 - 2grs + h(r^2 - s^2) - kA^2s + lA^2r + Ks(s^2 - 3r^2) + Lr(r^2 - 3s^2)$$
$$= 0 \tag{5.7, 1}$$

which enables, if for instance s is given, numerical evaluation of r and by this of A independently of the frequency variation α. The value α belonging to every triple A, r, s is given by one of the periodicity equations (5.1, 9) or by the equation

$$pr + aA^2 - g(r^2 - s^2) - 2hrs - 3kA^2r - 3lA^2s - Kr(r^2 - 3s^2)$$
$$+ Ls(s^2 - 3r^2) = 0 \tag{5.7, 2}$$

which follows from (5.1, 9) by multiplying $\begin{pmatrix} r \\ s \end{pmatrix}$ and is independent of damping and self-excitation.

In what follows we assume $K = L = 0$, that is, the third harmonic of non-linear parametric excitation vanishes (harmonic resonance) and for simplicity $l = 0$, that is, the first harmonic has no since term. By help of (5.1, 9), the non-linear terms in (5.7, 1), (5.7, 2) can be written

$$2krs = as + \beta r + gs - hr ,$$

$$k(r^2 - s^2) = k(3r^2 + s^2) - 2kA^2$$
$$= p + ar - \beta s - gr - hs - 2kA^2$$

and so be eliminated. This way the equations (5.7, 1), (5.7, 2) become linear in r, s:

$$\left. \begin{aligned} \varphi r + (\chi - \psi)\, s &= \varrho , \\ (\chi + \psi)\, r - \varphi s &= \sigma \end{aligned} \right\} \tag{5.7, 3}$$

where

$$\varphi = ha - g\beta , \qquad \chi = kp - ga - h\beta - 2k^2A^2 ,$$
$$\psi = g^2 + h^2 - k^2A^2 , \qquad \varrho = k\beta A^2 + 2hkA^2 - hp ,$$
$$\sigma = -kaA^2 - 2gkA^2 + gp .$$

Their solution is

$$(\varphi^2 + \chi^2 - \psi^2)\, r = \varphi\varrho + (\chi - \psi)\, \sigma ,$$
$$(\varphi^2 + \chi^2 - \psi^2)\, s = (\chi + \psi)\, \varrho - \varphi\sigma$$

if the left-hand side expression in parentheses does not vanish. Squaring and adding leads to the amplitude equation

$$(\varphi^2 + \chi^2 - \psi^2)^2\, A^2 = (\varphi^2 + \chi^2 + \psi^2)\, (\varrho^2 + \sigma^2) + 2\chi\psi(\varrho^2 - \sigma^2) - 4\varphi\psi\varrho\sigma \tag{5.7, 4}$$

which contains r and s only in form of the sum of the squares, A^2. It is of power four in a, that is, in the frequency variation, and can be solved numerically for different given values of A.

In case of the phase relation $g = 0$ between forced and parametric excitation, the equation (5.7, 4) is biquadratic in a with the amplitude formula

$$2h^2A^2(k^2A^2 - h^2) a^2 = -(\chi - \psi)^2 k^2A^4 + 2(\chi^2 - \psi^2) h^2A^2 - 4\psi\varrho hkA^2 - \varrho^2h^2$$
$$\pm \{(\chi - \psi)^4 k^4A^8 + 8\psi(\chi - \psi)^2 \varrho hk^3A^6 + 4(\chi - \psi)^2 (\psi^2 - \chi^2) h^2k^2A^6$$
$$+ 4(\chi^2 - \psi^2)^2 h^2k^2A^6 + 2(\chi - \psi)^2 \varrho^2h^2k^2A^4 + 16\psi(\psi^2 - \chi^2) \varrho h^3kA^4$$
$$- 4(\chi + \psi)^2 \varrho^2h^2k^2A^4 + 16\psi^2\varrho^2h^2k^2A^4 + 4(\psi^2 - \chi^2) \varrho^2h^4A^2$$
$$+ 4(\chi + \psi)^2 \gamma^2h^4A^2 + 8\psi\varrho^3h^3kA^2 + \varrho^4h^4\}^{1/2} .$$

In particular, for $h = 0$ (linear parametric excitation influences only Γ) the amplitude formula follows (with (5.1, 10), for $A \neq 0$, $A^2 \neq p/k$)

$$\alpha = eA^2 + fA^4 - \Gamma \pm (3kA^2 - p) \sqrt{\frac{1}{A^2} - \frac{(b + cA^2 + dA^4)^2}{(kA^2 - p)^2}} \qquad (5.7, 5)$$

or, if in addition $p = 0$ (no forced excitation influence)

$$\alpha = eA^2 + fA^4 - \Gamma \pm 3\sqrt{k^2A^2 - (b + cA^2 + dA^4)^2} . \qquad (5.7, 6)$$

Formula (5.4, 3) is a special case of (5.7, 5) for $k = \Gamma = 0$.

The condition $dA/d\alpha = 0$ for amplitude extreme values requires the radicand in (5.7, 5) respectively (5.7, 6) to be zero. Correspondingly, real values α appear if the radicand is non-negative, in case (5.7, 5) for

$$(kA^2 - p)^2 \geqq (b + cA^2 + dA^4)^2 A^2 . \qquad (5.7, 7)$$

If damping is linear, $c = d = 0$, the condition

$$(2k^2A^2 - 2kp - b^2)^2 \geqq b^2(b^2 + 4kp)$$

for real amplitudes follows, which holds (Fig. 5.7, 1) in case $b^2 + 4kp \leqq 0$ for all amplitudes, in the opposite case if $2k^2A^2$ lies outside of the interval defined by the positive values

$$b^2 + 2kp \pm b\sqrt{b^2 + 4kp} .$$

For additional non-linear damping, the interval of real amplitudes lies in the interior of that for linear damping and is bounded above because (5.7, 7) does not now hold for A sufficiently great. The maximum amplitude is the greatest value of A for which the equality sign holds in (5.7, 7).

If self-excitation occurs, for instance hard self-excitation $b > 0$, $c < 0$, $d > 0$, the interval of real amplitudes need not lie in the interior of that for $c = d = 0$, but because of (5.7, 7) it is bounded above as for non-linear damping.

In case (5.7, 6) of vanishing forced excitation, the condition for real α corresponding to (5.7, 7) reads

$$k^2A^2 \geqq (b + cA^2 + dA^4)^2 \qquad (5.7, 8)$$

which yields

$$A \geqq \left| \frac{b}{k} \right| \qquad (5.7, 9)$$

for $c = d = 0$ and

$$(2c^2A^2 + 2bc - k^2)^2 \leqq k^2(k^2 - 4bc) \tag{5.7, 10}$$

for $d = 0$ and therefore $c > 0$ (non-linear damping or van der Pol self-excitation). For linear damping the resonance curves coalesce only in the lower amplitude threshold value (5.7, 9), a maximum amplitude does not exist. For non-linear damping, (5.7, 10) shows that the resonance curves are limited, if linear damping vanishes ($b = 0$) to amplitudes from zero to $|k|/c$ and if $b > 0$ to an interval in the interior of this interval which completely vanishes for $4bc \geqq k^2$ (Fig. 5.7, 2).

The condition (5.7, 8) can be written with $d = 0$

$$|k|\, A \geqq b + cA^2\,, \quad \text{if} \quad A^2 \geqq -\frac{b}{c}$$

(especially for non-linear damping), from which follows

$$(2cA - |k|)^2 \leqq k^2 - 4bc \tag{5.7, 11}$$

and analogously

$$(2cA + |k|)^2 \geqq k^2 - 4bc\,, \quad \text{if} \quad A^2 < -\frac{b}{c} \tag{5.7, 12}$$

(only possible for self-excitation). For non-linear damping, α is real in the interval defined by the positive values

$$2cA = |k| \pm \sqrt{k^2 - 4bc}\,.$$

For van der Pol self-excitation (left-hand side of Fig. 5.7, 2), the relations (5.7, 11) and (5.7, 12) imply that α is real in the greatest amplitude interval built by the three positive terms

$$\frac{1}{2c}\sqrt{k^2 - 4bc} \pm \frac{|k|}{2c} \quad \text{and} \quad \sqrt{-\frac{b}{c}}\,.$$

The equation for the amplitude extreme values given by the equality sign in (5.7, 7) can be written

$$b = -cA^2 - dA^4 \pm \left(kA - \frac{p}{A}\right)\,. \tag{5.7, 13}$$

In case of linear damping, $b > 0$, $c = d = 0$, Fig. 5.7, 1 gives these amplitude extreme values for $p = 1$, $k = \mp 1$, but also for any p (> 0) and k if (5.7, 13) is transformed by

$$\tilde{A} = \sqrt{\frac{|k|}{p}}\, A\,, \quad \tilde{b} = \frac{b}{\sqrt{|k|\, p}}$$

to

$$\tilde{b} = \pm \left(\tilde{A} - \frac{\text{sign } k}{\tilde{A}}\right)\,.$$

For $k < 0$, no amplitude extreme values exist if damping is small, $b^2 \leqq 4\,|k|\, p$, whereas for greater damping and always for $k > 0$ we get a smaller maximum and a greater minimum, that is, an interval where no vibration amplitudes exist. This reveals the basic influence on the vibrations of the sign of k.

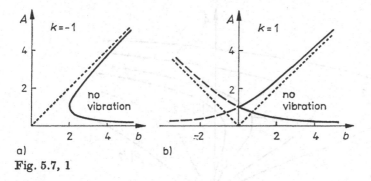

Fig. 5.7, 1

Fig. 5.7, 2 shows the amplitude extreme values (5.7, 13) for $c = 1$, $d = 0$ (that is non-linear damping for $b \geqq 0$, van der Pol self-excitation for $b < 0$), $p = 0$ (vanishing forced excitation) and different values of non-linear parametric excitation k. The corresponding minimum amplitude for linear damping, $c = 0$, is given by dotted lines whereas in the non-linear case the maximum and minimum amplitudes are given by the greater respectively (for $b \neq 0$ also positive) smaller value of the full-lined curve. Without self-excitation, for increasing linear damping the maximum amplitude decreases and the minimum amplitude increases, whereas for self-excitation, with increasing self-excitation coefficient the maximum as well as the minimum amplitude increase.

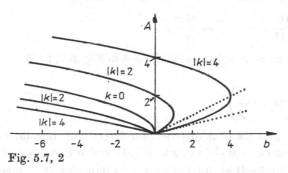

Fig. 5.7, 2

Fig. 5.7, 2 can be used for every $c > 0$ if for instance $\check{b} = b/c$ and $\check{k} = k/c$ are inserted instead of b, k. By transforming A, the value k also could be brought to ± 1 (compare (5.7, 15)), so that only one curve would do, but it is just the dependence on k that was to be illustrated.

For $d \neq 0$, (5.7, 13) with $p = 0$ can be transformed to the equation

$$\overline{b} = -\operatorname{sign} c \cdot \overline{A}^2 - \operatorname{sign} d \cdot \overline{A}^4 \pm \overline{k}\overline{A} \qquad (5.7, 14)$$

by setting

$$\overline{A} = \sqrt{\left|\frac{d}{c}\right|}\, A\,, \qquad \overline{b} = \frac{|d|}{c^2}b\,, \qquad \overline{k} = \sqrt{\left|\frac{d}{c^3}\right|}\, k\,.$$

Fig. 5.7, 3 illustrates (5.7, 14) for sign $c = 1$, sign $d = -1$, that is, for generalized van der Pol self-excitation (therefore only $\overline{b} < 0$ is of interest) and different values of $\overline{k}$. The interval of real vibration amplitudes enlarges with non-linear parametric excitation $|\overline{k}|$ and also for diminishing $|\overline{b}|$, for small $|\overline{k}|$ and $[\overline{b}]$, for instance for $|\overline{k}| = 0.2$, $\overline{b} = -0.08$ it splits into two such intervals.

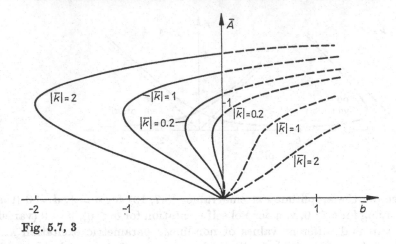

Fig. 5.7, 3

A hard self-excitation (5.1, 4) yields a corresponding behaviour; in Fig. 5.7, 3 only $\bar{b}$ has to be exchanged for $-\bar{b}$, therefore again the full lines are of interest, because now $\bar{b} > 0$.

The interaction of forced and non-linear parametric excitation with non-linear damping respectively self-excitation described by (5.7, 13) will be discussed now for van der Pol self-excitation respectively cubic damping, $c > 0$ and $d = 0$. Using

$$\check{A} = \frac{4c}{|k|}A, \qquad \check{b} = \frac{16c}{k^2}b, \qquad \check{p} = \frac{64c^2}{|k|^3}p, \tag{5.7, 15}$$

where the integers are chosen for comparison with $|k| = 4$ in Fig. 5.7, 2, (5.7, 13) now reads

$$\check{b} = -\check{A}^2 \pm \left(4\check{A} - \operatorname{sign} k \cdot \frac{\check{p}}{\check{A}}\right).$$

The extreme value behaviour is somewhat different for forced and non-linear parametric excitation having the same sign ($k > 0$, Fig. 5.7, 4) and for the opposite case $k < 0$ (Fig. 5.7, 5). For $k < 0$ and self-excitation as well as damping ($\check{b} < 0$ respectively $\check{b} \geqq 0$), the possibility of detached resonance curves, that is, of smaller amplitudes, decreases with increasing forced excitation $\check{p}$, whereas for $k > 0$ this possibility is nearly constant.

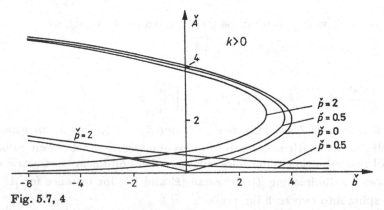

Fig. 5.7, 4

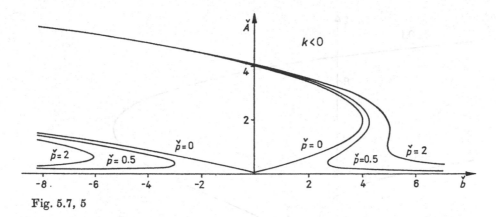

Fig. 5.7, 5

Up till now, in this section we have discussed the dependence of amplitude extremes on self-excitation respectively linear damping for different values of non-linear parametric excitation (Figures 5.7, 2 and 5.7, 3) and of forced excitation (Figures 5.7, 4 and 5.7, 5). To consider the dependence of amplitude extremes on forced excitation, we may write (5.7, 13) in the form

$$p = kA^2 \mp (bA + cA^3 + dA^5) \,.$$

Assuming $d = 0$ (following $c \geqq 0$) and transforming

$$\overset{\circ}{A} = \left| \frac{k}{b} \right| A \,, \qquad \overset{\circ}{p} = \frac{k}{b^2} p \,, \qquad \overset{\circ}{c} = \frac{b}{k^2} c \,,$$

so that $\overset{\circ}{p}$ has the sign of k and $\overset{\circ}{c}$ has the sign of b, we get

$$\overset{\circ}{p} = \overset{\circ}{A}{}^2 \pm (\overset{\circ}{A} + \overset{\circ}{c}\overset{\circ}{A}{}^3) \,.$$

For different non-negative values of $\overset{\circ}{c}$ (no self-excitation), Fig. 5.7, 6 gives the amplitude extreme values in dependence of forced excitation. Fig. 5.7, 7 gives the corresponding curves for van der Pol self-excitation, $\overset{\circ}{c} < 0$. It shows how increasing non-linear damping diminishes, while self-excitation as well as a negative sign of $\overset{\circ}{p}$ (different sign of forced and non-linear parametric excitation) augments the maximum amplitude.

As an example of *stability* investigations in connection with non-linear parametric excitation, consider the amplitude formula (5.7, 6). Differentiation by α shows that $dA/d\alpha \gtrless 0$ yields if

$$2(e + 2fA^2)\sqrt{k^2A^2 - (b + cA^2 + dA^4)^2}$$
$$\gtrless \pm 3[2(b + cA^2 + dA^4)(c + 2dA^2) - k^2] \tag{5.7, 16}$$

and the radicand is not zero. The first stability condition (5.2, 15) is independent of the special excitation at hand. Because

$$3kr = a \,, \qquad ks = -\beta \,, \qquad 3k^2rs = -a\beta \,,$$
$$3k^2(r^2 - s^2) = a^2 + 3\beta^2 - 6k^2A^2 \,,$$

the second stability condition (5.2, 16) leads with (5.1, 10), (5.7, 6) after a lengthy analysis, to the condition

$$\pm 2(e + 2fA^2)\sqrt{k^2A^2 - (b + cA^2 + dA^4)^2}$$
$$< 3[2(b + cA^2 + dA^4)(c + 2dA^2) - k^2] \,.$$

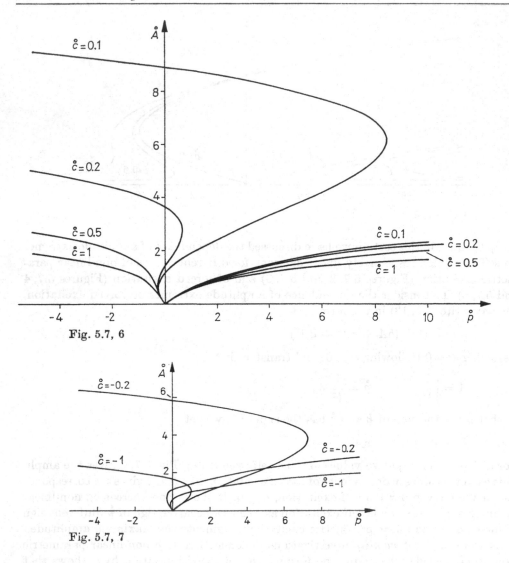

Fig. 5.7, 6

Fig. 5.7, 7

Comparison with (5.7, 15) gives stability for the parts of the resonance curve on the right-hand side of the backbone curve (upper sign) if $dA/d\alpha < 0$ and for the parts of the resonance curve on the left-hand side of the backbone curve (lower sign) if $dA/d\alpha > 0$. In other words, the upper parts of the resonance curves up to the points with vertical tangent are stable (if also the first stability condition holds), while the lower points are unstable.

5.8.   Non-linear parametric excitation. Subharmonic resonance

Now assume that with $L = k = l = 0$ a third harmonic component K of non-linear parametric excitation appears in the periodicity equations (5.1, 9). As in the case of harmonic resonance as a first step the second order terms in (5.7, 1), (5.7, 2) are transformed to linear ones, and these are eliminated as a second step, now the third order terms appearing in (5.7, 1), (5.7, 2) can be eliminated in three steps. We more easily obtain an amplitude equation from the equations

$$ps + 2ars + \beta(r^2 - s^2) - hA^2 + KA^2s = 0 \qquad (5.8, 1)$$

and

$$pr + a(r^2 - s^2) - 2\beta rs - gA^2 - KA^2r = 0 \qquad (5.8, 2)$$

which follow from (5.1, 9) by multiplication by $\begin{pmatrix} s \\ r \end{pmatrix}$ respectively $\begin{pmatrix} r \\ -s \end{pmatrix}$ and are only of second order in r, s. Now (5.1, 9) gives the equations

$$\left. \begin{array}{l} K(r^2 - s^2) = p + ar - \beta s - gr - hs\,, \\ 2Krs = -as - \beta r - gs + hr \end{array} \right\} \qquad (5.8, 3)$$

by which (5.8, 1), (5.8, 2) yield linear equations of the form (5.7, 3) where now

$$\varphi = ha - g\beta\,, \qquad \chi = Kp - ga - h\beta\,, \qquad \psi = a^2 + \beta^2 - K^2A^2\,,$$

$$\varrho = kKA^2 - p\beta\,, \qquad \sigma = gKA^2 - pa\,.$$

The amplitude equation (5.7, 4) is now of power eight in a and so too in the frequency variation α. In the special case $g = 0$ it is of power four in a^2, even if additionally $h = 0$ or $p = 0$. Only for $g = h = p = 0$ (no influence of forced *and* linear parametric excitation) a simple amplitude formula yields,

$$\psi = 0\,,$$

that is

$$\alpha = eA^2 + fA^4 - \Gamma \pm \sqrt{K^2A^2 - (b + cA^2 + dA^4)^2}\,,$$

which can be considered in the same way as (5.7, 6). The condition $dA/d\alpha \gtrless 0$ corresponds to

$$2(e + 2fA^2)\sqrt{K^2A^2 - (b + cA^2 + dA^4)^2}$$
$$\gtrless \pm [2(b + cA^2 + dA^4)(c + 2dA^2) - K^2] \qquad (5.8, 4)$$

when the radicand is not zero. The first stability condition is again (5.2, 15). In the second stability condition (5.2, 16), $r^2 - s^2$ and rs can be eliminated by (5.8, 3) from which it follows that

$$K^2(b + 2cA^2 + 3dA^4)^2$$
$$> [(e + 2fA^2)(ar - \beta s) - (c + 2dA^2)(\beta r + as) + 2K^2r]^2$$
$$+ [(e + 2fA^2)(\beta r + as) + (c + 2dA^2)(ar - \beta s) + 2K^2s]^2$$
$$- K^2[a - (e + 2fA^2)A^2]^2\,, \qquad (5.8, 5)$$

but now (5.1, 9) does not yield formulae for r and s. In the case in hand this does not matter because (5.8, 5) can be shown to depend on r, s only in form of the sum of squares A^2. Using (5.1, 10), the stability condition (5.8, 5) can be reduced to (5.8, 4) so that it coincides on the right-hand side of the backbone curve with the condition $dA/d\alpha < 0$, on the left-hand side with $dA/d\alpha > 0$, from which it follows that there is stability for the upper parts of the resonance curves up to the points with vertical tangent and instability for the corresponding lower parts of the resonance curves.

Other results on non-linear parametric excitation can be found in METTLER (1965), SCHMIDT (1969a, 1975), and TONDL (1978b).

6. Vibrations of systems with many degrees of freedom

6.1. Single and combination resonances

Vibrations of systems with many degrees of freedom show such complex behaviour that it seems impossible to develop a general theory for them. The analysis in this section should be understood not as part of such a theory but only as a kind of general framework for the investigation of certain classes of vibrations with many degrees of freedom. We shall discuss later and in more detail the vibrations arising in two typical problems of many degrees of freedom, the vibrations in gear drives (Sections 6.3 to 6.) and the autoparametric vibrations (Chapter 12).

First we formulate the non-linear integro-differential equations for a system of coupled differential equations

$$y_i'' + \lambda_i y_i = \Phi_i \qquad (i = 1, 2, \dots, N) \tag{6.1, 1}$$

where dashes denote derivatives with respect to a dimensionless time τ and

$$\Phi_i = \Phi_i(y_i, y_i', \varepsilon_p, \tau) = \Phi_i[\tau]$$

are given by power series in y_i, y_i' $(i = 1, 2, \dots, N)$ and parameters ε_p $(p = 1, 2, \dots, P)$, which may also depend on time.

Restricting attention for simplicity to continuous solutions with continuous first and second derivatives, we can prove the following theorem (compare SCHMIDT (1961, 1975)): Every periodic solution of the differential equations (6.1, 1) is a solution of the integro-differential equations

$$y_i(\tau) = \int_0^{2\pi} G_i(\tau, \sigma)\ \Phi_i[\sigma]\ \mathrm{d}\sigma + \delta_{\lambda_i}^{n_i^2}(r_i \cos n_i\tau + s_i \sin n_i\tau) \tag{6.1, 2}$$

where

$$G_i(\tau, \sigma) = \frac{1}{\pi}\left[\frac{1}{2\lambda_i}\right] + \sum_{\nu=1}^{\infty} \frac{\cos \nu(\tau - \sigma)}{\vartheta\lambda_i - \nu^2}\bigg]$$

are the corresponding generalized Green's functions, δ_μ^ν is the Kronecker symbol, and

$$\delta\vartheta = \vartheta_\mu^\nu = 1 - \delta_\mu^\nu$$

so that the denominator does not vanish. The bifurcation parameters r_i and s_i appearing in the resonance case

$$\lambda_i = n_i^2$$

are to be determined by the bifurcation or periodicity equations

$$r_i = \frac{1}{\pi}\int_0^{2\pi} y_i(\tau) \cos n_i\tau\ \mathrm{d}\tau\ , \qquad s_i = \frac{1}{\pi}\int_0^{2\pi} y_i(\tau) \sin n_i\tau\ \mathrm{d}\tau \tag{6.1, 3}$$

which are equivalent to

$$\int_0^{2\pi} \Phi_i[\tau] \cos n_i\tau \, d\tau = \int_0^{2\pi} \Phi_i[\tau] \sin n_i\tau \, d\tau = 0 \, .$$ (4.1 6,)

The solutions of (6.1, 2) and (6.1, 3) respectively (6.1, 4) can be found by the method of successive approximations based upon the following equations for the approximative solutions $y_{ik}(\tau)$, $k = 1, 2, 3, \ldots$

$$y_{ik}(\tau) = \int_0^{2\pi} G(\tau, \sigma) \, \Phi_{i,k-1}[\sigma] \, d\sigma + \delta_{\lambda_i}^{n_i^2}(r_i \cos n_i\tau + s_i \sin n_i\tau)$$

where

$$\Phi_{i0}[\tau] = \Phi_i(0, 0, \varepsilon_\rho; \tau) \, , \qquad \Phi_{ik}([\tau] = \Phi_i(y_{jk}, y_{jk}', \varepsilon_p; \tau) \, , \qquad k = 1, 2, 3, \ldots$$

The convergence of the method can be proved if the parameters are smaller than certain upper bounds. Second derivatives y_j'' in Φ_i can be handled in the same way, for the proof of convergence they can be substituted successively by means of the differential equations.

Vibrations with many degrees of freedom can very often be described by the equations

$$\ddot{x}_i + \Omega_i^2 x_i = \sum_{j=0}^{\infty} (P_{ij} \cos j\omega t + Q_{ij} \sin j\omega t) - \sum_j B_{ij}\dot{x}_1$$

$$- \sum_{jkl} C_{ijkl}x_jx_k\dot{x}_l - \sum_{jklpq} D_{ijklpq}x_jx_kx_l x_p\dot{x}_q$$

$$- \sum_{jkl} E_{ijkl}x_jx_kx_l - \sum_{jklpq} F_{ijklpq}x_jx_kx_lx_px_q$$

$$- \sum_j \sum_{k=1}^{\infty} (G_{ijk} \cos k\omega t + H_{ijk} \sin k\omega t) x_1$$ (6.1, 5)

where, corresponding to (5.1, 1), the coefficients P_{ij}, Q_{ij} denote general periodic forced excitations, B_{ij} linear and C_{ijkl}, D_{ijklpq} non-linear damping respectively self-excitation, E_{ijkl}, F_{ijklpq} non-linear restoring forces and G_{ijk}, H_{ijk} general periodic (linear) parametric excitations. Summations run, if not indicated otherwise, from 1 to N.

Consider first a single (at most) resonance of the system, when only one (say, the first) of the values Ω_i/ω is approximately an integer n,

$$\lambda_i = \frac{\Omega_i^2}{\omega_0^2} = n^2 \quad \text{for} \quad i = 1$$ (6.1, 6)

with

$$\omega = \omega_0(1 + \alpha)$$

where

$$\alpha = \frac{\omega - \omega_0}{\omega_0}$$ (6.1, 7)

is the frequency variation. With a dimensionless time

$$\tau = \omega t \, , \qquad = \frac{d}{d\tau} \, , \qquad x_i(t) = x_i\left(\frac{\tau}{\omega}\right) = y_i(\tau) \, ,$$

(6.1, 5) can be written, divided by ω_0^2,

$$y_i'' + \lambda_i y_i = \frac{1}{\omega_0^2} \sum_{j=0}^{\infty} (P_{ij} \cos j\tau + Q_{ij} \sin j\tau) - \alpha(2 + \alpha)\, y_i''$$

$$- \frac{1+\alpha}{\omega_0} \sum_j B_{ij} y_j' - \frac{1+\alpha}{\omega_0} \sum_{jkl} C_{ijkl} y_j y_k y_i'$$

$$- \frac{1+\alpha}{\omega_0} \sum_{jklpq} D_{ijklpq} y_j y_k y_l y_p y_q' - \frac{1}{\omega_0^2} \sum_{jkl} E_{ijkl} y_j y_k y_l$$

$$- \frac{1}{\omega_0^2} \sum_{jklpq} F_{ijklpq} y_j y_k y_l y_p y_q$$

$$- \frac{1}{\omega_0^2} \sum_j \sum_{k=1}^{\infty} (G_{ijk} \cos k\tau + H_{ijk} \sin k\tau)\, y_j . \tag{6.1, 8}$$

The convergence of the iteration method demands the coefficients P_{ij}, Q_{ij}, α, B_{ij}, G_{ijk}, H_{ijk} to be sufficiently small.

The *non-resonance* case, when no λ_i is the square of an integer, is included; then the first approximative solution of the integral equation method is

$$y_{i10} = \frac{1}{\omega} \sum_{j=0}^{\infty} \frac{P_{ij} \cos j\tau + Q_{ij} \sin j\tau}{\vartheta \lambda_i - j^2} .$$

All higher approximations contain only small terms multiplied by P_{ij}, Q_{ij}, so that, as for single-degree-of-freedom systems, non-linear restoring forces, damping, self-excitation and parametric excitation influence the solution only to a smaller order of magnitude than forced excitation.

A *single resonance* (6.1, 6) leads to the first approximation

$$y_{i1} = \delta_i^1 (r \cos n\tau + s \sin n\tau) + y_{i10} .$$

The second and parts of the third approximation, corresponding to (5.1, 8), yield the periodicity equations

$$\begin{pmatrix} p \\ q \end{pmatrix} + a \begin{pmatrix} r \\ s \end{pmatrix} - \beta \begin{pmatrix} s \\ -r \end{pmatrix} - g \begin{pmatrix} r \\ -s \end{pmatrix} - h \begin{pmatrix} s \\ r \end{pmatrix} = 0 \tag{6.1, 9}$$

where in the abbreviations (5.1, 10), (5.1, 11)

instead of $\qquad P_\nu \quad B \quad C \quad D \qquad E \qquad F \qquad G_\nu \quad H_\nu \quad K_\nu \quad L_\nu$

now substitute $P_{1\nu} \quad B_{11} \quad C_{1111} \quad D_{111111} \quad E_{1111} \quad F_{111111} \quad G_{11\nu} \quad H_{11\nu} \quad 0 \qquad 0$

and

$$q = \frac{Q_{1n}}{2n^2 \omega_0^2} .$$

The solution of (6.1, 9) is

$$(a^2 + \beta^2 - g^2 - h^2)\, r = -(a + g)\, p - (\beta + h)\, q ,$$
$$(a^2 + \beta^2 - g^2 - h^2)\, s = (\beta - h)\, p - (a - g)\, q .$$

Squaring and adding gives the amplitude equation

$$(a^2 + \beta^2 - g^2 - h^2)^2\, A^2 = (a^2 + \beta^2 + g^2 + h^2)\, (p^2 + q^2)$$
$$+ 2(ag - \beta h)\, (p^2 - q^2) + 4(ah + \beta g)\, pq \tag{6.1, 10}$$

which is of power four in a, that is, in the frequency variation α. The resonance curves can be found by numerically solving for A or α. Particularly for vanishing forced excitation ($p = q = 0$), we get $A = 0$ or

$$a = \pm \sqrt{g^2 + h^2 - \beta^2} \, ,$$

that is

$$\alpha = eA^2 + fA^4 - \Gamma \pm \sqrt{g^2 + h^2 - \beta^2} \, . \tag{6.1, 11}$$

The general investigation of systems with one degree of freedom in Chapter 5 based on the amplitude equation (5.1, 14) can be transmitted to equation (6.1, 10). We will do this when necessary in what follows.

We now take into consideration a *two-fold resonance*, when two (say, the first and the second) of the values Ω_i/ω are approximately integers,

$$\Lambda_1 \approx n_1^2 \, (= n^2) \, , \qquad \Lambda_2 \approx n_2^2 \, (< n^2) \tag{6.1, 12}$$

with the notation

$$\Lambda_i = \frac{\Omega_i^2}{\omega^2} \, . \tag{6.1, 13}$$

Instead of (6.1, 12) we write

$$\Lambda_i = \lambda_i(1 - 2\alpha_i) \, , \qquad \lambda_i = n_i^2 \qquad (i = 1, 2) \tag{6.1, 14}$$

with small parameters α_1, α_2. The other parameters α_i ($i \geqq 3$) are not used and therefore assumed equal to zero. Using (6.1, 14), the integers $m_\pm = n \pm n_2$ can be written

$$m_\pm = n \pm n_2 = \sqrt{\lambda_1} \pm \sqrt{\lambda_2} = \sqrt{\frac{\Lambda_1}{1 - 2\alpha_1}} \pm \sqrt{\frac{\Lambda_2}{1 - 2\alpha_2}}$$

or, developing the square roots,

$$m_\pm = \sqrt{\Lambda_1} \pm \sqrt{\Lambda_2} + n\alpha_1 \pm n_2\alpha_2 + O(\alpha_1^2, \alpha_2^2) \, . \tag{6.1, 15}$$

On the other hand, the approximative equation $m_\pm \approx \sqrt{\Lambda_1} \pm \sqrt{\Lambda_2}$ can be made exact when the variable excitation frequency ω in (6.1, 13) is replaced by a suitably chosen value ω_0,

$$m_\pm = (\sqrt{\Lambda_1} \pm \sqrt{\Lambda_2}) \frac{\omega}{\omega_0} \, , \qquad \text{i.e.} \qquad \frac{\omega_0}{\omega} m_\pm = \sqrt{\Lambda_1} \pm \sqrt{\Lambda_2} \, .$$

Subtracting from (6.1, 15) yields, up to terms $O(\alpha^2, \alpha_i^2)$, the relation

$$m_\pm \alpha = n\alpha_1 \pm n_2\alpha_2 \tag{6.1, 16}$$

between the frequency variation (6.1, 7) and the parameters α_i.

The dimensionless differential equations, divided by ω^2, now have on the right-hand side

$$2\lambda_i \alpha_i y_i \quad \text{instead of} \quad -\alpha(2 + \alpha) y_i'' \tag{6.1, 17}$$

and the factors

$$\frac{1}{\omega^2}, \frac{1}{\omega} \quad \text{instead of} \quad \frac{1}{\omega_0^2}, \frac{1 + \alpha}{\omega_0} \tag{6.1, 18}$$

which are $1/\omega_0^2, 1/\omega_0$ up to $O(\alpha)$.

The first approximative solution is now

$$y_{i1} = \delta_i^1(r \cos n\tau + s \sin n\tau) + \delta_i^2(r_2 \cos n_2\tau + s_2 \sin n_2\tau) + y_{i10} \,.$$

Evaluating the second approximation and inserting into the periodicity equations, we get, if we assume for simplicity

$$D_{ijklpq} = F_{ijklpq} = H_{ijk} = 0 \,, \qquad G_{ijk} \neq 0 \quad \text{for only one } j$$

and exclude the internal resonance $3n = n_2$ (compare Section 6.6) the four equations

$$\begin{pmatrix} P_{1n} \\ Q_{1n} \end{pmatrix} + a \begin{pmatrix} r \\ s \end{pmatrix} - \beta \begin{pmatrix} s \\ -r \end{pmatrix} - \delta_j^{2n} \frac{G_{11j}}{2} \begin{pmatrix} r \\ -s \end{pmatrix} - \delta_j^{n+n_2} \frac{G_{12j}}{2} \begin{pmatrix} r_2 \\ -s_2 \end{pmatrix}$$

$$- \delta_j^{n-n_2} \frac{G_{12j}}{2} \begin{pmatrix} r_2 \\ s_2 \end{pmatrix} = 0 \,, \tag{6.1, 19}$$

$$\begin{pmatrix} P_{2n_2} \\ Q_{2n_2} \end{pmatrix} + a_2 \begin{pmatrix} r_2 \\ s_2 \end{pmatrix} - \beta_2 \begin{pmatrix} s_2 \\ -r_2 \end{pmatrix} - \delta_j^{2n_2} \frac{G_{22j}}{2} \begin{pmatrix} r_2 \\ -s_2 \end{pmatrix} - \delta_j^{n+n_2} \frac{G_{21j}}{2} \begin{pmatrix} r \\ -s \end{pmatrix}$$

$$- \delta_j^{n-n_2} \frac{G_{21j}}{2} \begin{pmatrix} r \\ s \end{pmatrix} = 0 \tag{6.1, 20}$$

with the abbreviations

$$\left. \begin{aligned} a &= 2n^2\omega^2\alpha_1 - \tfrac{3}{4} E_{1111}A^2 - E_{12}A_2^2 \,, \\ a_2 &= 2n_2^2\omega^2\alpha_2 - E_{21}A^2 - \tfrac{3}{4} E_{2222}A_2^2 \,, \end{aligned} \right\} \tag{6.1, 21}$$

$$\beta = n\omega(B_{11} + \tfrac{1}{4} C_{1111}A^2 + \tfrac{1}{2} C_{1122}A_2^2) \,,$$

$$\beta_2 = n_2\omega(B_{22} + \tfrac{1}{2} C_{2211}A^2 + \tfrac{1}{4} C_{2222}A_2^2)$$

where

$$A_2 = \sqrt{r_2^2 + s_2^2}$$

is the partial amplitude of the resonance part of y_{21} and

$$E_{\mu\nu} = \tfrac{1}{2} \left(E_{\mu\mu\nu\nu} + E_{\mu\nu\mu\nu} + E_{\mu\nu\nu\mu} \right) \,.$$

In what follows, we assume $P_{1n} = Q_{1n} = P_{2n_2} = Q_{2n_2} = 0$, that is, neglect the influence of forced excitation and consider first the summed type *combination resonance*

$$j = n + n_2 \,.$$

We first seek a relation between A and A_2 only. Multiplication of (6.1, 19), (6.1, 20) with $\begin{pmatrix} -s \\ r \end{pmatrix}$ and $\begin{pmatrix} -s_2 \\ r_2 \end{pmatrix}$ respectively gives

$$\beta A^2 + \tfrac{1}{2} G_{12j}(rs_2 + sr_2) = 0 \,,$$

$$\beta_2 A_2^2 + \tfrac{1}{2} G_{21j}(rs_2 + sr_2) = 0$$

from which it follows that

$$\beta G_{21j}A^2 = \beta_2 G_{12j}A_2^2 \tag{6.1, 22}$$

a relation (because of the dependence of $\beta_1\beta_2$ on A, A_2 biquadratic) between A and A_2, independent of the frequency parameters α_1, α_2.

In order to find, for A and A_2 separately, formulae depending on the frequency variation, we use the determinant condition

$$\begin{vmatrix} a & -\beta & -\frac{1}{2}G_{12j} & 0 \\ \beta & a & 0 & \frac{1}{2}G_{12j} \\ -\frac{1}{2}G_{21j} & 0 & a_2 & -\beta_2 \\ 0 & \frac{1}{2}G_{21j} & \beta_2 & a_2 \end{vmatrix}$$

$$= (aa_2 + \beta\beta_2 - \tfrac{1}{4}G_{12j}G_{21j})^2 + (a\beta_2 - a_2\beta)^2 = 0$$

for non-vanishing solutions r, s, r_2, s_2, which yields (for $\beta\beta_2 \neq 0$)

$$a = \pm \sqrt{\frac{\beta}{4\beta_2}G_{12j}G_{21j} - \beta^2}, \qquad a_2 = \pm \sqrt{\frac{\beta_2}{4\beta}G_{12j}G_{21j} - \beta_2^2}. \tag{6.1, 23}$$

In these equations, in the normal (damping) case where $\beta\beta_2 > 0$, equal signs before the roots correspond with one another because multiplication of (6.1, 19) and (6.1, 20) with $\binom{r}{s}$ respectively $\binom{r_2}{s_2}$ leads to

$$\frac{a}{a_2} = \frac{\beta}{\beta_2}.$$

For $j = n + n_2$, equation (6.1, 16) reads

$$j\alpha = n\alpha_1 + n_2\alpha_2;$$

using (6.1, 21), (6.1, 20), we find the amplitude frequency formula

$$2jnn_2\omega^2\alpha = \left(\frac{3n_2}{4}E_{1111} + nE_{21}\right)A^2 + \left(n_2E_{12} + \frac{3n}{4}E_{2222}\right)A_2^2$$

$$\pm (n_2\beta + n\beta_2)\sqrt{\frac{G_{12j}G_{21j}}{4\beta\beta_2} - 1}, \tag{6.1, 24}$$

which determines, together with (6.1, 22), the partial amplitudes A_1, A_2 in dependence of the frequency variation α.

If we assume a *difference type* combination resonance

$$j = n - n_2, \tag{6.1, 25}$$

an analogous analysis leads to the amplitude formulae

$$2jnn_2\omega^2\alpha = \left(\frac{3n_2}{4}E_{1111} - nE_{21}\right)A^2 + \left(n_2E_{12} - \frac{3n}{4}E_{2222}\right)A_2^2$$

$$\pm (n_2\beta + n\beta_2)\sqrt{-\frac{G_{12j}G_{21j}}{4\beta\beta_2} - 1} \tag{6.1, 26}$$

and

$$\beta G_{21j}A^2 = -\beta_2 G_{12j}A_2^2. \tag{6.1, 27}$$

Compare the resonance amplitudes given by (6.1, 24), (6.1, 22) respectively (6.1, 26), (6.1, 27) with the amplitudes (6.1, 11). The order of magnitude of A^2 is in every case equal to the parametric excitation, consequently it is not smaller than linear damping. In the case of linear damping only, all the resonance curves are equidistant parabolae, while in the case of non-linear damping they are curves — with the same starting-points $A = 0$ — which coalesce for

$$\beta^2 = g^2 + h^2$$

in case of single resonance and for

$$4\beta\beta_2 = \pm\, G_{12j}G_{21j}$$

in case (6.1, 24), (6.1, 22) or (6.1, 26), (6.1, 27) of combination resonance.

An example of a difference resonance (6.1, 25) with $n = 8$, $n_2 = 3$, $\omega = 1$, $E_{iiii} = E_{12} = 0$,

$$\beta = 4\beta_2 = 0.04 + 0.4A^2 + 0.2A_2^2 \tag{6.1, 28}$$

and $G_{12j} = G_{21j} = 0.1$ is given in Fig. 6.1, 1 for $E_{21} = 1$ and in Fig. 6. 1, 2 for $E_{21} = 3$. Because β and β_2 are proportional independently of A and A_1, these amplitudes are also proportional (dashed line in Fig. 6.1, 3). If in contrast

$$\beta = 0.04 , \tag{6.1, 29}$$

the proportionality does not hold (full line in Fig. 6.1, 3). The corresponding resonance curves for $E_{21} = 1$ in Fig. 6.1, 4 and for $E_{21} = 3$ in Fig. 6.1, 5 show that the partial

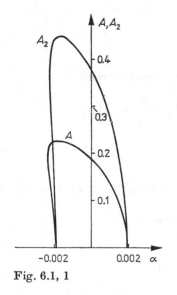

Fig. 6.1, 1

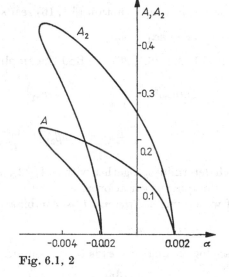

Fig. 6.1, 2

amplitude A grows much more quickly than A_2. If $E_{21} = 0$, but $E_{2222} = 1/3$ and (6.1, 28), we get the resonance curve of Fig. 6.1, 1, and for $E_{2222} = 1$ Fig. 6.1, 2 holds; whereas for the damping (6.1, 29) the resonance curves are not given by Figures 6.1, 4 and 6.1, 5 but, with a much steeper backbone curve, by Fig. 6.1, 6 for $E_{2222} = 1/3$ and by Fig. 6.1, 7 for $E_{2222} = 1$. It shows how the restoring force coefficients significantly influence the form of the resonance curves whereas the maximum amplitudes are, as (6.1, 24), (6.1, 22) respectively (6.1, 26), (6.1, 27) show quite generally, independent of them.

A detailed discussion of further combination resonances can be found in SCHMIDT (1969, 1975) and EVAN-IWANOWSKI (1976).

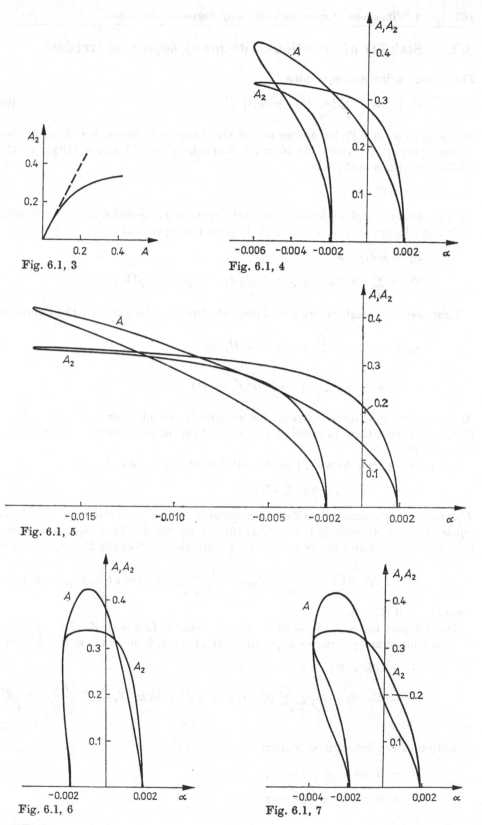

Fig. 6.1, 3

Fig. 6.1, 4

Fig. 6.1, 5

Fig. 6.1, 6

Fig. 6.1, 7

6.2.   Stability of vibrations with many degrees of freedom

The linear variational equations

$$z_i'' + \lambda_i z_i = \sum_j [u_{ij}(\tau) z_j + v_{ij}(\tau) z_j'] \tag{6.2, 1}$$

of the system (6.1, 1) have, because of the Floquet theorem, which holds also for systems (6.2, 1) (compare for instance METTLER (1949), MALKIN (1952) or CESARI (1963)), solutions of the form

$$z_i = e^{\varrho \tau} Z_i \tag{6.2, 2}$$

with generally complex characteristic exponents ϱ and periodic functions Z_i with the period 2π. Inserting (6.2, 2) into (6.2, 1) gives the equations

$$Z_i'' + \lambda_i Z_i = \Phi_i \,,$$

$$\Phi_i = \sum_j \{[u_{ij}(\tau) + \varrho v_{ij}(\tau) - \varrho^2] Z_j + [v_{ij}(\tau) - 2\varrho] Z_j'\} \,.$$

First assume a single resonance of (say) the first equation and use the Fourier series

$$u_{11}(\tau) = u_0 + 2 \sum_{j-1}^{\infty} (u_j \cos j\tau + U_j \sin j\tau) \,,$$

$$v_{11}(\tau) = v_0 + 2 \sum_{j=1}^{\infty} (v_j \cos j\tau + V_j \sin j\tau) \,.$$

As in Section 2.4, the conditions for asymptotic stability prove to be (2.4, 4) and (2.4, 5), whereas there is instability if at least one of the inequalities holds with the opposite sign.

Let us investigate in what follows a double resonance (6.1, 14)

$$\lambda_1 = n^2 \,, \qquad \lambda_2 = n_2^2 (< n^2) \,.$$

The linear variational equations corresponding either to the linearized differential equations (6.1, 8) with (6.1, 17), (6.1, 18), or to the non-linear equations and the solution $y_i \equiv 0$ (which exists if the forced excitation coefficients P_{ij}, Q_{ij} are zero), are

$$z_i'' + \lambda_i z_i = 2\lambda_i \alpha_i z_i - \frac{1}{\omega} \sum_j B_{ij} z_j' - \frac{1}{\omega^2} \sum_j \sum_{k=1}^{\infty} (G_{ijk} \cos k\tau + H_{ijk} \sin k\tau) z_j$$

where $\alpha_i = 0$ $(i \geqq 3)$.

The Floquet theorem yields (6.2, 2), with periodic functions Z_i. As in Section 2.4 we can omit ϱ^2 and ϱv_{ij} in the approximation at hand, from which it now follows that

$$\left.\begin{aligned} Z_i'' + \lambda_i Z_i &= \Phi_i \,, \\ \Phi_i &= 2\lambda_i \alpha_i Z_i - \frac{1}{\omega^2} \sum_j \sum_{k=1}^{\infty} (G_{ijk} \cos k\tau + H_{ijk} \sin k\tau) Z_j - \sum_j \left(\frac{B_{ij}}{\omega} + 2\varrho\right) Z_j' \,. \end{aligned}\right\} \tag{6.2, 3}$$

Insertion of the first approximation

$$Z_1 = R \cos n\tau + S \sin n\tau \,,$$

$$Z_2 = R_2 \cos n\tau + S_2 \sin n\tau$$

of the solution in (6.2, 3), (6.1, 8) gives, by aid of

$$2(G \cos k\tau + H \sin k\tau)(R \cos n\tau + S \sin n\tau)$$
$$= (GR - HS) \cos (k + n)\tau + (HR + GS) \sin (k + n)\tau$$
$$+ (GR + HS) \cos (k - n)\tau + (HR - GS) \sin (k - n)\tau$$

and similar formulae, the four periodicity equations

$$2n^2\alpha_1 \binom{R}{S} - n\left(\frac{B_{11}}{\omega} + 2\varrho\right)\binom{S}{-R}$$
$$- \frac{1}{2\omega^2}\left[G_{11,2n}\binom{R}{-S} + H_{11,2n}\binom{S}{R} + G_{12,n+n_2}\binom{R_2}{-S_2}\right.$$
$$\left. + H_{12,n+n_2}\binom{S_2}{R_2} + G_{12,n-n_2}\binom{R_2}{S_2} - H_{12,n-n_2}\binom{S_2}{-R_2}\right] = 0, \quad (6.2, 4)$$

and

$$2n_2^2\alpha_2 \binom{R_2}{S_2} - n_2\left(\frac{B_{22}}{\omega} + 2\varrho\right)\binom{S_2}{-R_2}$$
$$- \frac{1}{2\omega^2}\left[G_{22,2n_2}\binom{R_2}{-S_2} + H_{22,2n_2}\binom{S_2}{R_2} + G_{21,n+n_2}\binom{R}{-S}\right.$$
$$\left. + H_{21,n+n_2}\binom{S}{R} + G_{21,n-n_2}\binom{R}{S} + H_{21,n-n_2}\binom{S}{-R}\right] = 0. \quad (6.2, 5)$$

We now investigate a summed type combination resonance, assuming

$$G_{11,2n} = H_{11,2n} = G_{22,2n_2} = H_{22,2n_2} = G_{ij,n-n_2} = H_{ij,n-n_2} = 0$$

and for simplicity

$$H_{ij,n+n_2} = 0$$

for $i = 1$, $j = 2$ and $i = 2$, $j = 1$, and using the abbreviations

$$\Gamma_1 = \frac{1}{2\omega^2}G_{12,n+n_2}, \qquad \Gamma_2 = \frac{1}{2\omega^2}G_{21,n+n_2}.$$

The determinant condition for non-vanishing solutions R, S, R_2, S_2 of (6.2, 4), (6.2, 5) leads to the equation

$$\left[4n^4\alpha_1^2 + n^2\left(\frac{B_{11}}{\omega} + 2\varrho\right)^2\right]\left[4n_2^4\alpha_2^2 + n_2^2\left(\frac{B_{22}}{\omega} + 2\varrho\right)^2\right] + \Gamma_1^2\Gamma_2^2$$
$$- 8n^2n_2^2\alpha_1\alpha_2\Gamma_1\Gamma_2 - 2nn_2\left(\frac{B_{11}}{\omega} + 2\varrho\right)\left(\frac{B_{22}}{\omega} + 2\varrho\right)\Gamma_1\Gamma_2 = 0. \quad (6.2, 6)$$

It is not possible to determine, corresponding to a single resonance, ϱ as a function of the frequency variation α because of the unknown parameters α_1, α_2. In what follows we have first to determine these parameters. By scalar multiplication of (6.2, 4) with $\binom{R}{S}$ we get

$$2n^2\alpha_1(R^2 + S^2) = \Gamma_1(RR_2 - SS_2),$$

multiplying by $\begin{pmatrix} S \\ -R \end{pmatrix}$ gives, on the other hand,

$$-n\left(\frac{B_{11}}{\omega} + 2\varrho\right)(R^2 + S^2) = \Gamma_1(SR_2 + RS_2).$$

Squaring these two equations and adding yields the relation

$$\left[4n^4\alpha_1^2 + n^2\left(\frac{B_{11}}{\omega} + 2\varrho\right)^2\right](R^2 + S^2)^2 = \Gamma_1^2(R^2 + S^2)(R_2^2 + S_2^2). \qquad (6.2, 7)$$

Correspondingly, multiplying (6.2, 5) by $\begin{pmatrix} R_2 \\ S_2 \end{pmatrix}$ respectively by $\begin{pmatrix} S_2 \\ -R_2 \end{pmatrix}$, squaring and adding leads to

$$\left[4n_2^4\alpha_2^2 + n_2^2\left(\frac{B_{22}}{\omega} + 2\varrho\right)^2\right](R_2^2 + S_2^2)^2 = \Gamma_2^2(R^2 + S^2)(R_2^2 + S_2^2). \qquad (6.2, 8)$$

By multiplication of (6.2, 7) and (6.2, 8) we get

$$\left[4n^4\alpha_1^2 + n^2\left(\frac{B_{11}}{\omega} + 2\varrho\right)^2\right]\left[4n_2^4\alpha_2^2 + n_2^2\left(\frac{B_{22}}{\omega} + 2\varrho\right)^2\right] = \Gamma_1^2\Gamma_2^2, \qquad (6.2, 9)$$

noting that neither of the sums of squares on the right-hand side vanishes because, following (6.2, 4), (6.2, 5), the vanishing of one sum would imply the vanishing of the other one.

Just as for the relation (6.2, 6), the relation (6.2, 9) does not contain the quantities R, S, R_1, S_1. But these relations contain both the parameters α_1, α_2.

We need separate relations for α_1 and α_2. To begin with, (6.2, 6) simplifies to

$$4n^2n_2^2\alpha_1\alpha_2 + nn_2\left(\frac{B_{11}}{\omega} + 2\varrho\right)\left(\frac{B_{22}}{\omega} + 2\varrho\right) = \Gamma_1\Gamma_2 \qquad (6.2, 10)$$

by inserting (6.2, 9) and dividing through $2\Gamma_1\Gamma_2$. Subtracting the squared equation (6.2, 10) from (6.2, 9) yields

$$n_2\alpha_2\left(\frac{B_{22}}{\omega} + 2\varrho\right) = n\alpha_1\left(\frac{B_{11}}{\omega} + 2\varrho\right), \qquad (6.2, 11)$$

a relation between α_1 and α_2 only. Introducing this relation into equation (6.2, 10), multiplied by $\left(\frac{B_{11}}{\omega} + 2\varrho\right)$, gives

$$n^3n_2\alpha_1^2\left(\frac{B_{22}}{\omega} + 2\varrho\right) = \left(\frac{B_{11}}{\omega} + 2\varrho\right)\Gamma_1\Gamma_2 - nn_2\left(\frac{B_{11}}{\omega} + 2\varrho\right)^2\left(\frac{B_{22}}{\omega} + 2\varrho\right).$$

This equation determines the parameter α_1 from which follows by means of (6.2, 11) the parameter α_2.

For the summed type combination resonance chosen, the relation (6.1, 16)

$$(n + n_2)\alpha = n\alpha_1 + n_2\alpha_2$$

between the frequency variation α and the parameters α_1, α_2 holds. With (6.2, 11) there follows

$$(n + n_2)\alpha\left(\frac{B_{11}}{\omega} + 2\varrho\right) = n\alpha_1\left(\frac{B_{11}}{\omega} + \frac{B_{22}}{\omega} + 4\varrho\right)$$

and by squaring, multiplying with $\left(\dfrac{B_{22}}{\omega} + 2\varrho\right)$ and introducing (6.2, 12)

$$(n + n_2)^2 \alpha^2 \left(\frac{B_{11}}{\omega} + 2\varrho\right)\left(\frac{B_{22}}{\omega} + 2\varrho\right)$$
$$= \left(\frac{B_{11}}{\omega} + \frac{B_{22}}{\omega} + 4\varrho\right)^2 \left[\frac{\Gamma_1\Gamma_2}{nn_2} - \left(\frac{B_{11}}{\omega} + 2\varrho\right)\left(\frac{B_{22}}{\omega} + 2\varrho\right)\right].$$

Solving this equation gives the formula

$$4\varrho = -\frac{1}{\omega}(B_{11} + B_{22}) \underset{(-)}{+} \sqrt{T \pm \sqrt{T^2 + (n + n_2)^2 \frac{\alpha^2}{\omega^2}(B_{11} - B_{22})^2}}$$

where

$$T = \frac{2\Gamma_1\Gamma_2}{nn_2} + \frac{1}{2\omega^2}(B_{11} - B_{22})^2 - \frac{1}{2}(n + n_2)^2 \alpha^2.$$

From this formula we conclude: The solutions are asymptotically stable if and only if the conditions

$$B_{11} + B_{22} > 0$$

and

$$\frac{1}{\omega^2}(B_{11} + B_{22})^2 > T + \sqrt{T^2 + (n + n_2)^2 \frac{\alpha^2}{\omega^2}(B_{11} - B_{22})^2},$$

that is,

$$\frac{1}{\omega^2}(B_{11} + B_{22})^2 - T > \sqrt{T^2 + (n + n_2)^2 \frac{\alpha^2}{\omega^2}(B_{11} - B_{22})^2}$$

hold. The opposite sign in at least one of these conditions leads to instability.

The first condition is met if damping does not vanish. The second condition can, for $B_{11} > 0$ and $B_{22} > 0$, be expressed in the form

$$(n + n_2)|\alpha| > \frac{1}{\omega}(B_{11} + B_{22})\sqrt{\frac{\omega^2\Gamma_1\Gamma_2}{nn_2 B_{11}B_{22}} - 1}. \tag{6.2, 13}$$

An analogous condition, only with the opposite sign before the first part of the radicand, can be found for difference type combination resonance.

Because of the product of damping terms in the denominator, the stability condition (6.2, 13) can cause an enlargement of the instability regions when one of the damping terms increases, that is, a destabilizing effect of damping.

The stability condition (6.2, 13) was found by SCHMIDT and WEIDENHAMMER (1961), independently once more by MASSA (1967), and generalized by VALEEV (1963) and SCHMIDT (1967a, 1967b) with methods different from the one used here (compare, for the application of the integral equation method, SCHMIDT (1973)). Combined analytical, numerical and experimental investigations of combination vibrations and their stability are given by BENZ (1965), BECKER (1972), where the above effect is fully verified by experiment, and SCHMIEG (1976).

6.3.   Vibrations in one-stage gear drives

As a first example of coupled vibrations in systems with many degrees of freedom, we shall investigate vibrations in gear drives. Consider a pair of meshing spur gears (Fig. 6.3, 1). The vibrations of one-stage and in many cases also of multi-stage gear drives can be modelled in this way because other gear stages are often of minor influence on the vibrations of one stage, compare BOSCH (1965), HORTEL (1968, 1969, 1970).

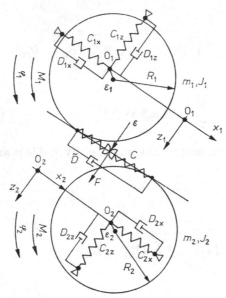

Fig. 6.3, 1

The i-th $(i = 1, 2)$ gear, on which acts a moment M_i, has rotating mass m_i and is elastically supported in two orthogonal directions, deflections in which are denoted by x_i, z_i. The spring stiffness and damping coefficients are C_{ix}, C_{iz} (combining the stiffness of the bearings, the oil film and the shafts) respectively D_{ix}, D_{iz}. The meshing of the teeth of the two wheels is modelled by the torsional stiffness C, the damping $\tilde{D}$, the tooth error ε which describes deviations from the ideal form of teeth and mounting, and the friction F which takes into consideration the sliding forces between the meshing teeth and also causes additional moments M_{iF}. Eccentricities of the wheels are denoted by ε_i, the moments of inertia relating to the rotational axes O_i by J_i, a phase angle between the torsional angles φ_i of the wheels by Δ and time derivatives by dots.

The relative deflection of the meshing teeth in direction of the path of contact can be shown (HORTEL (1968)) to be

$$x_3 = x_1 - x_2 + \varrho_1\varphi_1 + \varrho_2\varphi_2 + \varepsilon + \varepsilon_1 \sin \varphi_1 - \varepsilon_2 \sin (\Delta - \varphi_2) \qquad (6.3, 1)$$

where

$$\varrho_1 = R_1 + \varepsilon_1 \cos \varphi_1, \qquad \varrho_2 = R_2 - \varepsilon_2 \cos (\Delta - \varphi_2)$$

and R_i are the pitch radii of the two wheels.

The kinetic energy of the system is

$$T = \tfrac{1}{2}\left[J_1\dot\varphi_1^2 + J_2\dot\varphi_2^2 + m_1(\dot x_1^2 + \dot z_1^2) + m_2(\dot x_2^2 + \dot z_2^2)\right],$$ (6.3, 2)

the potential energy function

$$V = \tfrac{1}{2}\left(Cx_3^2 + C_{1x}x_1^2 + C_{2x}x_2^2 + C_{1z}z_1^2 + C_{2z}z_2^2\right),$$ (6.3, 3)

the dissipation function

$$W = \tfrac{1}{2}\left(\tilde D\dot x_3^2 + D_{1x}\dot x_1^2 + D_{2x}\dot x_2^2 + D_{1z}\dot z_1^2 + D_{2z}\dot z_2^2\right),$$ (6.3, 4)

the virtual work of the remaining generalized forces

$$(M_1 - M_{1F})\,\delta\varphi_1 + (M_2 + M_{2F})\,\delta\varphi_2 + F\delta z_1 + F\delta z_2\,.$$

Lagrange's equations are, using the Lagrangian $L = T - V$,

$$\frac{\mathrm d}{\mathrm dt}\left(\frac{\partial L}{\partial\dot\varphi_i}\right) - \frac{\partial L}{\partial\varphi_i} + \frac{\partial W}{\partial\dot\varphi_i} = M_i + (-1)^i M_{iF}\,,$$

$$\frac{\mathrm d}{\mathrm dt}\left(\frac{\partial L}{\partial\dot x_i}\right) - \frac{\partial L}{\partial x_i} + \frac{\partial W}{\partial\dot x_i} = 0\,, \qquad (i = 1,2)$$

$$\frac{\mathrm d}{\mathrm dt}\left(\frac{\partial L}{\partial\dot z_i}\right) - \frac{\partial L}{\partial z_i} + \frac{\partial W}{\partial\dot z_i} = F\,.$$

Insertion of (6.3, 2), (6.3, 3), (6.3, 4) yields taking into consideration

$$\varrho_i \approx R_i = \text{const}\,, \qquad J_i \approx \text{const}$$

and

$$\varepsilon_i\,, \frac{\partial\varepsilon}{\partial\varphi_i} \ll \varrho_i\,,$$

the equations

$$J_1\ddot\varphi_1 + C\varrho_1 x_3 + \frac{1}{2}\frac{\partial C}{\partial\varphi_1}x_3^2 + \tilde D\varrho_1\dot x_3 = M_1 - M_{1F}\,,$$

$$J_2\ddot\varphi_2 + C\varrho_2 x_3 + \frac{1}{2}\frac{\partial C}{\partial\varphi_2}x_3^2 + \tilde D\varrho_2\dot x_3 = M_2 + M_{2F}\,,$$

$$\left.\begin{array}{l} m_1\ddot x_1 + C_{1x}x_1 + Cx_3 + \dfrac{1}{2}\dfrac{\mathrm dC_{1x}}{\mathrm dx_1}x_1^2 + \dfrac{1}{2}\dfrac{\partial C}{\partial x_3}x_3^2 + D_{1x}\dot x_1 + \tilde D\dot x_3 = 0\,, \\[2mm] m_2\ddot x_2 + C_{2x}x_2 - Cx_3 + \dfrac{1}{2}\dfrac{\mathrm dC_{2x}}{\mathrm dx_2}x_2^2 - \dfrac{1}{2}\dfrac{\partial C}{\partial x_3}x_3^2 + D_{2x}\dot x_2 - \tilde D\dot x_3 = 0\,, \end{array}\right\} \quad (6.3, 5)$$

$$m_1\ddot z_1 + C_{1z}z_1 + \frac{1}{2}\frac{\mathrm dC_{1z}}{\mathrm dz_1}z_1^2 + D_{1z}\dot z_1 = F\,,$$

$$m_2\ddot z_2 + C_{2z}z_2 + \frac{1}{2}\frac{\mathrm dC_{2z}}{\mathrm dz_2}z_2^2 + D_{2z}\dot z_2 = F\,.$$

The last two equations are not coupled with the other ones, they describe forced vibrations z_1, z_2 and can be solved easily. We therefore confine ourselves to the remaining coupled equations. Differentiating the equation (6.3, 1) for the relative tooth deflection, taking into consideration

$$\varphi_1 \approx \Omega_1 t , \qquad \varphi_2 \approx \Omega_2 t$$

and inserting the first two equations, we get

$$\ddot{x}_3 + \frac{C}{\mu} x_3 + \frac{1}{2}\left(\frac{\varrho_1}{J_1}\frac{\partial C}{\partial \varphi_1} + \frac{\varrho_2}{J_2}\frac{\partial C}{\partial \varphi_2}\right)x_3^2 + \frac{\tilde{D}}{\mu}\dot{x}_3 - \ddot{x}_1 + \ddot{x}_2$$

$$= \frac{\varrho_1}{J_1}(M_1 - M_{1F}) + \frac{\varrho_2}{J_2}(M_2 + M_{2F}) + \ddot{\varepsilon} - \omega_1^2 \varepsilon_1 \sin \Omega_1 t + \omega_2^2 \varepsilon_2 \sin (\varDelta - \Omega_2 t)$$

$$(6.3,6)$$

where

$$\mu = \frac{J_1 J_2}{\varrho_1^2 J_2 + \varrho_2^2 J_1}$$

is the reduced mass.

The torsional stiffness C substantially depends on the number of meshing tooth pairs, it is periodic in time with the meshing frequency $\omega = \Omega_1 \zeta_1 = \Omega_2 \zeta_2$ where ζ_i is the number of teeth on the i-th wheel. It depends further on the relative tooth deflection; we assume therefore (compare HORTEL (1968), LINKE (1970)

$$C = \sum_{j=0}^{\infty}(\gamma_j \cos j\omega t + \bar{\gamma}_j \sin j\omega t) + \sum_{j=0}^{\infty}(\varkappa_j \cos j\omega t + \bar{\varkappa}_j \sin j\omega t)x_3 + \bar{c}x_3^2 \quad (6.3,7)$$

where $\bar{c} = $ const. The other stiffness coefficients as well as the damping coefficient are also assumed to depend (quadratically) on the deflection but not to depend on time:

$$C_{ix} = c_{ix} + \bar{c}_{ix}x_i^2 , \qquad \tilde{D} = \delta + \bar{\delta}x_3^2 , \qquad D_{ix} = \delta_i + \bar{\delta}_i x_i^2 \qquad (i = 1, 2)$$

and corresponding equations for C_{iz}, D_{iz}.

Inserting these expressions into (6.3, 5), (6.3., 6) and introducing a transformed time

$$\tau = \omega t , \qquad ' = \frac{\mathrm{d}}{\mathrm{d}\tau} , \qquad x_i(t) = y_i(\tau) \qquad (i = 1, 2, 3)$$

yields three equations of the form

$$y_i'' + \lambda_i y_i = -\alpha_i(2 + \alpha_i)\, y_i'' - (1 + \alpha_i)\,(B_i + D_i y_i^2)\, y_i' - (1 + \alpha_i)\,(\bar{B}_i + \bar{D}_i y_3^2)\, y_3'$$

$$- E_i y_i^3 - F_i y_3^3 - \sum_{j=0}^{\infty}(G_{ij} \cos j\tau + H_{ij} \sin j\tau)\, y_3$$

$$- \sum_{j=0}^{\infty}(K_{ij} \cos j\tau + L_{ij} \sin j\tau)\, y_3^2 \qquad (i = 1, 2) ,$$

$$y_3'' + \lambda_3 y_3 = \sum_{j=0}^{\infty}(P_j \cos j\tau + Q_j \sin j\tau) - \alpha_3(2 + \alpha_3)\, y_3'' + (1 + \alpha_3)^2\,(y_1'' - y_2'')$$

$$- (1 + \alpha_3)\,(B + D y_3^2)\, y_3' - E y_3^3 - \sum_{j=1}^{\infty}(G_j \cos j\tau + H_j \sin j\tau)\, y_3$$

$$+ (G_{10} - G_{20})\, y_3 - \sum_{j=0}^{\infty}(K_j \cos j\tau + L_j \sin j\tau)\, y_3^2 \qquad (6.3, 8)$$

for the lateral and the torsional vibrations where

$$H_{i0} = L_{i0} = L_0 = 0 \,,$$

the frequency parameters

$$\alpha_i = \frac{\omega - \omega_i}{\omega_i} \qquad (i = 1, 2, 3)$$

have been introduced describing the distance of ω from fixed frequencies ω_i. The abbreviations

$$\lambda_i = \frac{c_{ix}}{m_i \omega_i^2} \,, \qquad \lambda_3 = \frac{\gamma_0}{\mu \omega_3^2} + G_{10} - G_{20} \approx \frac{\gamma_0}{\mu \omega_3^2} \left(1 + \frac{\mu}{m_1} + \frac{\mu}{m_2} \right),$$

$$B_i = \frac{\delta_i}{m_i \omega_i} \,, \qquad \bar{B}_i = (-1)^{i-1} \frac{\delta}{m_i \omega_i} \,, \qquad D_i = \frac{\bar{\delta}_i}{m_i \omega_i} \,,$$

$$\bar{D}_i = (-1)^{i-1} \frac{\bar{\delta}}{m_i \omega_i} \,, \qquad B = \frac{\delta}{\mu \omega_3} \,, \qquad D = \frac{\bar{\delta}}{\mu \omega_3} \,,$$

$$E = \frac{\bar{c}}{\mu \omega_3^2} \,, \qquad E_i = \frac{2\bar{c}_{ix}}{m_i \omega_i^2} \,, \qquad F_j = (-1)^{i-1} \frac{2\bar{c}}{m_i \omega_i^2} \,,$$

$$G_{ij} = (-1)^{i-1} \frac{\gamma_j}{m_i \omega_i^2} \,, \qquad H_{ij} = (-1)^{i-1} \frac{\bar{\gamma}_j}{m_i \omega_i^2} \,, \qquad G_j = \frac{\gamma_j}{\mu \omega_3^2} \,,$$

$$H_j = \frac{\bar{\gamma}_j}{\mu \omega_3^2} \,, \qquad K_{ij} = (-1)^{i-1} \frac{3\varkappa_j}{2m_i \omega_i^2} \,,$$

$$K_j = \frac{\varkappa_j}{\mu \omega_3^2} + \frac{1 + \alpha_3}{2\omega_3} \left(\frac{\varrho_1}{J_1 \omega_1} + \frac{\varrho_2}{J_2 \omega_2} \right) j\bar{\gamma}_j \,, \qquad L_{ij} = (-1)^{i-1} \frac{3\bar{\varkappa}_j}{2m_i \omega_i^2} \,,$$

$$L_j = \frac{\bar{\varkappa}_j}{\mu \omega_3^2} - \frac{1 + \alpha_3}{2\omega_3} \left(\frac{\varrho_1}{J_1 \omega_1} + \frac{\varrho_2}{J_2 \omega_2} \right) j\gamma_j$$

for $i = 1, 2$ have been used, and the right-hand side of (6.3, 6) containing the tooth error ε and the eccentricities ε_i has been written (compare HORTEL (1968)) in the form

$$\omega_3^2 \sum_{j=0}^{\infty} (P_j \cos j\tau + Q_j \sin j\tau) \,.$$

On the right-hand sides of these equations, the static prestress P_0 predominates. We can take this into consideration by a suitable choice of the approximations. In the *non-resonance* case (λ_i not square of an integer for $i = 1, 2, 3$), we therefore use as a first approximation

$$y_{i1} = y_{i10} = 0 \qquad (i = 1, 2) \,,$$

$$y_{31} = y_{310} = \frac{P_0}{\lambda_3} \,.$$

For the second approximation we get

$$y_{i20} = -\frac{F_i P_0^3}{\lambda_i \lambda_3^3} - \frac{P_0}{\lambda_3} \sum_{j=0}^{\infty} \frac{G_{ij} \cos j\tau + H_{ij} \sin j\tau}{\vartheta \lambda_i - j^2} - \frac{P_0^2}{\lambda_3^2} \sum_{j=0}^{\infty} \frac{K_{ij} \cos j\tau + L_{ij} \sin j\tau}{\vartheta \lambda_i - j^2}$$

$$(i = 1, 2)$$

where now $\vartheta = 1$, and by inserting this approximation

$$y_{320} = \sum_{j=0}^{\infty} \frac{P_j \cos j\tau + Q_j \sin j\tau}{\vartheta \lambda_3 - j^2} + (G_{10} - G_{20}) \frac{P_0}{\lambda_3^2} - \frac{EP_0^3}{\lambda_3^4}$$

$$- \frac{P_0}{\lambda_3} \sum_{j=1}^{\infty} \frac{1}{\vartheta \lambda_3 - j^2} \left[G_j \cos j\tau + H_j \sin j\tau \right.$$

$$+ j^2 \sum_{i=1}^{2} (-1)^i \frac{G_{ij} \cos j\tau + H_{ij} \sin j\tau}{\vartheta \lambda_i - j^2}$$

$$+ \frac{P_0}{\lambda_3} (K_j \cos j\tau + L_j \sin j\tau)$$

$$\left. + \frac{j^2 P_0}{\lambda_3} \sum_{i=1}^{2} (-1)^i \frac{K_{ij} \cos j\tau + L_{ij} \sin j\tau}{\vartheta \lambda_i - j^2} \right].$$

This shows that in addition to the predominant static prestress, the time-dependent input, output and tooth error as well as the tooth-pair stiffness have a major, and damping has a minor, influence on the torsional vibrations, whereas only the static prestress and the tooth-pair stiffness have a significant influence on the lateral vibrations.

6.4.   Torsional gear resonance

In this section we investigate a single resonance of the third (torsional) equation (6.3, 8):

$$\lambda_3 = n^2 , \qquad \lambda_i \neq n_i^2 \qquad (i = 1, 2; n, n_i \text{ integer}) .$$

Designating $\alpha_3 = \alpha$ as the frequency variation and setting $\alpha_1 = \alpha_2 = 0$, the first approximation is now

$$y_{31} = r \cos n\tau + s \sin n\tau + y_{310} ,$$

$$y_{i1} = 0 \qquad (i = 1, 2) .$$

For the second approximation we find

$$y_{i2} = y_{i20} - \frac{n \overline{B}_i}{\vartheta \lambda_i - n^2} (s \cos n\tau - r \sin n\tau)$$

$$- \frac{\overline{D}_i}{\vartheta \lambda_i - n^2} \left(\frac{n}{4} A^2 + \frac{P_0^2}{n^3} \right) (s \cos n\tau - r \sin n\tau)$$

$$- \frac{\overline{D}_i P_0}{n(\vartheta \lambda_i - 4n^2)} [2rs \cos 2n\tau + (s^2 - r^2) \sin 2n\tau]$$

$$- \frac{n \overline{D}_i}{4(\vartheta \lambda_i - 9n^2)} [s(3r^2 - s^2) \cos 3n\tau + r(3s^2 - r^2) \sin 3n\tau]$$

$$- \frac{3F_i P_0 A^2}{2n^2 \lambda_i} - \frac{3F_i}{\vartheta \lambda_i - n^2} \left(\frac{A^2}{4} + \frac{P_0^2}{n^4} \right) (r \cos n\tau + s \sin n\tau)$$

$$- \frac{3F_i P_0}{2n^2(\vartheta\lambda_i - 4n^2)} [(r^2 - s^2) \cos 2n\tau + 2rs \sin 2n\tau]$$

$$- \frac{1}{4} \frac{F_i}{\vartheta\lambda_i - 9n^2} [r(r^2 - 3s^2) \cos 3n\tau + s(3r^2 - s^2) \sin 3n\tau]$$

$$- \frac{1}{2} \sum_{j=0}^{\infty} \frac{(G_{ij}r + H_{ij}s) \cos (j - n)\tau + (H_{ij}r - G_{ij}s) \sin (j - n)\tau}{\vartheta\lambda_i - (j - n)^2}$$

$$- \frac{1}{2} \sum_{j=0}^{\infty} \frac{(G_{ij}r - H_{ij}s) \cos (j + n)\tau + (H_{ij}r + G_{ij}s) \sin (j + n)\tau}{\vartheta\lambda_i - (j + n)^2}$$

$$- \frac{P_0}{n^2} \sum_{j=0}^{\infty} \frac{G_{ij} \cos j\tau}{\vartheta\lambda_i - j^2} - \frac{A^2}{2} \sum_{j=0}^{\infty} \frac{K_{ij} \cos j\tau + L_{ij} \sin j\tau}{\vartheta\lambda_i - j^2}$$

$$- \frac{P_0}{n^2} \sum_{j=0}^{\infty} \frac{(K_{ij}r + L_{ij}s) \cos (j - n)\tau + (L_{ij}r - K_{ij}s) \sin (j - n)\tau}{\vartheta\lambda_i - (j - n)^2}$$

$$- \frac{P_0}{n^2} \sum_{j=0}^{\infty} \frac{(K_{ij}r - L_{ij}s) \cos (j + n)\tau + (L_{ij}r + K_{ij}s) \sin (j + n)\tau}{\vartheta\lambda_i - (j + n)^2}$$

$$- \frac{1}{4} \sum_{j=0}^{\infty} \frac{[K_{ij}(r^2 - s^2) + 2L_{ij}rs] \cos (j - 2n)\tau + [L_{ij}(r^2 - s^2) - 2K_{ij}rs] \sin (j - 2n)\tau}{\vartheta\lambda_i - (j - 2n)^2}$$

$$- \frac{1}{4} \sum_{j=0}^{\infty} \frac{[K_{ij}(r^2 - s^2) - 2L_{ij}rs] \cos (j + 2n)\tau + [L_{ij}(r^2 - s^2) + 2K_{ij}rs] \sin (j + 2n)\tau}{\vartheta\lambda_i - (j + 2n)^2}$$

where

$$A = \sqrt{r^2 + s^2},$$

the amplitude of the resonance part of the first approximative solution, is simply termed amplitude. The real torsional amplitude consists (approximately) of A, the constant static prestress y_{310} and the amplitude $\mathfrak{A}$ of y_{320}, the latter depending mainly on the variable tooth stiffness and the tooth errors. When $A + \mathfrak{A}$ is greater than the constant prestress, the tooth faces separate and the underlying equations are no longer valid.

The second approximation y_{32} has been determined by introducing y_{i2} and y_{31}, it cannot be given here because of its great length. It leads to the two periodicity equations

$$\begin{pmatrix} p \\ q \end{pmatrix} + a \begin{pmatrix} r \\ s \end{pmatrix} - \beta \begin{pmatrix} s \\ -r \end{pmatrix} - g \begin{pmatrix} r \\ -s \end{pmatrix} - h \begin{pmatrix} s \\ r \end{pmatrix}$$
$$- k \begin{pmatrix} 3r^2 + s^2 \\ 2rs \end{pmatrix} - l \begin{pmatrix} 2rs \\ r^2 + 3s^2 \end{pmatrix} - K \begin{pmatrix} r^2 - s^2 \\ -2rs \end{pmatrix} - L \begin{pmatrix} 2rs \\ r^2 - s^2 \end{pmatrix} = 0 \qquad (6.4, 1)$$

generalizing (5.1, 9) where now

$$\begin{pmatrix} p \\ q \end{pmatrix} = \begin{pmatrix} P_n \\ Q_n \end{pmatrix} - \frac{P_0}{n^2} \begin{pmatrix} G_n \\ H_n \end{pmatrix} - P_0 \sum_{i=1}^{2} \frac{(-1)^i}{\vartheta\lambda_i - n^2} \begin{pmatrix} G_{in} \\ H_{in} \end{pmatrix}$$
$$- \frac{P_0}{n^4} \begin{pmatrix} K_n \\ L_n \end{pmatrix} - \frac{P_0^2}{n^2} \sum_{i=1}^{2} \frac{(-1)^i}{\vartheta\lambda_i - n^2} \begin{pmatrix} K_{in} \\ L_{in} \end{pmatrix},$$

$$a = 2n^2\alpha - 3\left(\frac{E}{n^2} - \frac{F_1}{\vartheta\lambda_1 - n^2} + \frac{F_2}{\vartheta\lambda_2 - n^2}\right)\left(\frac{n^2 A^2}{4} + \frac{P_0^2}{n^2}\right)$$

$$\left. + \left(\frac{\vartheta\lambda_1 G_{10}}{\vartheta\lambda_1 - n^2} - \frac{\vartheta\lambda_2 G_{20}}{\vartheta\lambda_2 - n^2}\right) - 2P_0\left(\frac{K_0}{n^2} - \frac{K_{10}}{\vartheta\lambda_1 - n^2} + \frac{K_{20}}{\vartheta\lambda_2 - n^2}\right), \right\} \quad (6.4, 2)$$

$$\beta = n\left[B + B_1 - B_2 + (D + D_1 - D_2)\left(\frac{A^2}{4} + \frac{P_0^2}{n^4}\right)\right],$$

$$\binom{g}{h} = \frac{1}{2}\binom{G_{2n}}{H_{2n}} + \frac{n^2}{2}\sum_{i=1}^{2}\frac{(-1)^i}{\vartheta\lambda_i - n^2}\binom{G_{i,2n}}{H_{i,2n}} + \frac{P_0}{n^2}\binom{K_{2n}}{L_{2n}}$$

$$+ P_0\sum_{i=1}^{2}\frac{(-1)^i}{\vartheta\lambda_i - n^2}\binom{K_{i,2n}}{L_{i,2n}},$$

$$\binom{k}{l} = \frac{1}{4}\binom{K_n}{L_n} + \frac{n^2}{4}\sum_{i=1}^{2}\frac{(-1)^i}{\vartheta\lambda_i - n^2}\binom{K_{in}}{L_{in}},$$

$$\binom{K}{L} = \frac{1}{4}\binom{K_{3n}}{L_{3n}} + \frac{n^2}{4}\sum_{i=1}^{2}\frac{(-1)^i}{\vartheta\lambda_i - n^2}\binom{K_{i,3n}}{L_{i,3n}}.$$

Assume a torsional stiffness (6.3, 7) such that the coefficients of non-linear parametric excitation appearing in (6.1, 4) are zero,

$$k = l = K = L = 0.$$

(The influence of non-linear parametric excitation is discussed by HORTEL and SCHMIDT (1979, 1981).) Then the periodicity equations (6.4, 1) are of the form (6.1, 9). The resulting amplitude equation (6.1, 10) represents a simple formula for A as dependent on a if for the coefficients of nonlinear damping, the equation

$$D + D_1 - D_2 = 0 \qquad (6.4, 3)$$

holds, and especially if nonlinear damping vanishes.

We shall now give an example of how the results of Chapter 5 can be transmitted to the gear vibration problem at hand.

In order to find the maximum curves, we differentiate (6.1, 10) by a and put $dA/da = 0$:

$$2A^2 a^3 + 2(\beta^2 - g^2 - h^2) A^2 a - (p^2 + q^2) a = g(p^2 - q^2) + 2hpq \qquad (6.4, 4)$$

Eliminating a from (6.1, 10) and (6.4, 4), we get the extreme values of A which depend only on the parameters in β, g, h, p, q, not on the restoring force parameters contained in a.

Assume the special phase relation

$$g(p^2 - q^2) + 2hpq = 0,$$

that is,

$$g = \frac{2pq}{q^2 - p^2} \quad \text{if} \quad q \neq \pm p \quad \text{and} \quad h = 0 \quad \text{if} \quad q = \pm p$$

between the parametric excitation by variable stiffness (represented by g, h) and the forced excitation, mainly by tooth errors (represented by p, q). Then the amplitude equation (6.1, 10) is biquadratic in a,

$$(a^2 + \beta^2 - \gamma^2)^2 A^2 = [a^2 + (\beta - \gamma)^2]\pi^2 \qquad (6.4, 5)$$

if we abbreviate

$$\pi = \sqrt{p^2 + q^2}$$

and

$$\gamma = \frac{p^2 + q^2}{p^2 - q^2} h \quad \text{if} \quad q \neq \pm p \quad \text{resp.} \quad \gamma = \mp g \quad \text{if} \quad q = \pm p$$

so that

$$\gamma^2 = g^2 + h^2 .$$

Solving (6.4, 5) for a yields

$$a^2 = \frac{\pi^2}{2A^2} - \beta^2 + \gamma^2 \pm \pi \sqrt{\frac{\pi^2}{4A^4} - \frac{2(\beta - \gamma)\gamma}{A^2}} .$$

The extreme value equation (6.3, 12) now leads to $a_0 = 0$ and

$$a_{1,2} = \pm \sqrt{\gamma^2 - \beta^2 + \frac{\pi^2}{2A^2}} , \tag{6.4, 6}$$

(6.4, 5) gives for the corresponding amplitude extremes

$$(\beta + \gamma)^2 A_0^2 = \pi^2 \tag{6.4, 7}$$

and

$$8\gamma(\beta - \gamma) A_{1,2}^2 = \pi^2 . \tag{6.4, 8}$$

Insertion of (6.4, 8) into (6.4, 6) shows that $a_{1,2}$ is real if

$$\frac{\beta}{3} \leq \gamma \leq \beta , \quad \text{that ist, if} \quad \gamma \leq \beta \leq 3\gamma .$$

In this domain, $A_1 = A_2$ is the maximum, A_0 a (relative) minimum amplitude, whereas A_0 is the maximum amplitude for $\gamma < \beta/3$.

6.5. Combination gear resonances

We now investigate a double torsional-lateral resonance:

$$\lambda_3 = n^2 , \quad \lambda_1 = n_1^2 , \quad \lambda_2 \neq n_2^2 \quad (n, n_1, n_2 \text{ integer}) .$$

For the first approximation we get

$$y_{11} = r_1 \cos n_1\tau + s_1 \sin n_1\tau , \quad y_{21} = 0$$

and through that,

$$y_{31} = r \cos n\tau + s \sin n\tau + y_{310} + N(r_1 \cos n_1\tau + s_1 \sin n_1\tau)$$

where

$$N = \frac{n_1^2}{n_1^2 - \vartheta n^2} .$$

Assuming for simplicity a cosine-shaped torsional stiffness (6.3, 7),

$$\bar{\gamma}_j = \varkappa_j = \bar{\varkappa}_j = \bar{c} = 0 ,$$

from which it follows that

$$E = F_i = H_j = H_{ij} = K_j = K_{ij} = L_{ij} = 0 .$$

We can now find the additional terms

$$
y_{i2,\,\mathrm{add}} = -\delta_i^1 \cdot 2\alpha_1(r_1 \cos n_1\tau + s_1 \sin n_1\tau) + \delta_i^1 \frac{B_1}{n_1}(s_1 \cos n_1\tau - r_1 \sin n_1\tau)
$$

$$
+ \delta_i^1 \frac{D_1}{4n_1} A_1^2(s_1 \cos n_1\tau - r_1 \sin n_1\tau)
$$

$$
+ \delta_i^1 \frac{D_1}{32n_1}[s_1(3r_1^2 - s_1^2)\cos 3n_1\tau + r_1(3s_1^2 - r_1^2)\sin 3n_1\tau]
$$

$$
- \frac{n_1 N \overline{B}_i}{\vartheta\lambda_i - n_1^2}(s_1 \cos n_1\tau - r_1 \sin n_1\tau) - \frac{nN^2 \overline{D}_i}{2(\vartheta\lambda_i - n^2)} A_1^2(s \cos n\tau - r \sin n\tau)
$$

$$
- \frac{n_1 N \overline{D}_i}{4(\vartheta\lambda_i - n_1^2)}\left(2A^2 + N^2 A_1^1 + \frac{4P_0^2}{n^4}\right)(s_1 \cos n_1\tau - r_1 \sin n_1\tau)
$$

$$
- \frac{n_1 N^2 \overline{D}_i P_0}{n^2(\lambda_i - 4n_1^2)}[2r_1 s_1 \cos 2n_1\tau + (s_1^2 - r_1^2)\sin 2n_1\tau]
$$

$$
- \frac{n_1 N^3 \overline{D}_i}{4(\lambda_i - 9n_1^2)}[s_1(3r_1^2 - s_1^2)\cos 3n_1\tau + r_1(3s_1^2 - r_1^2)\sin 3n_1\tau]
$$

$$
- \frac{(n - n_1)\,N \overline{D}_i P_0}{n^2[\vartheta\lambda_i - (n - n_1)^2]}[(sr_1 - rs_1)\cos(n - n_1)\tau - (ss_1 + rr_1)\sin(n - n_1)\tau]
$$

$$
- \frac{(n + n_1)\,N \overline{D}_i P_0}{n^2[\lambda_i - (n + n_1)^2]}[(sr_1 + rs_1)\cos(n + n_1)\tau + (ss_1 - rr_1)\sin(n + n_1)\tau]
$$

$$
- \frac{(n - 2n_1)\,N^2 \overline{D}_i}{4[\vartheta\lambda_i - (n - 2n_1)^2]}\{[s(r_1^2 - s_1^2) - 2rr_1 s_1]\cos(n - 2n_1)\tau
$$
$$
+ [r(s_1^2 - r_1^2) - 2sr_1 s_1]\sin(n - 2n_1)\tau\}
$$

$$
- \frac{(n + 2n_1)\,N^2 \overline{D}_i}{4[\lambda_i - (n + 2n_1)^2]}\{[s(r_1^2 - s_1^2) + 2rr_1 s_1]\cos(n + 2n_1)\tau
$$
$$
+ [r(s_1^2 - r_1^2) + 2sr_1 s_1]\sin(n + 2n_1)\tau\}
$$

$$
- \frac{(n_1 - 2n)\,N \overline{D}_i}{4[\vartheta\lambda_i - (n_1 - 2n)^2]}\{[s_1(r^2 - s^2) - 2r_1 rs]\cos(n_1 - 2n)\tau
$$
$$
+ [r_1(s^2 - r^2) - 2s_1 rs]\sin(n_1 - 2n)\tau\}
$$

$$
- \frac{(n_1 + 2n)\,N \overline{D}_i}{4[\lambda_i - (n_1 + 2n)^2]}\{[s_1(r^2 - s^2) + 2r_1 rs]\cos(n_1 + 2n)\tau
$$
$$
+ [r_1(s^2 - r^2) + 2s_1 rs]\sin(n_1 + 2n)\tau\}
$$

$$
+ \delta_i^1 \frac{3E_1}{4n_1^2} A_1^2(r_1 \cos n_1\tau + s_1 \sin n_1\tau)
$$

$$
+ \delta_i^1 \frac{E_1}{32n_1^2}[r_1(r_1^2 - 3s_1^2)\cos 3n_1\tau + s_1(3r_1^2 - s_1^2)\sin 3n_1\tau]
$$

$$
- \frac{N}{2}\sum_{j=0}^{\infty} G_{ij}\left[\frac{r_1 \cos(n_1 + j)\tau + s_1 \sin(n_1 + j)\tau}{\vartheta\lambda_i - (n_1 + j)^2}\right.
$$
$$
\left. + \frac{r_1 \cos(n_1 - j)\tau + s_1 \sin(n_1 - j)\tau}{\vartheta\lambda_i - (n_1 - j)^2}\right]
$$

where

$$A_1 = \sqrt{r_1^2 + s_1^2}$$

is the partial amplitude of (the first approximation of) y_1. Insertion of y_{12} and $y_{12,\,\mathrm{add}}$ into the periodicity equations yields

$$\binom{p_1}{0} + a_1 \binom{r_1}{s_1} - \beta_1 \binom{s_1}{-r_1} - \beta_2 \binom{s}{-r} - g_1 \binom{r_1}{-s_1} - g_2 \binom{r}{-s} - h_2 \binom{r}{s}$$

$$+ \delta_{n_1}^n \frac{n}{4} \overline{D}_1 \left[\binom{(r^2 - s^2)s_1 - 2rsr_1}{(r^2 - s^2)r_1 + 2rss_1} + \binom{s(r_1^2 - s_1^2) - 2rr_1 s_1}{r(r_1^2 - s_1^2) + 2sr_1 s_1} \right]$$

$$- \delta_{2n_1}^n \frac{P_0}{6n} \overline{D}_1 \binom{rs_1 - sr_1}{rr_1 + ss_1} + \delta_{3n_1}^n \frac{n}{768} \overline{D}_1 \binom{-s(r_1^2 - s_1^2) + 2rr_1 s_1}{r(r_1^2 - s_1^2) + 2sr_1 s_1}$$

$$- \delta_{n_1}^{2n} \frac{P_0}{n} \overline{D}_1 \binom{2rs}{s^2 - r^2} - \delta_{n_1}^{3n} \frac{n}{4} \overline{D}_1 \binom{s(3r^2 - s^2)}{r(3s^2 - r^2)} = 0 \qquad (6.5,\,1)$$

where

$$p_1 = -\frac{P_0}{n^2} G_{1n_1}, \qquad a_1 = 2n_1^2 \alpha_1 - \frac{3}{4} E_1 A_1^2 - N G_{10},$$

$$\beta_1 = n_1 \left(B_1 + N\overline{B}_1 + \frac{1}{4} D_1 A_1^2 + \frac{N^3}{4} \overline{D}_1 A_1^2 + \frac{N}{2} \overline{D}_1 A_1^2 + \frac{N}{n^4} P_0^2 \overline{D}_1 \right),$$

$$\beta_2 = \delta_{n_1}^n n \left(\overline{B}_1 + \frac{1}{2} \overline{D}_1 A_1^2 + \frac{1}{4} \overline{D}_1 A^2 + \frac{P_0^2}{n^4} \overline{D}_1 \right), \qquad \left. \begin{array}{c} \\ \\ \\ \\ \\ \\ \\ \end{array} \right\} \quad (6.5,\,2)$$

$$g_1 = \frac{N}{2} G_{1,\,2n_1}, \qquad g_2 = \frac{1}{2} G_{1,\,n+n_1},$$

$$h_2 = \frac{1}{2} G_{1,\,n-n_1} + \frac{1}{2} G_{1,\,n_1-n}.$$

The second approximation y_{32}, which results from the insertion of y_{i1}, y_{i2}, $y_{i2,\,\mathrm{add}}$ and y_{31} (with α_3 instead of α) into the right-hand side of (6.3, 8), leads, after a lengthy analysis, to the additional terms

$$q_{\mathrm{add}} = -\frac{N^2}{2} L_n A_1^2, \qquad \beta_{\mathrm{add}} = \frac{nN^2}{2} \overline{D} A_1^2 \qquad (6.5,\,3)$$

and

$$- \bar{a} \binom{r_1}{s_1} - \bar{\beta} \binom{s_1}{-r_1} - \bar{g} \binom{r_1}{-s_1} + l_1 \binom{s_1}{r_1} - l_2 \binom{rs_1}{ss_1} - l_3 \binom{rs_1 + sr_1}{rr_1 - ss_1}$$

$$+ l_4 \binom{rs_1 - rs_1}{rr_1 + ss_1} - l_5 \binom{2r_1 s_1}{r_1^2 - s_1^2} - l_6 \binom{2r_1 s_1}{s_1^2 - r_1^2} - \delta_{n_1}^{3n} \frac{9n\overline{D}}{32} \binom{(r^2 - s^2)s_1 - 2rsr_1}{(s^2 - r^2)r_1 - 2rss_1}$$

$$+ \delta_{n_1}^n \frac{n\overline{D}}{4} \binom{(r^2 - s^2)s_1 - 2rsr_1}{(r^2 - s^2)r_1 + 2rss_1} + \delta_{n_1}^n \frac{n\overline{D}}{4} \binom{s(r_1^2 - s_1^2) - 2rr_1 s_1}{r(r_1^2 - s_1^2) + 2sr_1 s_1}$$

$$+ \delta_{3n_1}^n \frac{n}{32} \left(\frac{\overline{D}}{192} - 3D_1 \right) \binom{s_1(3r_1^2 - s_1^2)}{r_1(3s_1^2 - r_1^2)} - \delta_{3n_1}^n \frac{9E_1}{32} \binom{r_1(r_1^2 - 3s_1^2)}{s_1(3r_1^2 - s_1^2)} \qquad (6.5,\,4)$$

on the left-hand side of (6.4, 1) where

$$
\begin{aligned}
\bar{a} &= \frac{N}{2}(\overline{G}_{n-n_1} + G_{n_1-n}) + \delta_{n_1}^n(1 - 2\alpha_3 - 2\alpha_1)\,n^2 + \delta_{n_1}^n(G_{20} - G_{10}) \\
&\quad + \delta_{n_1}^n\frac{3}{4}E_1A_1^2\,, \\[2ex]
\bar{\beta} &= \delta_{n_1}^n n\left[B + B_1 + \sum_{i=1}^{2}\frac{(-1)^i\,n^2\overline{B}_i}{\vartheta\lambda_i - n^2} + \frac{D_1}{4}A_1^2\right. \\
&\quad\left. + \frac{\overline{D}}{4}\left(2A^2 + A_1^2 + \frac{4P_0^2}{n^4}\right)\right] + \frac{NP_0}{n^2}(L_{n_1-n} - L_{n-n_1})\,, \\[2ex]
\bar{g} &= \frac{N}{2}\overline{G}_{n+n_1}\,, \qquad l_1 = L_{n+n_1}\,, \qquad l_2 = NL_{n_1}\,, \\[2ex]
l_3 &= \frac{N}{2}L_{2n+n_1}\,, \qquad l_4 = \frac{N}{2}(L_{n_1-2n} - L_{2n-n_1}) + \delta_{n_1}^{2n}\frac{4P_0\overline{D}}{3n}\,, \\[2ex]
l_5 &= \frac{N^2}{4}L_{n+2n_1}\,, \qquad l_6 = \frac{N^2}{4}(L_{2n_1-n} - L_{n-2n_1}) + \delta_{2n_1}^n\frac{P_0\overline{D}}{18n}
\end{aligned}
\right\} \tag{6.5, 5}
$$

with

$$
\overline{D} = D + \sum_{i=1}^{2}\frac{(-1)^i\,n^2\overline{D}_i}{\vartheta\lambda_i - n^2}\,, \qquad \overline{G}_\nu = G_\nu + \sum_{i=1}^{2}\frac{(-1)^i\,n^2 G_{i\nu}}{\vartheta\lambda_i - n^2}\,.
$$

The equations (6.5, 1) coven the case of a simple lateral resonance $\lambda_1 = n_1^2$, $\lambda_2 \neq n_2^2$, $\lambda_3 \neq n^2$ of the gear drive. We have to set $r = s = A = 0$ in (6.5, 1) and find the amplitude formula

$$
(a_1^2 + \beta_1^2 - g_1^2)^2\,A_1^2 = [(a_1 + g_1)^2 + \beta_1^2]\,p_1^2\,,
$$

which yields for $p_1 = 0$ especially $A_1 = 0$ or

$$
a_1 = \pm\sqrt{g_1^2 - \beta_1^2}\,,
$$

that is,

$$
2n_1^2\alpha_1 = \tfrac{3}{4}E_1A_1^2 + \nu G_{10} \pm \sqrt{g_1^2 - \beta_1^2}\,.
$$

Now we exclude the influence of an internal resonance in the formulae found by assuming

$$
\nu n_1 \neq n\,, \qquad n_1 \neq \nu n \qquad (\nu = 1, 2, 3)\,. \tag{6.5, 6}
$$

In order to find closed amplitude formulae, we further neglect by assuming

$$
p = q = p_1 = g = \bar{a} = g_1 = h_2 = 0
$$

certain components of forced and (linear)parametric excitation and by assuming

$$
h = \bar{\beta} = l_1 = l_2 = l_3 = l_4 = l_5 = l_6 = 0
$$

the influence of non-linear parametric excitation. Then the periodicity equations simplify to

$$
\left.
\begin{array}{l}
a\begin{pmatrix} r \\ s \end{pmatrix} - \beta \begin{pmatrix} s \\ -r \end{pmatrix} - \bar{g}\begin{pmatrix} r_1 \\ -s_1 \end{pmatrix} = 0 \,, \\[3mm]
a_1\begin{pmatrix} r_1 \\ s_1 \end{pmatrix} - \beta_1 \begin{pmatrix} s_1 \\ -r_1 \end{pmatrix} - g_2\begin{pmatrix} r \\ -s \end{pmatrix} = 0 \,,
\end{array}
\right\}
\tag{6.5, 7}
$$

the coefficients being given by (6.4, 2), (6.5, 3), (6.5, 2), (6.5, 5).

As in Section 6.1, (6.5, 7) leads because of $(n + n_1)\alpha = n\alpha_3 + n_1\alpha_1$ to the formula

$$
2nn_1(n + n_1)\alpha = \frac{3n}{4} E_1 A_1^2 + n^2 \left(\frac{\nu G_{10}}{n} - \frac{n_1 G_{10}}{\vartheta\lambda_1 - n^2} + \frac{n_1 G_{20}}{\lambda_2 - n^2} \right)
$$

$$
\pm (n_1\beta + n\beta_1) \sqrt{ \frac{\nu}{4\beta\beta_1} \overline{G}_{n+n_1} G_{1,\,n+n_1} - 1 }
$$

for the frequency variation as dependent on the two partial amplitudes and to the relation

$$
\beta G_{1,\,n+n_1} A^2 = \nu\beta\overline{G}_{n+n_1} A_1^2
$$

between the partial amplitudes.

If $\bar{g} = g_2 = 0$ holds instead of $\bar{a} = h_2 = 0$, a similar analysis yields because of $(n - n_1)\alpha = n\alpha_3 - n_1\alpha_1$ the formulae

$$
2nn_1(n - n_1)\alpha = - \frac{3n}{4} E_1 A_1^2 - n^2 \left(\frac{\nu G_{10}}{n} + \frac{n_1 G_{10}}{\vartheta\lambda_1 - n^2} - \frac{n_1 G_{20}}{\lambda_2 - n^2} \right)
$$

$$
\pm (n_1\beta + n\beta_1) \sqrt{ - \frac{\nu}{4\beta\beta_1} \overline{G}_{|n-n_1|} G_{1,\,|n-n_1|} - 1 }
$$

and

$$
\beta G_{1,\,|n-n_1|} A^2 = -\nu\beta_1 G_{|n-n_1|} A_1^2 \,.
$$

In all these formulae, β_{add} from (6.5, 3) has to be taken into consideration.

If for example we set $n = 8$, $n_1 = 3$, $\beta = 4\beta_1 = 0.04 + 0.4A^2 + 0.2A_1^2$, $G_{15} = 0.1$, $\overline{G}_5 = -\dfrac{1}{10N} = \dfrac{55}{90}$ and $G_{10} = G_{20} = 0$ we get the resonance curves of Fig. 6.1, 1 for $E_1 = 1/3$ and of Fig. 6.1, 2 for $E_1 = 1$, where A_1 has to be written instead of A_2. Non-vanishing values of G_{10} and G_{20} only shift the resonance curve in direction α. If in contrast $\beta = 0.04$, the resonance curves are give by Fig. 6.1, 6 for $E_1 = 1/3$ and by Fig. 6.1, 7 for $E_1 = 1$.

The amplitudes found are of the same order of magnitude as in case of single torsional resonance where for instance formula (6.1, 11) yields for

$$
g = \tfrac{1}{2}\overline{G}_{2n} = 0.2 \,, \qquad h = 0 \,, \qquad \beta = 0.04 + 1.2A^2
$$

the maximum amplitude $A_{\text{max}} = 0.365$.

6.6.   Internal resonances in gear drives

We now investigate the internal resonances excluded by (6.5, 6) in the last section, beginning by assuming

$$n_1 = n .$$

In contrast to single torsional resonance ($r_1 = s_1 = 0$), the parameters r_1, s_1 need not be small in comparison with r, s, and a first approximation of the periodicity equations (6.4, 1), (6.5, 3), (6.5, 4) is

$$\begin{pmatrix} p \\ q \end{pmatrix} - n^2 \begin{pmatrix} r_1 \\ s_1 \end{pmatrix} = 0$$

that is

$$r_1 = \frac{p}{n^2}, \qquad s_1 = \frac{q}{n^2}, \qquad A_1 = \frac{\sqrt{p^2 + q^2}}{n^2} . \tag{6.6, 1}$$

Assuming a linear tooth damping, $D = 0$ (that is, also $\overline{D}_i = \overline{D} = 0$), the periodicity equations (6.5, 1) read

$$\beta_2 \begin{pmatrix} s \\ -r \end{pmatrix} + g_2 \begin{pmatrix} r \\ -s \end{pmatrix} + h_2 \begin{pmatrix} r \\ s \end{pmatrix} = \begin{pmatrix} P \\ Q \end{pmatrix} \tag{6.6, 2}$$

where

$$P = \frac{a_1 p}{n^2} - \frac{\beta_1 q}{n^2} - \frac{g_1 p}{n^2} + p_1 , \qquad Q = \frac{a_1 q}{n^2} + \frac{\beta_1 p}{n^2} + \frac{g_1 q}{n^2} . \tag{6.6, 3}$$

The solution

$$\left.\begin{aligned} (\beta_2^2 - g_2^2 + h_2^2)\, r &= (h_2 - g_2)\, P - \beta_2 Q \quad (\equiv R) , \\ (\beta_2^2 - g_2^2 + h_2^2)\, s &= \beta_2 P + (g_2 + h_2)\, Q \quad (\equiv S) \end{aligned}\right\} \tag{6.6, 4}$$

leads to the amplitude formula

$$(\beta_2^2 - g_2^2 + h_2^2)^2\, A^2 = [\beta_2^2 + (g_2 - h_2)^2]\, P^2 + [\beta_2^2 + (g_2 + h_2)^2]\, Q^2 + 4\beta_2 g_2 PQ$$

$$(\equiv R^2 + S^2) , \tag{6.6, 5}$$

where the coefficients are independent of A, and the right-hand side is quadratic in a_1, that is, in the frequency parameter α_1 (and in A_1^2 which is given directly by (6.6, 1)). Because the constant tooth stiffness component γ_0 in (6.3, 7) is greater than the varying components, the inequality $g_2 < h_2$ holds, so that the factor of A^2 in (6.6, 5) does not vanish, and A has a positive minimum. In the approximation at hand, the resonance amplitudes A and A_1 are independent of a, that is, of the detuning which is defined as the small distance between the two resonances,

$$\alpha_1 - \alpha \approx \frac{\omega_3 - \omega_1}{\omega_1} .$$

In second approximation, we have to use in (6.6, 2), instead of (6.6, 3),

$$P = a_1 r_1 - \beta_1 s_1 - g_1 r_1 + p_1 ,$$

$$Q = a_1 s_1 + \beta_1 r_1 + g_1 s_1 ,$$

so that (6.6, 4) is now

$$\left.\begin{array}{l}(\beta_2^2 - g_2^2 + h_2^2)\, r = o_1 r_1 + o_2 s_1 + (h_2 - g_2)\, p_1\,, \\[2mm] (\beta_2^2 - g_2^2 + h_2^2)\, s = o_3 r_1 + o_4 s_1 + \beta_2 p_1\end{array}\right\} \qquad (6.6,\,6)$$

where for abbreviation

$$o_1 = -\beta_1 \beta_2 - (a_1 - g_1)\,(g_2 - h_2)\,,$$
$$o_2 = -\beta_2 (a_1 + g_1) + \beta_1 (g_2 - h_2)\,,$$
$$o_3 = \beta_2 (a_1 - g_1) + \beta_1 (g_2 + h_2)\,,$$
$$o_4 = -\beta_1 \beta_2 + (a_1 + g_1)\,(g_2 + h_2)\,.$$

Inserting (6.6, 6) into the periodicity equations (6.4, 1), (6.5, 4) and neglecting by assuming $l_\nu = 0$ ($\nu = 1, 2, \ldots, 6$) the influence of non-linear parametric excitation, we get two equations

$$\left.\begin{array}{l}o_5 r_1 + o_6 s_1 = o_9\,, \\[2mm] o_7 r_1 + o_8 s_1 = o_{10}\end{array}\right\} \qquad (6.6,\,7)$$

with

$$o_5 = (a - g)\, o_1 - (\beta + h)\, o_3 - (\bar a + \bar g)\,(\beta_2^2 - g_2^2 + h_2^2)\,,$$
$$o_6 = (a - g)\, o_2 - (\beta + h)\, o_4 - \bar\beta (\beta_2^2 - g_2^2 + h_2^2)\,,$$
$$o_7 = (\beta - h)\, o_1 + (a + g)\, o_3 + \bar\beta (\beta_2^2 - g_2^2 + h_2^2)\,,$$
$$o_8 = (\beta - h)\, o_2 + (a + g)\, o_4 - (\bar a - \bar g)\,(\beta_2^2 - g_2^2 + h_2^2)\,,$$
$$o_9 = -(\beta_2^2 - g_2^2 + h_2^2)\, p + (a - g)\,(g_2 - h_2)\, p_1 + (\beta + h)\, \beta_2 p_1\,,$$
$$o_{10} = -(\beta_2^2 - g_2^2 + h_2^3)\, q + (\beta - h)\,(g_2 - h_2)\, p_1 - (a + g)\, \beta_2 p_1$$

for r_1, s_1 only, with coefficients depending on A_1, A_2, α and α_1. The solution of (6.6, 7) is

$$\left.\begin{array}{l}(o_5 o_8 - o_6 o_7)\, r_1 = o_8 o_9 - o_6 o_{10}\,, \\[2mm] (o_5 o_8 - o_6 o_7)\, s_1 = o_5 o_{10} - o_7 o_9\,.\end{array}\right\} \qquad (6.6,\,8)$$

Squaring and adding gives the amplitude formula

$$(o_5 o_8 - o_6 o_7)^2\, A_1^2 = (o_7^2 + o_8^2)\, o_9^2 + (o_5^2 + o_6^2)\, o_{10}^2 - 2(o_5 o_7 + o_6 o_8)\, o_9 o_{10}\,. \qquad (6.6,\,9)$$

A corresponding formula for the torsional resonance amplitude A results from (6.6, 6) by introducing (6.6, 8), squaring and adding:

$$(\beta_2^2 - g_2^2 + h_2^2)^2\,(o_5 o_8 - o_6 o_7)^2\, A^2$$
$$= [\beta_2^2 + (g_2 - h_2)^2]\,[(a_1 - g_1)\,(o_8 o_9 - o_6 o_{10}) - \beta_1 (o_5 o_{10} - o_7 o_9) + (o_5 o_8 - o_6 o_7)\, p_1]^2$$
$$+ [\beta_2^2 + (g_2 + h_2)^2]\,[\beta_1 (o_8 o_9 - o_6 o_{10}) + (a_1 + g_1)\,(o_5 o_{10} - o_7 o_9)]^2$$
$$+ 4\beta_2 g_2 [(a_1 - g_1)\,(o_8 o_9 - o_6 o_{10}) - \beta_1 (o_5 o_{10} - o_7 o_9) + (o_5 o_8 - o_6 o_7)\, p_1]$$
$$\times [\beta_1 (o_8 o_9 - o_6 o_{10}) + (a_1 + g_1)\,(o_5 o_{10} - o_7 o_9)]\,. \qquad (6.6,\,10)$$

In contrast to (6.6, 5) where only the right-hand side depends on α_1 and A_1 is directly given by (6.6, 1), the coefficients o_1 to o_4 now depend linearly on β_1 or a_1, that is, on $D_1 A_1^2$, $E_1 A_1^2$ or α_1, whereas the coefficients o_9, o_{10} depend linearly on β or a, that is, on DA^2, $D_i A^2$ or α, and the coefficients o_5 to o_8 contain the product of both these groups of terms. Thus (6.6, 9), (6.6, 10) are coupled non-linear equations in A^2, A_1^2, α and α_1. They can be solved numerically for A^2 and A_1^2 as dependent on frequency variation and detuning. If $D = D_i = E_1 = 0$ holds (linear damping and restoring forces), they simplify because then all the coefficients are independent of A, A_1 so that the amplitudes can be found immediately for every given value of frequency variation and detuning.

Fig. 6.6, 1 gives an example of the torsional amplitudes (6.6, 5) (full line) and the lateral amplitudes (6.6, 1) (dotted line) for $n = n_1 = 3, \beta_1 = 0.1, \beta_2 = 0.2, P_3 = 0.001$, $Q_3 = 0$, $G_{i0} = 0$ and $2\alpha_1 \overline{G}_3 \ll G_{13}$ where we write

$$\overline{\alpha} = \alpha_1 - \frac{P_0 G_{13}}{18 P_3}$$

(an amplitude maximum would appear if a non-linear damping were considered), and for comparison by dashed line the corresponding amplitude (6.6, 16) for single torsional resonance and $P_0 = 0.2$, $\beta = 0.1$, $\overline{G}_3 = 0.04$.

As a second internal resonance excluded in Section 6.5 by (6.5, 6), we consider now the internal resonance

$$n = 3n_1 .$$

Assuming also in this case $D = 0$ (from where follows $\overline{D}_i = \overline{D} = 0$), a linear tooth damping, the periodicity equations (6.4, 1) because $\beta_2 = 0$, are

$$\left. \begin{aligned} (h_2 + g_2)\, r &= (a_1 - g_1)\, r_1 - \beta_1 s_1 + p_1 , \\ (h_2 - g_2)\, s &= \beta_1 r_1 + (a_1 + g_1)\, s_1 . \end{aligned} \right\} \qquad (6.6, 11)$$

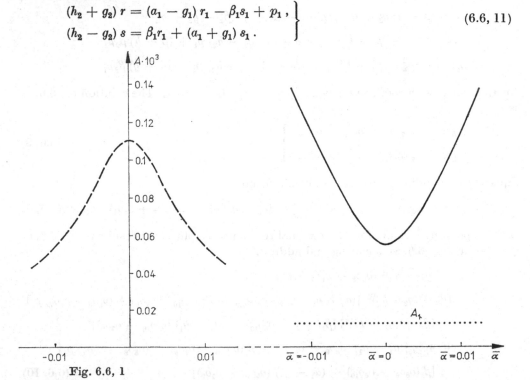

Fig. 6.6, 1

Inserting into (6.4, 1), (6.5, 4) and assuming $l_\nu = 0$ ($\nu = 1, 2, \dots , 6$), gives two equations for r_1, s_1, the coefficients of which depend on A, A_1, α, α_1:

$$t r_1 + u s_1 = x , \qquad v r_1 + w s_1 = y \qquad\qquad (6.6, 12)$$

where

$$t = (a - g)(a_1 - g_1)(h_2 - g_2) - \beta\beta_1(h_2 + g_2) - (\bar{a} + \bar{g})(h_2^2 - g_2^2) ,$$

$$u = -(a - g)\beta_1(h_2 - g_2) - (a_1 + g_1)\beta(h_2 + g_2) ,$$

$$v = (a + g)\beta_1(h_2 + g_2) + (a_1 - g_1)\beta(h_2 - g_2) ,$$

$$w = (a + g)(a_1 + g_1)(h_2 + g_2) - \beta\beta_1(h_2 - g_2) - (\bar{a} - \bar{g})(h_2^2 - g_2^2) ,$$

$$x = -(h_2^2 - g_2^2) p - (a - g)(h_2 - g_2) p_1 + \frac{3n}{32} D_1(h_2^2 - g_2^2) s_1(3r_1^2 - s_1^2)$$
$$+ \frac{9}{32} E_1(h_2^2 - g_2^2) r_1(r_1^2 - 3s_1^2) ,$$

$$y = -(h_2^2 - g_2^2) q - \beta(h_2 - g_2) p_1 + \frac{3n}{32} D_1(h_2^2 - g_2^2) r_1(3s_1^2 - r_1^2)$$
$$+ \frac{9}{32} E_1(h_2^2 - g_2^2) s_1(3r_1^2 - s_1^2) .$$

The equations (6.6, 12) are of third order and have to be solved numerically for given coefficients (that is, also for given α, α_1 and A^2). The corresponding values of r and s have to be determined numerically from (6.6, 11). If $D_i = 0$ (linear damping), the equations (6.6, 12) do not depend on A^2. If further $E_1 = 0$ (linear stiffness of the bearings), the system (6.6, 12) is linear in r_1, s_1 with the solution

$$(tw - uv) r_1 = wx - uy ,$$

$$(tw - uv) s_1 = ty - vx$$

which leads to the amplitude formula

$$(tw - uv)^2 A_1^2 = (v^2 + w^2) x^2 + (t^2 + u^2) y^2 - 2(tv + uw) xy . \qquad (6.6, 13)$$

The small-letter terms are now independent of r_i, s_i, A, A_1.

A second formula for A can be found for $\bar{D}_1 = 0$ by solving (6.5, 1) for r_1, s_1:

$$(\beta_1^2 + a_1^2 - g_1^2) r_1 = (a_1 + g_1)(h_2 + g_2) r + \beta_1(h_2 - g_2) s - (a_1 + g_1) p_1 ,$$

$$(\beta_1^2 + a_1^2 - g_1^2) s_1 = -\beta_1(h_2 + g_2) r + (a_1 - g_1)(h_2 - g_2) s + \beta_1 p_1 .$$

Insertion into (6.4, 1), (6.5, 4) yields, if again $D_i = E_1 = 0$,

$$T r + U s = X , \qquad V r + W s = y \qquad\qquad (6\;6, 14)$$

with

$$T = (a - g)(\beta_1^2 + a_1^2 - g_1^2) - (\bar{a} + \bar{g})(a_1 + g_1)(h_2 + g_2) ,$$

$$U = -\beta(\beta_1^2 + a_1^2 - g_1^2) - (\bar{a} + \bar{g})\beta_1(h_2 - g_2) ,$$

$$V = \beta(\beta_1^2 + a_1^2 - g_1^2) + (\bar{a} - \bar{g})\beta_1(h_2 + g_2) ,$$

$$W = (a + g)(\beta_1^2 + a_1^2 - g_1^2) - (\bar{a} - \bar{g})(a_1 - g_1)(h_2 - g_2) ,$$

$$X = -(\beta_1^2 + a_1^2 - g_1^2) p - (\bar{a} + \bar{g})(a_1 + g_1) p_1 ,$$

$$Y = -(\beta_1^2 + a_1^2 - g_1^2) q + (\bar{a} - \bar{g}) \beta_1 p_1 .$$

The solution

$$(TW - UV)\, r = WX - UY \,,$$

$$(TW - UV)\, s = TY - VX$$

of (6.6, 14) gives the amplitude formula

$$(TW - UV)^2\, A^2 = (V^2 + W^2)\, X^2 + (T^2 + U^2)\, Y^2 - 2(TV + UW)\, XY \,.$$

$$(6.6, 15)$$

The torsional resonance amplitude found shall now be compared with the amplitude (6.1, 10) for single torsional resonance wdich tends to infinity, in the frame of the linearized equations, for the parametric excitation reaching the threshold value, $g^2 + h^2 = \beta^2$ (if we choose $a = 0$ and for instance $g = 0$, $q \neq 0$), which is approximately

$$A^2 = \frac{p^2 + q^2}{a^2 + \beta^2} \tag{6.6, 16}$$

for small parametric excitation, $g, h \ll \beta$ in modulus.

Formula (6.6, 15) gives for small parametric excitation,

$$\beta_1 \gg g, g_1, g_2, h_2, p_1 \quad \text{in modulus}, \tag{6.6, 17}$$

approximately

$$A^2 = \frac{(a_1^2 + \beta_1^2)^2\, p^2 + [(a_1^2 + \beta_1^2)\, q + (\bar{g} - \bar{a})\, \beta_1 p_1]^2}{(a^2 + \beta^2)\, (a_1^2 + \beta_1^2)^2} \tag{6.6, 18}$$

which is for every a_1 approximately equal to (6.6, 16) if in modulus

$$(\bar{g} - \bar{a})\, p_1 \ll \beta_1 \sqrt{p^2 + q^2} \,,$$

that is

$$G_{1n_1}(\overline{G}_{4n_1} - \overline{G}_{2n_1}) \ll \frac{2n^2}{N}\, \frac{\beta_1 \sqrt{p^2 + q^2}}{P_0} \,,$$

otherwise (for smaller forced excitation) the torsional amplitude A can become greater than for single resonance. Fig. 6.6, 2 shows an example of the amplitudes (6.6, 18) for $n = 3$, $n_1 = 1$, $\alpha_1 = \alpha$ (that is vanishing detuning), $P_0 = 0.2$, $P_3 = 0.001$, $Q_3 = 0$, $\beta = \beta_1 = 0.1$, $G_{11} = \overline{G}_2 = \overline{G}_3 = -\overline{G}_4 = 0.04$, $G_{i0} = 0$ in comparison with the corresponding single resonance (6.6, 16) for $n = 3$, $P_0 = 0.2$, $P_3 = 0.001$, $Q_3 = 0$, $\beta = 0.1$, $\overline{G}_3 = 0.04$, $G_{i0} = 0$ (dashed curve).

For the corresponding lateral resonance amplitude A_1, equation (6.6, 13) and (6.6, 17) lead to the approximate formula

$$[(a^2 + \beta^2)\, (a_1^2 + \beta_1^2)\, g_2^2 - (a^2 - \beta^2)\, (a_1^2 - \beta_1^2)\, h_2^2]^2\, A_1^2$$

$$= (a_1^2 + \beta_1^2)\, [(a^2 + \beta^2)\, (g_2^2 + h_2^2)\, (x^2 + y^2) + 2(a^2 - \beta^2)\, g_2 h_2 (x^2 - y^2)$$

$$+ 8 a\beta g_2 h_2 xy]$$

where

$$x = (g_2^2 - h_2^2)\, p + a(g_2 - h_2)\, p_1 \,,$$

$$y = (g_2^2 - h_2^2)\, q + \beta(g_2 - h_2)\, p_1 \,.$$

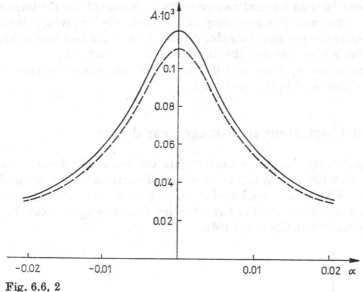

Fig. 6.6, 2

If we choose α such that $a = 0$, we get

$$\beta^2[(a_1^2 + \beta_1^2)\, g_2^2 + (a_1^2 - \beta_1^2)\, h_2^2]^2\, A_1^2$$
$$= (a_1^2 + \beta_1^2)\, (g_2 - h_2)^2\, \{(g_2^2 - h_2^2)^2\, p^2 + (g_2 + h_2)^2\, [(g_2 + h_2)\, q + \beta p_1]^2\}$$

$$(6.6, 19)$$

from which it follows, for instance, for $a_1 = 0$ and $\beta p_1 \leqq (g_2 + h_2)\, q$ because of (6.6, 17),

$$A_1 \ll A\,,$$

corresponding with $A_1 = 0$ for single torsional resonance; but for $\beta p_1 \gg (g_2 + h_2)\, q$ or if the expression in brackets on the left-hand side of (6.6, 19) vanishes,

$$a_1^2(g_2^2 + h_2^2) = \beta_1^2(h_2^2 - g_2^2)\ (\neq 0)\,,$$

the lateral resonance amplitude can become greater.

If we dispense with the condition (6.6, 17) for parametric excitation being small compared with damping, we can find threshold conditions for finite torsional amplitudes A. For instance if $a = a_1 = \beta_1 = q = 0$, formula (6.6, 15) reads

$$[\beta^2 g_1^2 + \bar{g}^2 h_2^2 - (g g_1 - \bar{g} g_2)^2]^2\, A^2 = [\beta^2 g_1^2 + (g g_1 + \bar{g} h_2 - \bar{g} g_2)^2]\, (g_1 p - \bar{g} p_1)^2\,.$$

If the bracket on the left-hand side vanishes (corresponding to the threshold equation $\beta^2 = g^2 + h^2$ in case of single resonance), the right-hand side is

$$2[(g g_1 - \bar{g} g_2)^2 + \bar{g} h_2(g g_1 - \bar{g} g_2)]\, (g_1 p - \bar{g} p_1)^2\,.$$

Because

$$\bar{g} h_2(g g_1 - \bar{g} g_2) = \frac{N^2}{16}\, \bar{G}_{4n_1} G_{1,\, 2n_1}\, (\bar{G}_{6n_1} G_{1,\, 2n_1} - \bar{G}_{4n_1} G_{1,\, 4n_1})$$

can be positive, the right-hand side need not vanish simultaneously.

It has been shown that an internal resonance $3n_1 = n$ can enlarge the torsional and lateral vibrations. The excitation and damping parameters being given, it is therefore necessary to evaluate the resonance amplitudes A, A_1 as dependent on the frequency variation and detuning by means of the formulae (6.6, 15), (6.6, 13).

The internal resonance $n_1 = 3n$, which is of minor importance for practical gear vibrations, can be discussed in the same way.

6.7.   Torsional vibrations in N-stage gear drives

In this and the following sections, we investigate the torsional vibrations in an N-stage gear drive, sketched in Fig. 6.7, 1. An essential contribution to this problem is due to MOLERUS (1963) who investigated N-stage gear drives with rigid shafts, in particular, the stability question. For the following results compare SCHMIDT (1984), SCHMIDT and SCHULZ (1983), SCHULZ (1986).

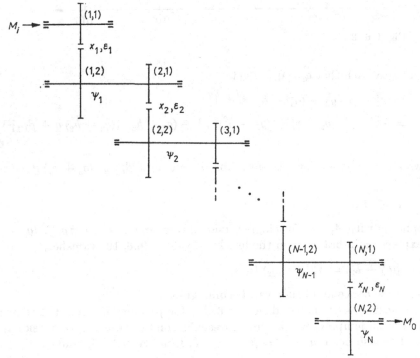

Fig. 6.7, 1

We denote by $\varphi_{nk}^{\text{abs}}$ $(n = 1, 2, \ldots, N; k = 1, 2)$ the torsional angle of the k-th wheel of the n-th stage, which can be divided into a stationary part with the constant angular velocity Ω_{nk} of the ideal rotation of perfect wheels, where $\Omega_{n1} = \Omega_{n-1,2}$, and small supplementary terms φ_{nk},

$$\varphi_{nk}^{\text{abs}} = \Omega_{nk}t + \varphi_{nk} . \tag{6.7, 1}$$

The latter terms express the deflection of the real rotation from ideal rotation and are connected with the elastic displacements x_n of the teeth of the n-th stage and the tooth error functions ε_n — characterizing the joint deviation of the teeth of the n-th stage from the ideal form — by the relation

$$\varrho_{n1}\varphi_{n1} - \varrho_{n2}\varphi_{n2} = x_n - \varepsilon_n \qquad (6.7, 2)$$

where ϱ_{nk} is the radius of the base circle of the wheel (n, k). The relative deflection of the wheels of the n-th stage from ideal rotation is denoted by

$$\psi_n = \varphi_{n2} - \varphi_{n+1,1} . \qquad (6.7, 3)$$

The equations (6.7, 2) and (6.7, 3) yield

$$\varphi_{n1} = \frac{\varphi_{11}}{i_{1,n-1}} + \sum_{j=1}^{n-1} \frac{1}{i_{j+1,n-1}} \left(\frac{\varepsilon_j - x_j}{i_{j+1,n}\varrho_{j2}} - \psi_j \right) \qquad (6.7, 4)$$

and

$$\varphi_{n2} = \frac{\varphi_{11}}{i_{1,n}} + \sum_{j=1}^{n} \frac{\varepsilon_j - x_j}{i_{j+1,n}\varrho_{j2}} - \sum_{j=1}^{n-1} \frac{\psi_j}{i_{j+1,n}} \qquad (6.7, 5)$$

where

$$i_j = \frac{\Omega_{j1}}{\Omega_{j2}} = \frac{\varrho_{j2}}{\varrho_{j1}} = \frac{\zeta_{j2}}{\zeta_{j1}}$$

are the transmission ratios, ζ_{jk} being the number of teeth of the wheel (j, k), and

$$i_{j,k} = \prod_{\nu=j}^{k} i_\nu$$

with

$$i_{n+1,n} = 1 .$$

The inertia moment of the wheel (n, k) is denoted by J_{nk}, the tooth stiffness and shaft stiffness coefficients by C_n and $\overline{C}_n$ respectively, the tooth damping and shaft damping coefficients by D_n and $\overline{D}_n$ respectively.

The kinetic energy is

$$T = \sum_{n=1}^{N} \frac{1}{2} (J_{n1}\dot{\varphi}_{n1}^2 + J_{n2}\dot{\varphi}_{n2}^2) ,$$

the potential energy function

$$V = \frac{1}{2} \sum_{n=1}^{N} C_n x_n^2 + \frac{1}{2} \sum_{n=1}^{N-1} \overline{C}_n \psi_n^2 ,$$

the dissipation function

$$W = \frac{1}{2} \sum_{n=1}^{N} D_n \dot{x}_n^2 + \frac{1}{2} \sum_{n=1}^{N-1} \overline{D}_n \dot{\psi}_n^2$$

and the virtual work of the external input and output moment

$$M_i \delta\varphi_{11} = - M_0 \delta\varphi_{N2} .$$

We assume, as usual, the tooth error functions ε_n and the tooth stiffness coefficients C_n — which, strictly speaking, both depend periodically on the torsional angles (6.7, 1) — as periodic functions of time (we write

$$C_n = c_n + \gamma_n(t)$$

with the mean tooth stiffness c_n), the shaft stiffness and shaft damping coefficients $\overline{C}_n$ respectively $\overline{D}_n$ as constants, whereas the tooth damping coefficients D_n may depend quadratically on the elastic displacements x_n.

The Lagrange formalism for the generalized coordinates φ_{11}, x_n $(n = 1, 2, \dots, N)$ and ψ_n $(n = 1, 2, \dots, N - 1)$ leads, together with the equations (6.7, 4), (6.7, 5), to the equation

$$\sum_{n=1}^{N} \frac{J_{n1}}{i_{1,n-1}} \left[\frac{\ddot{\varphi}_{11}}{i_{1,n-1}} + \sum_{j=1}^{n-1} \frac{1}{i_{j+1,n-1}} \left(\frac{\ddot{\varepsilon}_j - \ddot{x}_j}{\varrho_{j2}} - \ddot{\psi}_j \right) \right]$$

$$+ \sum_{n=1}^{N} \frac{J_{n2}}{i_{1n}} \left(\frac{\ddot{\varphi}_{11}}{i_{1n}} + \sum_{j=1}^{n} \frac{\ddot{\varepsilon}_j - \ddot{x}_j}{i_{j+1,n}\varrho_{j2}} - \sum_{j=1}^{n-1} \frac{\ddot{\psi}_j}{i_{j+1,n}} \right) = M_i - \frac{M_0}{i_{1n}} \qquad (6.7, 6)$$

which corresponds to the law of moment of momentum for the whole system, and the equations

$$- \sum_{n=m+1}^{N} \frac{J_{n1}}{i_{m+1,n-1}} \left[\frac{\ddot{\varphi}_{11}}{i_{1,n-1}} + \sum_{j=1}^{n-1} \frac{1}{i_{j+1,n-1}} \left(\frac{\ddot{\varepsilon}_j - \ddot{x}_j}{\varrho_{j2}} - \ddot{\psi}_j \right) \right]$$

$$- \sum_{n=m}^{N} \frac{J_{n2}}{i_{m+1,n}} \left(\frac{\ddot{\psi}_{11}}{i_{1n}} + \sum_{j=1}^{n} \frac{\ddot{\varepsilon}_j - \ddot{x}_j}{i_{j+1,n}\varrho_{j2}} - \sum_{j=1}^{n-1} \frac{\ddot{\psi}_j}{i_{j+1,n}} \right) + C_m x_m + D_m \dot{x}_m = \frac{M_0}{i_{m+1,N}\varrho_{m2}}$$

$$(m = 1, 2, \dots, N) \qquad (6.7, 7)$$

and

$$- \sum_{n=m+1}^{N} \frac{J_{n1}}{i_{m+1,n-1}} \left[\frac{\ddot{\varphi}_{11}}{i_{1,n-1}} + \sum_{j=1}^{n-1} \frac{1}{i_{j+1,n-1}} \left(\frac{\ddot{\varepsilon}_j - \ddot{x}_j}{\varrho_{j2}} - \ddot{\psi}_j \right) \right]$$

$$- \sum_{n=m+1}^{N} \frac{J_{n2}}{i_{m+1,n}} \left(\frac{\ddot{\varphi}_{11}}{i_{1n}} + \sum_{j=1}^{n} \frac{\ddot{\varepsilon}_j - \ddot{x}_j}{i_{j+1,n}\varrho_{j2}} - \sum_{j=1}^{n-1} \frac{\ddot{\psi}_j}{i_{j+1,n}} \right) + \overline{C}_m \psi_m + \overline{D}_m \dot{\psi}_m = - \frac{M_0}{i_{m+1,N}}$$

$$(m = 1, 2, \dots, N - 1) \qquad (6.7, 8)$$

describing the elastic vibrations of the gear drive system. Eliminate $\ddot{\varphi}_{11}$ by means of (6.7, 6) and use the abbreviations

$$J_n = J_{n1} + J_{n-1,2} \quad (n = 2, 3, \dots, N),$$

$$J_1 = J_{11}, \qquad J_{N+1} = J_{N2},$$

$$\overline{J}_n = J_{n1} + \frac{1}{i_n^2} J_{n2} \quad (n = 1, 2, \dots, N),$$

$$I_0 = \sum_{n=0}^{N} \frac{1}{i_{1n}^2} J_{n+1},$$

$$I_j = \sum_{n=j}^{N} \frac{J_{n+1}}{i_{1n} i_{j+1,n}} \quad (j = 1, 2, \dots, N),$$

$$\overline{I}_j = \sum_{n=j+1}^{N} \frac{\overline{J}_n}{i_{1,n-1} i_{j+1,n-1}} \quad (j = 1, 2, \dots, N),$$

$$I_{mj} = \frac{1}{\varrho_{j2}\varrho_{m2}} \left(i_{1,\min(m,j)} I_{\max(m,j)} - \frac{1}{I_0} I_m I_j \right) \quad (m, j = 1, 2, \dots, N),$$

$$\overline{I}_{mj} = \frac{1}{\varrho_{m2}} \left(i_{1,\min(m,j)} \overline{I}_{\max(m,j)} - \frac{1}{I_0} I_m \overline{I}_j + \sum_{l=1}^{m-1} \delta_l^j J_{m2} \right)$$

$$(m = 1, 2, \dots, n; \ j = 1, 2, \dots, N - 1)$$

$$\left. \right\} \qquad (6.7, 9)$$

and

$$\bar{\bar{I}}_{mj} = i_{1,\min(m,j)}\bar{I}_{\max(m,j)} - \frac{1}{I_0}\bar{I}_m\bar{I}_j\,.$$

Then the differential equations (6.7, 7), (6.7, 8) can be written in the form

$$\left.\begin{aligned}
&\sum_{j=1}^{N} I_{mj}\ddot{x}_j + \sum_{j=1}^{N-1}\bar{I}_{mj}\ddot{\psi}_j + C_m x_m + D_m\dot{x}_m = P_m\,,\\
&P_m = \sum_{j=1}^{N-1} I_{mj}\ddot{\varepsilon}_j + \frac{1}{I_0} I_m\left(M_i - \frac{M_0}{i_{1N}}\right) + \frac{M_0}{i_{m+1,\,N\varrho m2}}\\
&\qquad\qquad (m = 1, 2, \dots, N)
\end{aligned}\right\} \quad (6.7, 10)$$

and

$$\left.\begin{aligned}
&\sum_{j=1}^{N}\bar{I}_{rj}\ddot{x}_j + \sum_{j=1}^{N-1}\bar{\bar{I}}_{rj}\ddot{\psi} + \bar{C}_r\psi_r + \bar{D}_r\dot{\psi}_r = Q_r\,,\\
&Q_r = \sum_{j=1}^{N}\bar{I}_{rj}\ddot{\varepsilon}_j + \frac{1}{I_0}\bar{I}_r\left(M_i - \frac{M_0}{i_{1N}}\right) - \frac{M_0}{i_{r+1,\,N}}\,.
\end{aligned}\right.\begin{aligned}(r = 1, 2, \dots, N-1)\end{aligned}\right\} (6.7, 11)$$

If the *shafts* can be assumed to be *rigid*, $\psi_j \equiv 0$ holds and the equations of motion reduce to (6.7, 10):

$$\sum_{j=1}^{N} I_{mj}\ddot{x}_j + C_m x_m + D_m\dot{x}_m = P_m\,.$$

In this case we can write

$$I_{mj} = \begin{cases} i_{1j}k_m K_j & \text{if } m \geq j\,,\\ i_{1m}k_j K_m & \text{if } m < j \end{cases} \tag{6.7, 12}$$

where

$$k_\nu = \frac{i_{1\nu}}{I_0}\sum_{\mu=\nu}^{N}\frac{J_{\mu+1}}{i_{1\mu}^2}\,, \qquad K_\nu = \sum_{\mu=0}^{\nu-1}\frac{J_{\mu+1}}{i_{1\mu}^2}\,.$$

As (6.7, 12) shows directly, the symmetry condition $I_{jm} = I_{mj}$ holds for the inertia matrix. If all transmission ratios are equal, $i_\nu = i$, we get

$$I_0 = J_1 + \frac{J_2}{i^2} + \dots + \frac{J_{N+1}}{i^{2N}}\,.$$

In the general case of rigid shafts, (6.7, 9), (6.7, 12) yield for a one-stage drive ($N = 1$)

$$I_0 = J_1 + \frac{J_2}{i_1^2}\,, \qquad k_1 = \frac{J_2}{i_1 J_0}\,, \qquad K_1 = J_1\,, \qquad J_{11} = i_1 J_1 k_1\,,$$

for a two-stage drive

$$I_0 = J_1 + \frac{J_2}{i_1^2} + \frac{J_3}{i_1^2 i_2^2}\,, \qquad k_1 = \frac{1}{i_1 I_0}\left(J_2 + \frac{J_3}{i_2^2}\right)\,, \qquad k_2 = \frac{J_3}{i_1 i_2 I_0}\,,$$

$$J_{11} = i_1 J_1 k_1\,, \qquad J_{12} = J_{21} = i_1 J_1 k_2\,, \qquad J_{22} = i_1 i_2\left(J_1 + \frac{J_2}{i_1^2}\right)k_2$$

and for a three-stage drive

$$I_0 = J_1 + \frac{J_2}{i_1^2} + \frac{J_3}{i_1^2 i_2^2} + \frac{J_4}{i_1^2 i_2^2 i_3^2}, \qquad k_1 = \frac{1}{i_1 I_0}\left(J_2 + \frac{J_3}{i_2^2} + \frac{J_4}{i_2^2 i_3^2}\right),$$

$$k_2 = \frac{1}{i_1 i_2 I_0}\left(J_3 + \frac{J_4}{i_3^2}\right), \qquad k_3 = \frac{J_4}{i_1 i_2 i_3 I_0},$$

$$J_{11} = i_1 J_1 k_1, \qquad J_{12} = J_{21} = i_1 J_1 k_2, \qquad J_{13} = J_{31} = i_1 J_1 k_3,$$

$$J_{22} = i_1 i_2 \left(J_1 + \frac{J_2}{i_1^2}\right) k_2, \qquad J_{23} = J_{32} = i_1 i_2 \left(J_1 + \frac{J_2}{i_1^2}\right) k_3,$$

$$J_{33} = i_1 i_2 i_3 \left(J_1 + \frac{J_2}{i_1^2} + \frac{J_3}{i_1^2 i_2^2}\right) k_3.$$

Corresponding to the method of Sections 6.3 to 6.6, we first assume a *weak coupling* of the different gear stages in the sense that we can solve iteratively the equations (6.7, 10), (6.7, 11) by taking into consideration in a first approximation, besides the constant components c_m, $\overline{C}_m$, only the diagonal elements of the inertia matrix of the left-hand sides. Therefore we can write these equations in the form

$$y_n'' + \lambda_n y_n = \sum_{\nu=0}^{\infty} (p_{n\nu} \cos \nu\tau + q_{n\nu} \sin \nu\tau) - \alpha_n(2 + \alpha_n)\, y_n''$$

$$- (1 + \alpha_n)^2 \sum_{\substack{\nu=1 \\ (\nu \neq n)}}^{2N-1} j_{n\nu} y_\nu'' - (1 + \alpha_n)(b_n + d_n y_n^2)\, y_n'$$

$$- \sum_{\nu=1}^{\infty} (g_{n\nu} \cos \nu\tau + h_{n\nu} \sin \nu\tau)\, y_n \qquad (n = 1, 2, \dots, 2N - 1).$$

$$(6.7, 13)$$

Here we have introduced a dimensionless time

$$\tau = \omega t$$

where ω is the greatest common divisor [3]) of all (circular) frequencies appearing in (6.7, 10), (6.7, 11). Differentiations by τ are denoted by dashes. For

$$n, \nu = 1, 2, \dots, 2N - 1,$$

$$m, \mu = 1, 2, \dots, N \quad \text{and} \quad r, \varrho = 1, 2, \dots, N - 1$$

use the vectors

$$(y_n(\tau)) = \begin{pmatrix} x_m(t) \\ \psi_r(t) \end{pmatrix}, \qquad (\lambda_n) = \begin{pmatrix} \dfrac{c_m}{I_{mm}\omega_m^2} \\ \dfrac{\overline{C}_r}{\overline{\overline{I}}_{rr}\omega_r^2} \end{pmatrix},$$

$$\left(\sum_{\nu=0}^{\infty} (p_{n\nu} \cos \nu\tau + q_{n\nu} \sin \nu\tau)\right) = \begin{pmatrix} \dfrac{P_m}{I_{mm}\omega_m^2} \\ \dfrac{Q_r}{\overline{\overline{I}}_{rr}\omega_r^2} \end{pmatrix},$$

---

[1]) Common divisors exist because the meshing frequencies appearing in C_m are proportional to the numbers of teeth and the frequencies appearing in the error terms P_m, Q_m are equal to, multiples or fractions of, these meshing frequencies.

where p_{n0} is the predominant static prestress,

$$(b_n + d_n y_n^2) = \begin{pmatrix} \dfrac{c_m}{I_{mm}\omega_m^2} \\[2ex] \dfrac{\bar{\bar{C}}_r}{\bar{\bar{I}}_{rr}\omega_r^2} \end{pmatrix},$$

$$\left(\sum_{\nu=1}^{\infty} (g_{n\nu} \cos \nu\tau + h_{n\nu} \sin \nu\tau) \right) = \begin{pmatrix} \dfrac{\gamma_m(t)}{I_{mm}\omega_m^2} \\[2ex] 0 \end{pmatrix}$$

and the matrix

$$(j_{n\nu}) = \begin{pmatrix} \dfrac{I_{m\mu}}{I_{mm}} & \dfrac{\bar{I}_{m\mu}}{I_{mm}} \\[2ex] \dfrac{\bar{I}_{r\varrho}}{\bar{\bar{I}}_{rr}} & \dfrac{\bar{\bar{I}}_{r\varrho}}{\bar{\bar{I}}_{rr}} \end{pmatrix}.$$

The frequency parameters

$$\alpha_n = \frac{\omega - \omega_n}{\omega_n}$$

are, as will be shown, related to the frequency variation and express the small distance of ω from fixed frequencies ω_n.

In the *non-resonance* case, the first approximation is

$$y_{n10} = \frac{p_{n0}}{\lambda_n} \quad (n = 1, 2, \ldots, 2N - 1),$$

from which follows the second approximation

$$y_{n20} = \sum_{\nu=0}^{\infty} \frac{p_{n\nu} \cos \nu\tau + q_{n\nu} \sin \nu\tau}{\vartheta\lambda_n - \nu^2} - \frac{p_{n0}}{\lambda_n} \sum_{\nu=1}^{\infty} (g_{n\nu} \cos \nu\tau + h_{n\nu} \sin \nu\tau).$$

This shows that, besides the predominant static prestress, the error and the variable stiffness terms (the latter multiplied by the static prestress) influence the vibration already in second approximation whereas damping as well as inertia coupling of the different gear stages by means of $j_{n\nu}$ $(\nu \neq n)$ influence the solution only in third approximation. The formulae for the third approximation can be derived easily.

In what follows, we investigate a *single resonance*

$$\lambda_k = K^2 \quad (K \text{ integer})$$

of the k-th stage, calling λ_k an eigenvalue (leading to an eigenfrequency ω_k) and $\alpha_k = \alpha$ the frequency variation and choose $\alpha_n = 0$ for $n \neq k$.

The first approximation is

$$y_{n1} = \delta_n^k(r \cos K\tau + s \sin K\tau) + y_{n10}.$$

Inserting this and the second approximation into the periodicity equations yields

$$\begin{pmatrix} p \\ q \end{pmatrix} + 2K^2\alpha \begin{pmatrix} r \\ s \end{pmatrix} - \beta \begin{pmatrix} s \\ -r \end{pmatrix} - g \begin{pmatrix} r \\ -s \end{pmatrix} - h \begin{pmatrix} s \\ r \end{pmatrix} = 0 \qquad (6.7, 14)$$

with the abbreviations

$$p = p_{kK} + \sum_{\substack{\nu=1 \\ (\nu \neq k)}}^{2N-1} l_{k\nu} p_{\nu K} - \frac{p_{k0} g_{kK}}{K^2},$$

$$q = q_{kK} + \sum_{\substack{\nu=1 \\ (\nu \neq k)}}^{2N-1} l_{k\nu} q_{\nu K} - \frac{p_{k0} h_{kK}}{K^2}, \qquad\qquad (6.7, 15)$$

$$\beta = K(b_k + \tfrac{3}{4} d_k A^2),$$

$$g = \tfrac{1}{2} g_{k,2K}, \qquad h = \tfrac{1}{2} h_{k,2K}$$

where

$$l_{k\nu} = \frac{j_{k\nu}}{\dfrac{\lambda_\nu}{K^2} - 1} \qquad\qquad (6.7, 16)$$

gives the coefficients of the secondary components in p, q and

$$A = \sqrt{r^2 + s^2}$$

is the amplitude of the resonance part of the first approximative solution, designated simply as amplitude. The real amplitude of the elastic displacement x_k consists approximately of A, the constant prestress y_{k10} and the amplitude $\mathfrak{A}$ of y_{k20}. When $A + \mathfrak{A}$ is greater than the prestress y_{k10}, the tooth flanks separate.

The parametric excitation with the coefficients $g_{k,2K}$, $h_{k,2K}$ of the periodic tooth stiffness is called stiffness excitation. The forced excitation p, q stemming mainly from the tooth errors (but containing also components of periodic tooth stiffness multiplied by static prestress) is called error excitation.

As in Section 6.1, the solution of (6.7, 14) leads to the amplitude equation

$$(4K^2\alpha^2 + \beta^2 - g^2 - h^2)^2 A^2 = (4K^2\alpha^2 + \beta^2 + g^2 + h^2)(p^2 + q^2)$$

$$+ 2(2K^2\alpha g - \beta h)(p^2 - q^2) + 4(2K^2\alpha h + \beta g) pq. \qquad (6.7, 17)$$

This equation can be treated in the same way as equation (6.1, 10). It is linear in A^2 if the non-linear damping coefficient d_k vanishes.

The backbone curve is now the coordinate axis $\alpha = 0$. Differentiating by α and setting $dA/d\alpha = 0$ yields

$$16K^6 A^2\alpha^3 + 4K^2(\beta^2 - g^2 - h^2) A^2\alpha - 2K^2(p^2 + q^2)\alpha$$

$$= g(p^2 - q^2) + 2hpq. \qquad\qquad (6.7, 18)$$

The amplitude extreme values result from (6.7, 17) and (6.7, 18) by eliminating α. They — as well as the resonance curves given by (6.7, 17) — depend only on the damping coefficients b_k, d_k and stiffness coefficients $g_{k,2K}$, $h_{k,2K}$, $g_{k,K}$, $h_{k,K}$ *of the k-th stage*, whereas they depend on the error excitation coefficients $p_{\nu K}$, $q_{\nu K}$ *of all stages*. A diminution of the amplitudes is possible by appropriately influencing these coefficients.

Transferring the analysis in Sections 5.1 to 5.3 shows how the vibration amplitudes depend on linear and non-linear damping as well as on stiffness and error excitation and the phase relation between them.

The above coefficients depend significantly on the number k of the stage where resonance occurs, on the order K of resonance and on the inertia moment I_{kk}. The other frequencies λ_ν, inertia moments $I_{k\nu}$ and by these the transmission ratios i_ν influence only the secondary components in p, q containing $p_{\nu K}$, $q_{\nu K}$ ($\nu \neq K$). This influence can be discussed using the formulae derived above.

In order to make such a discussion easier to follow, we assume in the rest of this section that the shafts are rigid, all transmission ratios are equal, $i_\nu = i$, and the inertial moments are proportional to i^4,

$$J_2 = (i^4 + 1) J_1, \dots, J_N = (i^4 + 1) J_1, \qquad J_{N+1} = i^4 J_1. \qquad (6.7, 19)$$

A two-stage gear drive yields

$$j_{12} = \frac{i^3}{i^4 + i^2 + 1}, \qquad j_{21} = \frac{i}{i^4 + i^2 + 1},$$

a three-stage drive leads to

$$j_{12} = \frac{i(i^4 + i^2 + 1)}{i^6 + i^4 + 2i^2 + 1}, \qquad j_{13} = \frac{i^4}{i_6 + i^4 + 2i^2 + 1},$$

$$j_{23} = \frac{i^3}{i^4 + i^2 + 1}, \qquad j_{21} = \frac{i}{i^4 + i^2 + 1},$$

$$j_{32} = \frac{i(i^4 + i^2 + 1)}{i^6 + 2i^4 + i^2 + 1}, \qquad j_{31} = \frac{i^2}{i^6 + 2i^4 + i^2 + 1}.$$

If for gearing down $i \gg 1$ holds, we get for two-stage drives and three-stage drives respectively j_{12}, j_{23}, $j_{32} = O(1/i)$, $j_{13} = O(1/i^2)$, $j_{21} = O(1/i^3)$, and $j_{31} = O(1/i^4)$. If for gearing up $i \ll 1$ holds, we find analogously j_{12}, j_{21}, $j_{32} = O(i)$, $j_{31} = O(i^2)$, $j_{23} = O(i^3)$, and $j_{13} = O(i^4)$. Already that suggests that the influence of the secondary components in p, q is small in comparison with that of p_{kK}, q_{kK}.

In order precisely to evaluate the influence of the secondary components, we have also to evaluate λ_ν/K^2. If we bear in mind that the quantities ω_n are approximately equal to ω, we get, approximately,

$$\frac{\lambda_n}{\lambda_k} = \frac{\lambda_n}{K^2} = \frac{c_n J_{kk}}{c_k J_{nn}},$$

that is, in the two-stage case for resonance of the first ($k = 1$) as well as of the second stage ($k = 2$)

$$\frac{\lambda_1}{\lambda_2} = \frac{i^2 c_1}{c_2}, \qquad (6.7, 20)$$

in the three-stage case

for $k = 1$ (resonance of the first stage)

$$\frac{\lambda_2}{\lambda_1} = \frac{i^6 + i^4 + 2i^2 + 1}{(i^4 + i^2 + 1)^2} \frac{c_2}{c_1}, \qquad \frac{\lambda_3}{\lambda_1} = \frac{i^6 + i^4 + 2i^2 + 1}{i^2(i^6 + 2i^4 + i^2 + 1)} \frac{c_3}{c_1},$$

for $k = 2$ (resonance of the second stage)

$$\frac{\lambda_1}{\lambda_2} = \frac{(i^4 + i^2 + 1)^2}{i^6 + i^4 + 2i^2 + 1} \frac{c_1}{c_2}, \qquad \frac{\lambda_3}{\lambda_2} = \frac{(i^4 + i^2 + 1)^2}{i^2(i^6 + 2i^4 + i^2 + 1)} \frac{c_3}{c_2},$$

for $k = 3$ (resonance of the third stage)

$$\frac{\lambda_1}{\lambda_3} = \frac{i^2(i^6 + 2i^4 + i^2 + 1)}{i^6 + i^4 + 2i^2 + 1} \frac{c_1}{c_3}, \qquad \frac{\lambda_2}{\lambda_3} = \frac{i^2(i^6 + 2i^4 + i^2 + 1)}{(i^4 + i^2 + 1)^2} \frac{c_2}{c_3}.$$

By means of these formulae, the eigenvalues λ_ν and coefficients $l_{k\nu}$ of the secondary components can be evaluated immediately. Here are some examples.

For a two-stage drive and resonance in the second stage, Fig. 6.7, 2 gives the secondary components for $c_2 = c_1$ (full line) and $c_2 = 1.5c_1$ (dashed line) as dependent on the

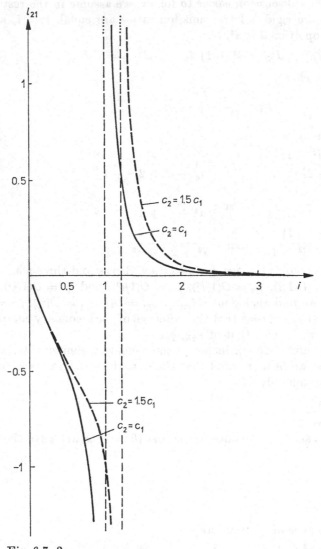

Fig. 6.7, 2

transmission ratio. It shows that they are very small for $i > 2$ (for instance $l_{21} = 0.004\,121$ for $i = 3$, $l_{21} = 0.000\,976\,8$ for $i = 4$ and $c_2 = c_1$). If resonance occurs in the first stage, i and c_2/c_1 have to be replaced by the inverse values.

For a three-stage drive, $i = 57/20$, $c_2 = 1.5c_1$ and resonance in the third stage, $l_{32} = -0.9034$ and $l_{31} = 0.002\,439$, whereas $l_{12} = -0.3883$ and $l_{13} = -0.1153$ if the resonance occurs in the first stage.

6.8.   Strong coupling between gear stages

In this section we consider a strong coupling between several or all, say $M \geq 2$, of the $2N - 1$ equations of motion (6.7, 13) in the sense that already in first approximation the coupling terms are taken into account, in other words, that the starting point is a transformation to principal axes. Without loss of generality we assume that the first M equations are strongly coupled. The equations (6.7, 13) are written in the form

$$y_n'' + \sum_{\substack{\mu=1 \\ (\mu \neq n)}} j_{n\mu} y_\mu'' + \lambda_n y_n = \Phi_n \qquad (n = 1, 2, \dots, M) , \qquad (6.8, 1)$$

$$\Phi_n = \sum_{\nu=0}^{\infty} (p_{n\nu} \cos \nu\tau + q_{n\nu} \sin \nu\tau) - \alpha_n(2 + \alpha_n) y_n''$$

$$- \alpha_n(2 + \alpha_n) \sum_{\substack{\mu=1 \\ (\mu \neq n)}}^{M} j_{n\mu} y_\mu'' - (1 + \alpha_n)^2 \sum_{\substack{\nu=M+1 \\ (\nu \neq n)}}^{2N-1} j_{n\nu} y_\nu''$$

$$- (1 + \alpha_n) (b_n + d_n y_n^2) y_n' - \sum_{\nu=1}^{\infty} (g_{n\nu} \cos \nu\tau + h_{n\nu} \sin \nu\tau) y_n ,$$

$$y_n'' + \lambda_n y_n = \sum_{\nu=0}^{\infty} (p_{n\nu} \cos \nu\tau + q_{n\nu} \sin \nu\tau) - \alpha_n(1 + \alpha_n) y_n''$$

$$- (1 + \alpha_n)^2 \sum_{\substack{\nu=1 \\ (\nu \neq n)}}^{2N-1} j_{n\nu} y_\nu'' - (1 + \alpha_n) (b_n + d_n y_n^2) y_n'$$

$$- \sum_{\nu=1}^{\infty} (g_{n\nu} \cos \nu\tau + h_{n\nu} \sin \nu\tau) y_n \qquad (n = M + 1, \dots, 2N - 1) .$$

The equations (6.8, 1) have to be decoupled by principal axes transformation, which amounts to multiplication by such factors $t_{\nu n}$ $(\nu, n = 1, 2, \dots, M)$, so that the equations given by summation are of the decoupled form

$$V_\nu(z_\nu'' + \Lambda_\nu z_\nu) = \sum_{n=1}^{M} t_{\nu n} \Phi_n ,$$

that is,

$$z_\nu'' + \Lambda_\nu z_\nu = \varphi_\nu , \qquad \varphi_\nu = \sum_{n=1}^{M} v_{\nu n} \Phi_n , \qquad v_{\nu n} = \frac{t_{\nu n}}{V_\nu} \qquad (\nu = 1, 2, \dots, M)$$

$$\qquad (6.8, 2)$$

with new coordinate functions

$$z_\nu = \sum_{n=1}^{M} T_{\nu n} y_n , \qquad (6.8, 3)$$

the original coordinates can be expressed as:

$$y_n = \sum_{\mu=1}^{M} U_{n\mu} z_\mu \qquad (n = 1, 2, \dots, M) . \qquad (6.8, 4)$$

The expressions φ_ν can — if (6.8, 2), (6.8, 4) are used and the frequency parameters α_ν ($\nu = 1, 2, \ldots, M$) are defined by means of Λ_ν instead of λ_ν — be shown to be

$$\varphi_\nu = \sum_{n=1}^{M} v_{\nu n} \sum_{m=0}^{\infty} (p_{nm} \cos m\tau + q_{nm} \sin m\tau)$$

$$-\alpha_\nu(2 + \alpha_\nu)\, z_\nu'' - (1 + \alpha_\nu)^2 \sum_{n=1}^{M} v_{\nu n} \sum_{\substack{m=M+1 \\ (m \neq n)}}^{2N-1} j_{nm} y_m'$$

$$- (1 + \alpha_\nu) \sum_{n=1}^{M} v_{\nu n} \left[b_n + d_n \left(\sum_{\varkappa=1}^{M} U_{n\varkappa} z_\varkappa\right)^2\right] \sum_{\mu=1}^{M} U_{n\mu} z_\mu'$$

$$- \sum_{n=1}^{M} v_{\nu n} \sum_{m=1}^{\infty} (g_{nm} \cos m\tau + h_{nm} \sin m\tau) \sum_{\mu=1}^{M} U_{n\mu} z_\mu\, .$$

The first approximative solution of (6.8, 2) is, in the non-resonance case,

$$z_{\nu 10} = \sum_{n=1}^{M} \frac{v_{\nu n} p_{n0}}{\Lambda_\nu}\, ,$$

while the second approximation is

$$z_{\nu 20} = \sum_{n=1}^{M} v_{\nu n} \sum_{m=0}^{\infty} \frac{p_{nm} \cos m\tau + q_{nm} \sin m\tau}{\vartheta \Lambda_\nu - m^2}$$

$$- \sum_{n=1}^{M} v_{\nu n} \sum_{m=1}^{\infty} \frac{g_{nm} \cos m\tau + h_{nm} \sin m\tau}{\vartheta \Lambda_\nu - m^2} \sum_{\mu=1}^{M} U_{n\mu} \sum_{\varkappa=1}^{M} \frac{v_{\mu\varkappa} p_{\varkappa 0}}{\Lambda_\mu}\, .$$

For a *single resonance*

$$\Lambda_k = K^2 \qquad (1 \leq k \leq M;\, K \text{ integer})$$

we get, putting $\alpha_k = \alpha$ as the frequency variation and choosing $\alpha_\nu = 0$ for $\nu \neq k$, the first approximation

$$z_{\nu 1} = \delta_\nu^k (r \cos K\tau + s \sin K\tau) + z_{\nu 10}\, .$$

The periodicity equations are with $z_{\nu 1}$ and the second approximation $z_{\nu 2}$ again of the form (6.7, 14) where now

$$\left. \begin{aligned}
p &= \sum_{n=1}^{M} v_{kn} \left(p_{nK} + \sum_{m=M+1}^{N} l_{nm} p_{mK} - \sum_{m, \mu=1}^{M} U_{nm} v_{m\mu} \frac{p_{\mu 0}}{\Lambda_m} g_{nK}\right), \\
q &= \sum_{n=1}^{M} v_{kn} \left(q_{nK} + \sum_{m=M+1}^{N} l_{nm} q_{mK} - \sum_{m, \mu=1}^{M} U_{nm} v_{m\mu} \frac{p_{\mu 0}}{\Lambda_m} h_{nK}\right), \\
\beta &= \sum_{n=1}^{M} K\left(w_n b_n + \tfrac{3}{4} W_n d_n A^2\right), \\
g &= \tfrac{1}{2} \sum_{n=1}^{M} w_n g_{n, 2K}\, , \qquad h = \tfrac{1}{2} \sum_{n=1}^{M} w_n h_{n, 2K}
\end{aligned} \right\} \qquad (6.8, 5)$$

with

$$w_n = v_{kn} U_{nk}\, , \qquad W_n = v_{kn} U_{nk}^3\, .$$

Using these terms, the amplitude equation (6.7, 17) and the ensuing discussion remain valid, only now the damping and stiffness coefficients of all M stages strongly coupled influence the resonance curves given by (6.7, 17) as well as the amplitude extreme values. The real influence of the different system parameters occuring in (6.8, 5) is revealed after determining the quantities Λ_k, $v_{\nu n}$, $U_{n\mu}$ by principal axes transformation.

We now take into special consideration a strong coupling of $M = 2$ gear stages. The analysis is simplified by choosing two of the values $t_{\nu n}$ (say, t_{12} and t_{22}) equal to 1 instead of normalizing z_ν. Thus we derive the formulae

$$2\frac{\lambda_1}{\lambda_2} j_{12}t_{\nu 1} = 1 - \frac{\lambda_1}{\lambda_2} \mp \sqrt{\left(1 - \frac{\lambda_1}{\lambda_2}\right)^2 + 4\frac{\lambda_1}{\lambda_2} j_{12}j_{21}},$$

$$T_{12} = T_{22} = 1, \qquad T_{\nu 1} = \frac{\lambda_1}{\lambda_2}t_{\nu 1}, \qquad V_\nu = 1 + j_{12}t_{\nu 1},$$

$$\Lambda_\nu = \frac{\lambda_2}{V_\nu}, \qquad v_{\nu 1} = \frac{t_{\nu 1}}{V_\nu}, \qquad v_{\nu 2} = \frac{1}{V_\nu},$$

$$U_{11} = -U_{12} = \frac{1}{T_{11} - T_{21}}, \qquad U_{21} = \frac{T_{21}}{T_{21} - T_{11}}, \qquad U_{22} = \frac{T_{11}}{T_{11} - T_{21}}$$

for $\nu = 1, 2$.

For a *two-stage gear drive* ($N = 2$), equal transmission ratios i, inertia moments (6.7, 19), $c_2 = 1.5c_1$, and resonance in the second equation (mainly in the second stage), using (6.7, 16), we get the following values:

i	l_{21}	v_{21}	v_{22}	w_2	W_2	w_1	W_1
1.5	0.3609	0.2443	0.9008	0.7216	0.4630	0.1195	0.02856
2	0.05714	0.05414	0.9794	0.9484	0.8895	0.01169	0.000532
3	0.00659	0.00656	0.9981	0.9957	0.9910	0.00038	0.0000014

These values have to be inserted into formulae (6.7, 15) for weak coupling respectively (6.8, 5) for strong coupling, which if for simplicity the last terms in p, q containing $p_{\mu 0}$ are neglected, can now be written:

weak coupling	strong coupling
$p = l_{21}p_{1K} + p_{2K}$,	$= v_{21}p_{1K} + v_{22}p_{2K}$,
$q = l_{21}q_{1K} + q_{2K}$,	$= v_{21}q_{1K} + v_{22}q_{2K}$,
$\dfrac{\beta}{K} = b_2 + \frac{3}{4}d_2A^2$,	$= w_2b_2 + \frac{3}{4}W_2d_2A^2 + w_1b_1 + \frac{3}{4}W_1d_1A^2$,
$2g = g_{2,2K}$,	$= w_2g_{2,2K} + w_1g_{1,2K}$,
$2h = h_{2,2K}$,	$= w_2h_{2,2K} + w_1h_{1,2K}$.

This shows that for transmission ratio $i = 3$, the influence of any (weak or strong) coupling on the left-hand side parameters determining maximum amplitude and response curve is smaller than 1 per cent. For $i = 2$ the influence of coupling at all (of the first stage) on error excitation p, q expressed by v_{21} is about 5.4 per cent (the latter being fairly well taken into consideration by l_{21}, that is by weak coupling), the influence of strong coupling on the error excitation in the second stage (compare v_{22}) is about 2 per cent. The influence of (strong) coupling on the stiffness excitation g, h as well as on the linear damping is (compare w_2) for $i = 2$ more than 5 per cent but the influence on the maximum amplitude is much smaller because only the quotient of stiffness excitation and linear damping determines the maximum amplitude, as follows from (5.1, 17), (5.3, 3), (5.3, 4). For $i = 1.5$ the corrections of weak and strong coupling are greater.

The corresponding values for the quotient of strong and weak coupling eigenvalues and of T_{21} and T_{11} — characterizing, because of (6.8, 3) and $T_{\nu 2} = 1$, the share of y_1 in the resonance coordinate z_2 and in z_1 respectively — are

i	Λ_1/λ_1	Λ_2/λ_2	T_{21}	T_{11}
1.5	1.1979	0.9008	0.4069	-1.6384
2	1.0595	0.9794	0.1474	-4.5227
3	1.0118	0.9981	0.0395	-16.8912

We see that the resonance in the second equation leads to pronounced vibrations not only in the second but also in the first stage. The eigenvalues, consequently also the eigenfrequencies, differ for $i = 3$ by about 1 per cent respectively 0.2 per cent from those found by the equations of weak coupling, for $i = 2$ by about 6 per cent respectively 2 per cent.

We have seen that the weak coupling model, which avoids the main axes transformation and therefore can lead to general and relatively simple formulae for more than two degrees of freedom, represents a far better approximation than the model with one degree of freedom.

6.9.   Application of computer algebra

Under certain simplifying assumptions, Sections 6.7 and 6.8 give results on the resonance vibrations of N-stage gear drives. If we dispense with these simplifications or investigate more complicated gear drives, the amount of analytical evaluation further increases. On the one hand, the basic differential equations of motion become much more complicated; on the other hand, often a complete transformation to principal axes cannot be avoided. Thus solving the problem analytically becomes very toilsome, and the possibility of mistakes is great.

The question arises if such complicated analytical evaluations can be performed by aid of computers. In fact, during the last two decades activities in computer-aided analysis have led to several different computer languages and program systems. Starting in general from certain problems of theoretical physics, computer languages have been developed which permit analytical evaluations by reducing them to algebraic ones and which are therefore termed computer algebra languages. Examples are the languages MACSYMA, REDUCE and FORMAC.

For certain types of multy-body systems, several programs have been developed which allow to find the differential equations of motion by help of computers. However up to now much less is known on the implementation of analytical approximation methods with computers although this is of highest importance for problems as those discussed in this chapter.

R. SCHULZ (1983, 1985) has developed a program system ASB (,,Automatisierte Schwingungsberechnung") for the automatic evaluation of vibrations by means of the computer algebra language FORMAC. This system combines the generation of differential equations of motion (taking into consideration especially gear drives) with the construction of iterative analytical solutions and the numerical evaluation of these iterative solutions.

The program system ASB consists of four parts which can be used also separately.

1. The form of the gear drive leads to formulae for the kinetic and potential energy function and the dissipation function of a linear model describing the torsional vibrations of the gear drive. It is possible to change the number of degrees of freedom as well as to introduce non-linearities and other modifications. Only this part of the system ASB is especially adapted to gear problems, the other parts are universally applicable.

2. From the terms found in the first part (or from corresponding terms given by the user), Lagrange's equations of motion are derived by means of the computer.

3. In order to solve vibration problems modelled by ordinary differential equations, the iterative determination of periodic solutions by help of the integral equation method widely used in this book has been realized in form of a FORMAC program. Thus the advantages of this analytical method is used for the automatic evaluation of vibrations. The results are firstly the eigenfrequencies and eigenfunctions, secondly (taking into consideration also the time-varying terms) the resonances of different kind, in the above case of gear drives the critical torsional frequencies, thirdly the iterative solutions in form of finite Fourier series and at last the periodicity equations, a system of algebraic equations which determine the bifurcation parameters occuring in the Fourier series. As has been shown in Chapter 5 and this chapter, the periodicity equations give manyfold general analytical information on the system behaviour; for further information by numerically evaluating the iterative solutions and the periodicity equations, PL/1 programs are generated from these Fourier series and periodicity equations.

4. In the last part the analytical results are evaluated numerically. The PL/1 programs generated in the third part are used for numerical solution of the periodicity equations and numerical evaluations of the Fourier series.

The influence of the stiffness of the shafts on the resonance vibrations of a two-stage gear drive has been investigated by means of the system ASB by R. SCHULZ. The parameters taken into consideration were

$$i_1 = i_2 = i \, ,$$

$$J_{12} = i^4 J_{11} \, , \qquad J_{21} = i^{\frac{1}{2}} J_{11} \, , \qquad J_{22} = i^{\frac{9}{2}} J_{11} \, ,$$

$$M_0 = -i^2 M_t \, ,$$

$$c_1 = c_2 = c \, ,$$

$$\gamma_1 = \gamma(\zeta_{11} \Omega_{11} t) \, , \qquad \gamma_2 = \gamma(\zeta_{21} \Omega_{21} t)$$

where

$$\gamma(\tau) = c(\hat{\gamma}_1 \cos \tau + \hat{\gamma}_2 \cos 2\tau) \, ,$$

further

$$D_1 = D_2 = d \, , \qquad \overline{D}_1 = 0 \, ,$$

$$\varrho_{11} = \varrho_{21} = \varrho_1 \, , \qquad \varrho_{12} = \varrho_{22} = \varrho_2 \, ,$$

$$\varepsilon_1 = \varepsilon_2 = 0 \, .$$

In the numerical evaluations the values

$$i = 3 , \qquad J_{11} = 0.1 \ [\text{kg m}^2] , \qquad c = 10^9 \ [\text{N m}^{-1}] ,$$

$$\hat{\gamma}_1 = 0.1 , \qquad \hat{\gamma}_2 = 0.02 , \qquad d = 2 \cdot 10^4 \ [\text{N sec m}^{-1}] ,$$

$$\varrho_1 = 0.1 \ [\text{m}] , \qquad \varrho_2 = 0.3 \ [\text{m}] , \qquad M_0 = 1000 \ [\text{N m}]$$

were chosen. The results for the resonances in the second stage are

$\overline{C}_1$ [Nm]	Ω_{11}	$\hat{x}_1$	$\hat{x}_2$
10^{11}	410.5	1.38	2.90
(approximately rigid shaft)			
10^8	392.1	1.51	3.00
$3 \cdot 10^7$	356.0	1.54	3.23

where Ω_{11} is the angular velocity of the wheel $(1, 1)$, which leads to the main resonance in the second gear stage, and

$$\hat{x}_i = \frac{x_{i,\,\text{max}}}{x_{i,\,\text{stat}}} ,$$

$x_{i,\,\text{max}}$ being the maximum value of the solution x_i and $x_{i,\,\text{stat}}$ being the mean value of the quasistatic solution,

$$x_{i,\,\text{stat}} = \frac{M_0}{\varrho_{11} c} .$$

It shows that the results of the simplified model with rigid shaft hold approximately if the rigidity of the shaft is taken into consideration.

The ASB system has also been used successfully for higher-stage gear drives (SCHULZ (1986)), for planetary gear drives (SCHULZ and FRIEDRICH (1985)) and for other problems (compare Section 12.7).

7. Investigation of stability in the large

7.1. Fundamental considerations

Many non-linear systems of engineering interest possess more than one stationary solution stable against small disturbances, that is, more than one attractor. Such solutions are termed *locally stable*. In practical applications usually only one of these is desirable; the others, for the most part, are unwelcome because they signal danger to safe and reliable operation of the device whose model is being analyzed. The problem is to establish the conditions which lead to a particular steady state, or alternatively, to examine the disturbances which are apt to cause one stationary state to change to another (for example, small-amplitude stationary vibration to large-amplitude motion, a non-oscillatory state to an oscillatory condition, etc.).

The solution to the first aspect of the problem, i.e. establishing the domains of the initial conditions which lead to different stationary solutions, is well known. These domains are called the domains of attraction of a particular solution. For systems which are directly described by a set of two first-order differential equations of the type (4.1, 1) or whose set of such equations is identical with the original second-order differential equation the problem is solved by means of phase portrait analysis outlined in Chapter 4. The differential equations of motion of one-degree-of-freedom systems excited harmonically by an external force or parametrically can be converted, by application of known procedures (van der Pol or Krylov and Bogoljubov methods) to a system of two first-order differential equations of the type (4.1, 1) which, however, are not identical with the original equation of motion. Consequently, an analysis made in the phase plane (see Section 4.4) is only approximate.

This chapter reviews some exact methods of stability-in-the-large investigations effected by means of digital or analogue computers. The best known of these is the stroboscopic method which has been found very useful for analyses of one-degree-of-freedom systems. The analyst obtains a sequence of points of the separatrix, which — on being connected — yield the boundaries of the domains of attraction of the various steady solutions. The TV scanning and the circular scanning method proposed and used by TONDL (1970a, b) (see also the Russian translation (1973b)) are applicable to systems with several degrees of freedom as well; their author used them in combination with the fast repetition procedure of analogue computers.

For systems determined by two initial conditions both the solution and its graphical representation are comparatively simple. Difficulties begin to arise with graphical representation of the boundaries of the domains of attraction when the system being examined is governed by more than three initial conditions; in four- and multi-dimensional spaces intuitive orientation is no longer unambiguous and charting can only be effected by means of sections. This is the reason the problem has been solved, almost exclusively, for systems with two initial conditions (the work of HAYASHI, especially his book (1964), constitutes without doubt the greatest contribution to its analysis). In the case of more complicated systems the initial conditions are subjected to restrictive assumptions (for example, are assumed to be such that only a single normal mode vibration can be initiated — see BENZ (1962), BECKER (1972)). Some systems governed by three or four initial conditions are analyzed in TONDL's monographs mentioned above.

A solution to the second problem, the investigation of the resistance of a particular steady solution to disturbances of a certain type or establishment of the disturbances which do not lead to a qualitative change of a steady solution is less well known. In the discussion which follows special attention is accorded to the class of problems in which the disturbances acting on the system satisfy definite assumptions.

In real systems, the actual disturbances cannot always be expressed in terms of the initial conditions. This is particularly true of disturbances which act on the system during a definite time interval. Consider a disturbance (an external pulse, a change of a system's parameter) expressed in terms of a function which is wholly determined by a small number of parameters (for example, two: the amplitude and the time of application of a sinusoidal pulse). If both the beginning of the disturbance action and the initial steady state of the system are known, we can determine the domain of different values of the parameters defining a certain type of disturbance, for which the transient solutions lead to the given steady state. Accordingly, the meaning of the term "domains of attraction" can also be extended to include, as the coordinates of the boundaries of the domains of attraction, the parameters which define a certain type of disturbance. When a disturbance is described by two parameters, the domain of attraction can be represented by a plane diagram irrespective of the number of initial conditions which govern the system being examined.

In the discussion which follows, the solution to this problem is restricted to cases for which the time interval between two successive disturbances is long enough to permit the analyst to assume that the transient motion of the system can be regarded as stationary before the next disturbance begins to take place. This assumption enables a solution to be obtained even in the case when the disturbance or the time of the beginning of its application is stochastic. The probability that a disturbance will lead to a particular stationary state can be determined using a finite number of deterministic solutions. This approach is similar to that of the Monte Carlo method.

Disturbances can be defined in various ways and modified to represent those occurring in real systems. A disturbance may be represented by a force pulse, for example, a sinusoidal (Fig. 7.1, 1 — light lines) or an oscillatory decaying pulse (Fig. 7.1, 1 — heavy lines), a step load, a constant load acting during a definite time interval, etc. A step change of an internal force or of a system parameter, such as a change of damping or the friction force, of the mass of the system or the stiffness of an elastic element, etc. can be counted among other types of disturbances. A change of a parameter (denoted Q) can be expressed in terms of a step change from Q_I to Q_{II} (Fig. 7.1, 2a) or as a step change which exists for a definite time interval (Fig. 7.1, 2b). The basic types of disturbances are sufficiently defined by two parameters. This fact is advantageous for graphical representation of the results, which no longer depends on the number of initial conditions of the system. The results can also be used in connection with other problems, for example, in considerations relating to the stochastic character of the disturbance or optimization of a particular parameter.

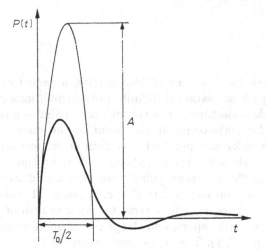

Fig. 7.1, 1

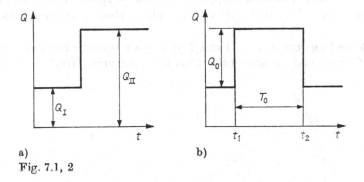

a)   b)

Fig. 7.1, 2

Modern analogue computers, particularly those provided with fast repetition and logic circuits, have been found very useful for analyses of the kind described. Although an analogue computer was employed in the solution of the examples presented in this chapter, the principles of the methods proposed here do not exclude the application of a fast digital computer.

7.2.   Methods of investigating stability in the large for disturbances in the initial conditions

(a) *Stroboscopic method*

Consider a system governed by the differential equation of motion

$$\ddot{y} + f(\dot{y}, y) = P \cos \omega t \tag{7.2, 1 a}$$

(external excitation), or by the equation

$$\ddot{y} + f(\dot{y}, y, \omega t) = 0 \tag{7.2, 1 b}$$

(parametric excitation) where the function $f(\dot{y}, y, \omega t)$ is a periodic function of time having a period of $2\pi/\omega$. Writing $\dot{y} = v$ one can convert (7.2, 1) to an equation of the form

$$\dot{v} = V(v, y, t),$$
$$\dot{y} = Y(v, y, t)$$

(7.2, 2)

where the functions V, Y are periodic functions of time having a period of $2\pi/\omega$. Trajectories obtained in the (v, y) phase plane for definite initial conditions do not, generally, obey the rule that a single trajectory passes through each point — although, as it is assumed, this rule holds for trajectories at any point of the space (v, y, t). Follow now the projection of the reference point of a certain trajectory onto the (v, y) plane during the time intervals spaced $2\pi/\omega$ apart, that is at times $t = 0$, $2\pi/\omega$, $4\pi/\omega$, ..., $n \cdot 2\pi/\omega$, as though the reference point were illuminated stroboscopically during those time intervals and projected onto the (v, y) plane. Alternatively, imagine the trajectories in the (v, y, t) space to be intersected by a system of planes parallel to the (v, y) plane and spaced $2\pi/\omega$ apart, and the points of intersection to be projected onto the (v, y) plane (Fig. 7.2, 1). The trajectories obtained by connecting the corresponding projections are termed trajectories in the stroboscopic plane (in the sense of the well-known Poincaré mapping); for the purpose of differentiation, their coordinates are denoted by $(v)^*$, $(y)^*$ and the phase plane is spoken of as stroboscopic.

Considering that the functions $V(v, y, t)$ and $Y(v, y, t)$ are periodic functions of time having a period of $2\pi/\omega$, and consequently, that the functions $V[(v)^*, (y)^*, 2n\pi/\omega]$

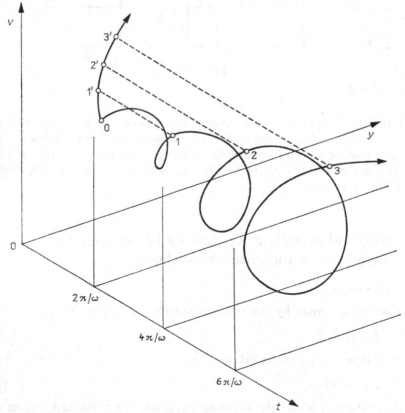

Fig. 7.2, 1

and $Y[(v)^*, (y)^*, 2n\pi/\omega]$ are invariant with respect to n, it can readily be shown that the stroboscopic trajectories are controlled by conditions of uniqueness which are wholly analogous to those applying to system (4.1, 1).

Rather than using approximate equations, the stroboscopic method solves directly (7.2, 1a) (or (7.2, 1b) or (7.2, 2)). Using the track-hold memory circuits records are made, from the solutions of $y(t)$ and $\dot{y}(t)$, of the values of $y(2n\pi/\omega)$, $\dot{y}(2n\pi/\omega)$ ($n = 0, 1, 2, ...$) or of the values of $y(t_0 + 2n\pi/\omega)$, $\dot{y}(t_0 + 2n\pi/\omega)$ when examining the separatrix or another important trajectory for the initial conditions at time $t = t_0$ (instead of at time $t = 0$). In practical applications the signals of $y(2n\pi/\omega)$ and $\dot{y}(2n\pi/\omega)$ are brought to the automatic plotter whose stylus jumps at intervals from point to point thus drawing a broken line; alternatively, the stylus is dipped at intervals by means of a relay and records only individual points. The procedure of drawing the phase portraits is analogous to that described in Chapter 4.

The procedure of obtaining the separatrix is again based on the idea of realizing the negative time. In the $(y, -\dot{y})$ plane the system

$$\ddot{y} + f(-\dot{y}, y) = P \cos \omega t \tag{7.2, 3}$$

(or possibly $f(\dot{y}, y, -\omega t)$ for parametric systems) has trajectories which are identical with those of the system (7.2, 1) in the $(y, \dot{y})$ plane; the reference point, however, moves in opposite direction.

When the points of the set thus drawn are not close enough for the purpose of drawing a trajectory, the procedure is repeated; the starting point is a point whose position can readily be determined from the first set. In selecting the starting point use is usually made of the fact that in the neighbourhood of a singular point (for example, a saddle), the points are spaced close together. Thus, for example, if in the first solution the analyst obtains a sequence of points 1, 2, ... where the first points are spaced close together (see Fig. 7.2, 2) he can easily determine the position of the first point 1' of the second sequence 1', 2', ... By repeating the solution he can obtain as many points for fitting the trajectory as he desires.

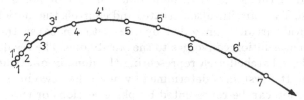

Fig. 7.2, 2

Compare now the stroboscopic method with the approximate solution effected, for example, by means of the van der Pol method; in the application of the latter, the solution is sought in the form

$$y = a \cos \omega t + b \sin \omega t \tag{7.2, 4}$$

where, for a non-stationary solution, the coefficients a, b are slowly varying functions of time and investigations concerning stability in the large are carried out using the phase portrait analysis in the (a, b) phase plane (see Section 4.4). The more closely the trajectories in the (a, b) phase plane approach those in the stroboscopic phase plane $((\dot{y}/\omega)^*, (y)^*)$, the better the approximation (7.2, 4) satisfies the solution of (7.2, 1) and the more perfect is the fulfilment of the assumption that $a(t)$, $b(t)$ are slowly varying functions of time, for then the relations

$$(y)^* = a , \qquad (\dot{y})^* = \omega b \tag{7.2, 5}$$

are satisfied to a higher degree. The relations (7.2, 5) were also the reason why the plane $((\dot{y}/\omega)^*, (y)^*)$ was used in place of the plane $((\dot{y})^*, (y)^*)$ in most examples presented here. Since the solution $y(t)$ is not identical with the approximate solution (7.2, 4) because it generally contains higher harmonic components (also in the case of steady solutions), the positions of the singular points in the (a, b) plane are not identical with those in the $((\dot{y}/\omega)^*, (y)^*)$ plane; consequently, the phase portrait obtained in the $((\dot{y}/\omega)^*, (y)^*)$ for $t = 0$ can no longer be made applicable to the initial conditions at $t = t_0$ by mere turning. For these delayed initial conditions (or for excitation $P \cos(\omega t + \varphi)$) — if an exact procedure is in order — the problem must be solved separately. For the Duffing equation the effect is analyzed in a paper by FIALA and TONDL (1974). The differences between the two procedures (the van der Pol and the stroboscopic method) grow larger with increasing non-linearity of the system being examined; qualitatively, however, the methods are in close agreement even for comparatively strong non-linearities.

(b) *Method of television scanning*

The method is based on repeated solution of the equation of motion (or of a set of such equations if the system is governed by more than two initial conditions) for initial conditions which — except for one which is being varied — are kept constant. In the specified interval of values the variable initial condition is changed in steps. After running through the whole interval of values, the other initial condition is changed by a certain value and the process of step-wise changing of the original variable condition is repeated. In this way the analyst scans a plane whose coordinates are the values of the two initial conditions — hence the term the *television (TV) scanning method*. Each transient solution is examined by means of a suitable criterion (which will be outlined at the end of this section) to determine whether or not it leads to a particular steady solution. TV scanning also actuates and controls the motion of the graph plotter stylus. Logical circuits controlled by the criterion cause the stylus to dip whenever the transient solution converges to the steady one. The result of this process can be, for example, a hatched area representing the domain of attraction of a particular steady solution. If a system is determined by more than two initial conditions, the domain of attraction can be represented by plane sections or three-dimensional surfaces. Fast repetitive procedure has been found expedient when an analogue computer is used in the solution. From the point of view of practical solving and presentation of results, the application of TV scanning is not restricted to systems with one degree of freedom excited externally or parametrically; the method has been found equally suitable for analyses of self-excited systems with several degrees of freedom.

(c) *Method of circular scanning*

This method differs from TV scanning by an auxiliary subroutine which is provided to generate a function whose reference point moves continuously in a circle. The circle, called the *circle of disturbances*, is centred at a definite point, for example, a singular point. At regular time intervals the values of the function (the coordinates of the reference point on the circle) are introduced to the problem being solved in fast repetition as the initial conditions. After a circle has been completed, the radius is changed automatically by ΔR (Fig. 7.2, 3). The analyst can maintain either a constant angular velocity of rotation on the individual circles (and equal numbers of points corresponding to the initial conditions of the various solutions in fast repetition on each circle) or a constant peripheral velocity of motion on the individual circles (and obtain a number of points whose spacing is kept constant and which grows larger

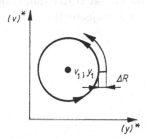

Fig. 7.2, 3

with increasing radius). Using a suitable criterion and the memory circuits he then decides whether or not a solution leads to the given steady solution. The circular scanning also actuates the stylus of the graph plotter whose motion is coordinately controlled by the circuits which determine whether or not a transient solution leads to the given steady solution.

The domains of attraction of a particular steady solution are obtained in a way similar to that used in TV scanning. The method of circular scanning, however, was proposed primarily to facilitate solution of the second problem, i.e. investigation of stability of a particular steady, locally stable solution against disturbances in initial conditions which are not fully determined. This problem will be dealt with in the next section.

Returning now to the question of the criterion for determining whether or not a transient solution leads to one or the other steady solution: consider the so-called integral criterion and its application to the following example. A steady solution with a large amplitude should be distinguished from that with a small amplitude.

Let the time interval T'' used for the criterion application be bounded by times t_1 and t_2, and denote by $\dot{x}(t)$ the velocity of a chosen coordinate which describes the motion of the system being examined. The integral monitored on the computer is

$$J = \int_{t_1}^{t_2} [\dot{x}(t)]^2 \, dt \quad \text{or} \quad J = \int_{t_1}^{t_2} |\dot{x}(t)| \, dt \, . \tag{7.2, 6}$$

Denote by J_1 the value of J for the large-amplitude steady solution and by J_2 that for the small-amplitude (or non-oscillatory) solution. It is found that

$$J_1 \gg J_2 \, .$$

Choosing a suitable value J_0 from the interval $J_2 < J_0 < J_1$ the transient response leads to the large-amplitude solution for $J > J_0$ and to the small-amplitude one for $J < J_0$. A detailed discussion of the various criteria may be found in the monographs by TONDL (1970a, 1970b, 1973a). Questions of analogue computer programming for the application of the methods and criteria outlined above are dealt with in a monograph by FIALA (1976).

7.3. Investigation of stability in the large for not-fully determined disturbances

This section will discuss the principle of solution effected by means of the circular scanning method. Consider first a one-degree-of-freedom system for which two stable steady solutions (one resonant, the other non-resonant) exist for a specified value of ω. In the stroboscopic phase plane they are represented by the stable foci 1 and 3, respectively (Fig. 7.3, 1). The separatrix s is formed by two stroboscopic trajectories leading to the saddle point 2; from this point issue two trajectories terminating at points 1 and 2.

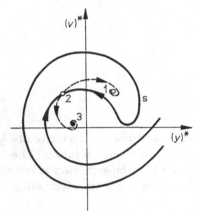

Fig. 7.3, 1

Consider the system to be in a definite steady state, for example, resonant vibration, to which corresponds focus 1 in the stroboscopic phase plane. The points of interest are the disturbance in the initial conditions from the steady state which will lead to the return to the original steady solution, and the disturbance which will lead to a qualitative change, i.e. to the transition to non-resonant vibration represented by focus 3.

In the coordinates $(y)^*$, $(v)^*$ a disturbance from the steady state is defined by

$$\Delta(y)^* = y(0) - y_1 ,$$

$$\Delta(v)^* = \frac{1}{\omega} \dot{y}(0) - v_1$$

where y_1, v_1 are the coordinates of the singular point 1. For $\Delta(y)^* = \Delta(v)^* = 0$, i.e. for $y(0) = y_1$, $\dot{y}(0) = \omega v_1$ no transient motion will arise and the system will immediately vibrate in resonant vibration. For fully determined disturbances in time $t = n(2\pi/\omega)$ $(n = 0, 1, 2, ...)$ the question can be answered directly by consulting the separatrix diagram shown in Fig. 7.3, 1.

So much for fully determined disturbances. Consider now the effect of disturbances from the steady state, which are no longer fully determined but contain a stochastic element introduced in their definition. Choose the initial conditions at time $t = 0$ such that it will make the distance between the starting point and the singular point (point 1 in the case being considered) in the phase plane equal to a given value, R. This means that the starting point is chosen quite freely, and so represents the stochastic element because it may lie anywhere on the circle with radius R centred at point 1 (Fig. 7.3, 2).

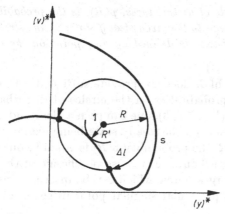

Fig. 7.3, 2

Definition 7.I. *A disturbance from the steady state represented in the stroboscopic phase plane (or Hayashi's plane)* [1] *by singular point 1 with coordinates y_1, v_1 is defined by the distance R between the starting point with coordinates $y(0)$, $v(0) \cdot \big(v(0) = 1/\omega\ \dot{y}(0)\big)$ and the singular point 1. The initial conditions are defined by any point of the circle described by the equation*

$$R^2 = [y(0) - y_1]^2 + [v(0) - v_1]^2 \tag{7.3, 1}$$

and termed the circle of disturbances.

The position of the starting point on the circle with radius R is, therefore, stochastic and it is assumed that the starting point has equal probability of occurring anywhere on the circle. Denoting by Δl the length of the arc of the circle of disturbances, which lies inside the domain of attraction relating to the singular point 1, the relation

$$p(R) = \frac{\Delta l}{2\pi R} \tag{7.3, 2a}$$

defines the probability that a solution with a disturbance in the initial conditions such that the starting point can lie anywhere on the circle with radius R centered at point 1 will again lead to the steady solution represented by the singular point 1. If n (a sufficiently large number) solutions, starting from points uniformly distributed on the circle of disturbances, are examined and k solutions are found to lead to the steady solution being investigated, the probability $p(R)$ is also given by the relation

$$p(R) = \frac{k}{n}. \tag{7.3, 2b}$$

This equation makes it possible to determine the probability $p(R)$ directly (by means of an analogue or a digital computer).

---

[1] The term Hayashi's plane (used for reason of briefness) denotes the phase plane of the system of two first-order differential equations, which is obtained by application of the van der Pol transformation (the (a, b) phase plane in Section 4.4). It was used by TONDL in his monograph (1973) because Professor Hayashi was the first who employed the phase portraits in this plane.

Definition 7.II. *Probability on the circle of disturbances, $p(R)$, is the probability that the disturbed solution at time $t = 0$ will lead to the given steady state; in the stroboscopic phase plane (Hayashi's plane) a disturbance is defined by any point on the circle of disturbances.*

Knowledge of $p(R)$ for a single value of R does not provide sufficient information. If, on the other hand, one knows the probabilities on the circles of disturbances of different radii R, i.e. the function $p(R)$ (Fig. 7.3, 3), one can gain a clear and comprehensive idea of the effect of disturbances from the steady state being examined. Consulting the diagram, we see that up to R' the probability $p(R)$ is equal to one, that is to say, so long as $R \leq R'$ arbitrary disturbances can cause no change of the steady state. The circle with radius R' is the largest circle which can be inscribed inside the respective domain of attraction from a particular singular point in the stroboscopic phase plane.

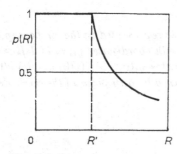

Fig. 7.3, 3

As mentioned above, all information so far presented relates to disturbances at time $t = 0$. Upson dividing the period $2\pi/\omega$ into n equal time intervals and determining $p_k(R)$ for the beginning of each interval, the mean value of the probabilities on the circle of disturbances is found to be defined by the formula

$$\bar{p}(R) = \frac{1}{n} \sum_{k=1}^{n} p_k(R) \,. \tag{7.3, 3}$$

This mean value makes it possible to eliminate the dependence on time.

Definition 7.III. *The probability on the circle of disturbances, $p(R)$, obtained from (7.3, 3) is the mean value of the probabilities on the circles of disturbances established for various instants uniformly distributed over the interval of a vibration period. It thus represents the probability on the circle of disturbances with which a solution with a disturbance defined on the circle of disturbances will lead to the steady solution being examined at any instant.*

The probability on the circle of disturbances can also be employed for determining disturbances defined in different ways (see the already quoted monograph by TONDL (1973 a)).

Since in practical cases the occurrence of small disturbances is much more frequent than that of large disturbances, and the magnitude of a disturbance can, for the most part, be restricted to a finite value, we can choose the density of occurrence of the disturbances and express it also in terms of the circle of disturbances. To that end introduce the function of the probability density of disturbances on the circle of disturbances, $f(R)$; the function can be bounded (see Fig. 7.3, 4) by the condition that the maximum disturbance will not extend beyond the circle of disturbances with radius R_0.

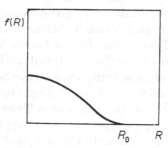

Fig. 7.3, 4

The function $f(R)$ satisfies the relation

$$\int_0^{R_0} f(R)\,\mathrm{d}R = 1 \qquad \begin{aligned} f(R) &\neq 0 \quad \text{for} \quad R < R_0\,, \\ f(R) &= 0 \quad \text{for} \quad R \geqq R_0\,. \end{aligned} \tag{7.3, 4}$$

If function $f(R)$ is applied to the probability function on the circle of disturbances $p(R)$ obtained as indicated above, we can, given the function of the probability density of the occurrence of the disturbances on the circle of disturbances, establish directly the probability with which the steady solution being examined will be attained by the effect of any disturbance at time $t = 0$.

Definition 7.IV. *If the specified function of the density of distribution of the occurrence of disturbances on the various circles of disturbances satisfies* (7.3, 4), *the probability P that any disturbance from the steady state being examined will lead to that state can be determined from the equation*

$$P = \int_0^{R_0} f(R)\,p(R)\,\mathrm{d}R \tag{7.3, 5a}$$

when the function $p(R)$ is known and the disturbances arise at time $t = 0$, or from the equation

$$P = \int_0^{R_0} f(R)\,\bar{p}(R)\,\mathrm{d}R \tag{7.3, 5b}$$

when the function $\bar{p}(R)$ is known and the disturbances arise at any time.

In this case the numerical value of the probability represents a measure of estimating the resistance of the steady state in question to disturbances.

To facilitate solution, it is possible to choose for the function $f(R)$ the linear function

$$f(R) = \frac{2}{R_0}\left(1 - \frac{R}{R_0}\right) \tag{7.3, 6}$$

and for R_0 a value which is common to a whole class of systems, for example, a multiple of the static deflection corresponding to the amplitude of the excitation force. This enables the analyst to compare resistances of certain solutions of various systems or to optimize a particular parameter.

As mentioned in the preceding section, solutions to problems of the sort discussed are expediently obtained by means of the circular scanning method, which makes it possible to determine not only the domains of attraction but the function $p(R)$ as well. The graph plotter stylus, which is controlled by circular scanning and the logic circuits deciding whether or not a transient solution leads to the given one, moves at a uniform rate and draws a vertical line whenever the solution in fast repetition converges to the steady solution. In the segments of the circle of disturbances in which the transient solutions fail to lead to the steady one, the stylus is stopped. If the circle of disturbances lies whole in the domain of attraction being examined, the stylus draws the entire prescribed length and the probability $p(R) = 1$. After completing a circle, the stylus is automatically reset by R and the process is repeated. The resulting diagram contains a series of vertical lines whose lengths represent, to an appropriate scale, the values of the probability $p(R)$; by fitting a curve to the end points of the line segments we obtain the dependence $p(R)$ (Fig. 7.3, 5).

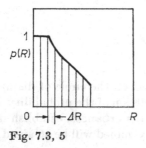

Fig. 7.3, 5

Consider now systems with several degrees of freedom, for example, a system excited by a periodic force with a period $T = 2\pi/\omega$ whose stable steady solutions are characterized by singular points in the n-th dimensional stroboscopic phase (or the n-dimensional Hayashi's) space. Just as in the case of one-degree-of-freedom systems the solution, although approximate, is independent of time. Proceeding in the way outlined for one-degree-of-freedom systems examine the effect of the disturbances from the given steady state replacing, logically, the circle of disturbances by the sphere of disturbance of radius R centered at a particular singular point.

Definition 7.V. *If* y_{01}, y_{02}, ... , y_{0n} *are the coordinates of the given singular point,* $y_1(0)$, $y_2(0)$, ... , $y_n(0)$ *the coordinates of the starting point, and* $x_k = y_k(0) - y_{0k}$ *(*$k = 1$, *2, ... , n*) *in the stroboscopic phase (Hayashi's) space, the sphere of disturbances is defined by the equation*

$$\sum_{k=1}^{n} x_k^2 = R^2 \, . \tag{7.3, 7}$$

If the analyst plans to use a computer for a practical solution, the points on this sphere must be replaced by a finite number of points. Since it is usually much easier to make such a replacement for curves than for surfaces, the analyst should proceed in two steps. First, the spherical surface is replaced by an exactly defined number of curves distributed as uniformly as possible; second, the curves are replaced by a finite number of points. Although this double replacement tends to make the procedure slightly less accurate, it enables the analyst quickly to form a qualitative opinion of the situation. The simplest substitute curves seem to be the principal circles of the sphere of disturbances defined by the quations

$$x_k^2 + x_j^2 = R^2 , \qquad (k, j = 1, 2, \ldots , n; k \neq j) , \atop x_s = 0 \qquad (s = 1, 2, \ldots , n; s \neq k, s \neq j). \qquad \Bigg\} \qquad (7.3, 8)$$

These equations represent the curves of intersection of the coordinate planes (x_k, x_j) and the *sphere of disturbances*. The points on the sphere surface are thus replaced by points lying on the principal circles whose number is $N = \binom{n}{2}$. This system of circles is termed the substitute sphere of disturbances.

Definition 7.VI. *The substitute sphere of disturbances. The substitute sphere of disturbances is the set of all starting points for a disturbed solution that lie on the principal circles of the sphere of disturbances defined by (7.3, 8).*

Assume that a starting point has the same probability of occurring at any point of the substitute sphere of disturbances.

Denote by $p_s(R) = p_{kj}(R)$ $\left(s = 1, 2, \ldots , N = \binom{n}{2}\right)$ the probability that a disturbed solution will lead to the given steady one where the disturbance is described by any of the points of the principal circle of disturbances. The mean value of all probabilities $p_s(R)$ is defined by the equation

$$p(R) = \frac{1}{N} \sum_{s=1}^{N} p_s(R) \qquad (7.3, 9)$$

and represents the probability that a disturbed solution will lead to the steady solution where the disturbance is described by any of the points of the substitute sphere of disturbances. To simplify the discussion the same notation is used for the function $p(R)$ as for the probability on the circle of disturbances of one-degree-of-freedom systems.

Definition 7.VIII. *The probability on the substitute sphere of disturbances $p(R)$ is the probability that a disturbed solution at time $t = 0$ (at any time when using Hayashi's space) will lead to the steady solution being examined. The starting point of the disturbed solution may lie at any point of the substitute sphere of disturbances. The value of $p(R)$ is the mean of the probabilities on all principal circles of the sphere of disturbances.*

If the analyst is using the stroboscopic phase space and wishes to eliminate the time from his considerations, he can — just as in the case of one-degree-of-freedom systems — divide the period of steady vibration into a number of equal parts, determine the functions $p_s(R)$ for the beginnings of the individual intervals and by establishing the mean values $\overline{p}_s(R)$ obtain the probability

$$\overline{p}(R) = \frac{1}{N} \sum_{s=1}^{N} \overline{p}_s(R) . \qquad (7.3, 10)$$

This is the probability that a disturbed solution at any time will lead to the steady solution where the disturbance is defined by the substitute sphere of disturbances.

Note: Although the use of Hayashi's space is more straightforward and eliminates the time dependence directly, the stroboscopic phase space approach is more rigorous.

A procedure similar to that just outlined can be used to determine the probability P that any disturbed solution will lead to the given steady solution if one knows the variation of function $p(R)$ (at time $t = 0$) or of function $\overline{p}(R)$ (at any time) and the function of the density of distribution of the probability of occurrence of the disturbances, $f(R)$. The resulting relation is formally analogous to (7.3, 5).

7.4.   Examples of investigations concerning stability in the large for disturbances in the initial conditions

Example I. For the purpose of comparison with the approximate procedure outlined in Section 4.4, an analysis will first be presented of the Duffing system described by equation (4.4, 1). Figs. 7.4, 1 and 7.4, 2 show the phase portraits in the stroboscopic

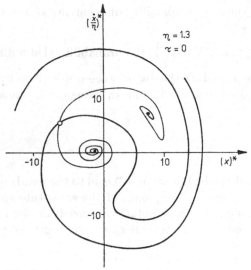

Fig. 7.4, 1

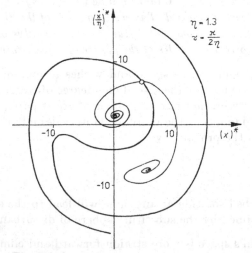

Fig. 7.4, 2

phase plane obtained for $\varkappa = 0.05$ and $\gamma = 10^{-2}$ (i.e. the same as those used in Section 4.4) and $\eta = 1.3$. The first portrait relates to the initial time $\tau = 0$, the second to $\tau = \pi/2\eta$; both show the effect of time of application of the disturbance. Unlike those obtained by application of the approximate method of Section 4.4, the two diagrams do not become one when turned through an angle of 90° relative to one another. Fig. 7.4, 3 shows the phase portrait obtained for $\eta = 1.5$ and $\tau = 0$. To facilitate comparison the separatrix established by application of the approximate method is shown in dotted line. It can be seen that although the qualitative agreement is close, the results differ quantitatively. The difference grows larger with increasing value of the non-linearity γ.

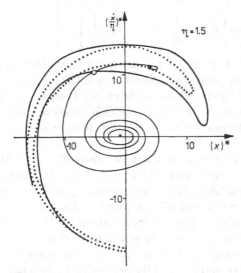

Fig. 7.4, 3

Example II. Consider a one-mass, two degrees-of-freedom (two-dimensional) system excited by a circulating vector whose motion is described by the differential equations

$$\left.\begin{aligned} m\ddot{x} + \varkappa_1\dot{x} + c_1 x + \beta_1 xy^2 &= m\varepsilon\omega^2\cos\omega t \, , \\ m\ddot{y} + \varkappa_2\dot{y} + c_2 y + \gamma_2 xy &= m\varepsilon\omega^2\sin\omega t \end{aligned}\right\} \tag{7.4, 1}$$

where $\varkappa_1$, $\varkappa_2$, c_1, c_2, β_1, γ_2, ε are positive constants. Assume that the system is tuned to the internal resonance for which the ratio of c_1/c_2 is close to 4. Introducing the relative deflection $u = x/\varepsilon$, $v = y/\varepsilon$, the notation $\omega_1^2 = c_1/m$, $\omega_2^2 = c_2/m$ (the partial frequencies of the linearized system without damping) and the substitution $\omega_1 t = \tau$, the equations of motion can be written in the dimensionless form

$$\left.\begin{aligned} u'' + D_1 u' + u + \beta_0 u v^2 &= \eta^2\cos\eta\tau \, , \\ v'' + D_2 v' + K v + \gamma_0 u v &= \eta^2\sin\eta\tau \end{aligned}\right\} \tag{7.4, 2}$$

where $\eta = \omega/\omega_1$, $D_1 = \varkappa_1/m\omega_1$, $D_2 = \varkappa_2/m\omega_1$, $K = \omega_2^2/\omega_1^2$, $\beta_0 = \beta_1\varepsilon^2/c_1$, $\gamma_0 = c_2/c_1\,\gamma_2\varepsilon/c_2$.

If $\eta \sim 1$, application of the van der Pol transformation (the solution is sought in the form

$$\left.\begin{aligned} u &= a\cos\eta\tau + b\sin\eta\tau \, , \\ v &= c\cos\tfrac{1}{2}\eta\tau + d\sin\tfrac{1}{2}\eta\tau \end{aligned}\right\} \tag{7.4, 3}$$

where the coefficients a, b, c, d are assumed to be slowly varying functions of time (constants for a steady solution)) enables equations (7.4, 2) to be converted to the following system of first-order equations:

$$\left.\begin{aligned}
a' &= -\frac{1}{2} D_1 a + \frac{1}{2\eta}\left[1 - \eta^2 + \frac{1}{2}\beta_0(c^2 + d^2)\right]b\,,\\[2mm]
b' &= -\frac{1}{2} D_1 b - \frac{1}{2\eta}\left[1 - \eta^2 + \frac{1}{2}\beta_0(c^2 + d^2)\right]a + \frac{1}{2}\eta\,,\\[2mm]
c' &= -\frac{1}{2} D_2 c + \frac{1}{\eta}\left\{\frac{1}{2}\gamma_0 bc + \left[K - \left(\frac{1}{2}\eta\right)^2 - \frac{1}{2}\gamma_0 a\right]d\right\}\,,\\[2mm]
d' &= -\frac{1}{2} D_2 d - \frac{1}{\eta}\left\{\frac{1}{2}\gamma_0 bd + \left[K - \left(\frac{1}{2}\eta\right)^2 + \frac{1}{2}\gamma_0 a\right]c\right\}\,.
\end{aligned}\right\} \qquad (7.4, 4)$$

In a definite interval of η, system (7.4, 4) can have the semi-trivial solution $a \neq 0$, $b \neq 0$, $c = d = 0$ as well as two stable and two unstable non-trivial solutions. In each of these pairs, the solutions differ by the phase angle rather than by the magnitude of the amplitude. Since a detailed analysis of the results is presented in a monograph by TONDL (1970a), only the information necessary for further discussion will be given below.

The results relating to stationary vibration in the case when $K = 0.3$, $D_1 = D_2 = 0.1$, $\gamma_0 = 3 \cdot 10^{-2}$ and $\beta_0 = 0.8 \cdot 10^{-2}$ are reviewed in the diagrams included in Fig. 7.4, 4. The diagrams show the dependence of the amplitudes $r_1 = (a^2 + b^2)^{1/2}$ and $r_2 = (c^2 + d^2)^{1/2}$ on η (r_{10} denotes the semi-trivial solution) and in the planes (a, b) and (c, d) the corresponding values for various η. The results obtained for the stable solutions are drawn in heavy lines, those for the unstable solutions in light lines.

The figure also includes the results of an analysis of the domains of attraction obtained by means of the method of TV scanning for fully determined disturbances and $\eta = 1.15$. The results refer to disturbances in the initial conditions of the solution defined by the stable singular point with coordinates a_i, b_i, c_i and d_i. The disturbances are marked $\alpha = a(0) - a_i$, $\beta = b(0) - b_i$, $\gamma = c(0) - c_i$ and $\delta = d(0) - d_i$. Since the space of the initial conditions is four dimensional, plane sections are taken of the separatrix surfaces with always two of the coordinates α, β, γ, δ made equal to zero (for example, $c(0) = c_i$, $d(0) = d_i$). The diagrams in Fig. 7.4, 4 relate to the case of $c = d = 0$ when the singular point belongs to the semi-trivial solution. In the diagrams of $r_1(\eta)$, $r_2(\eta)$ and in the planes (a, b), (c, d), the points corresponding to this solution are marked by numeral 1. The cross-hatched areas represent the regions of the initial conditions which lead to the semi-trivial solutions, the hatched areas those which lead to the non-trivial solutions (in the diagrams of $r_1(\eta)$, $r_2(\eta)$ and in the planes (a, b), (c, d) the non-trivial solutions which, as is recalled, are two differing only as to the phase angle, are marked by points 3a and 3b). Since $\binom{4}{2} = 6$, there ought to be six plane sections. As, however, all the initial conditions in the (α, β) plane lead to the semi-trivial solution, this plane is omitted. The diagrams shown in Fig. 7.4, 5 are similar to those just discussed except that the singular point 3a corresponds to one of the non-trivial stable solutions.

The results obtained in determination of the function $p(R)$ by means of the method of circular scanning are shown in Figs. 7.4, 6 and 7.4, 7. The first figure (corresponding to Fig. 7.4, 4) belongs to the case when the singular point represents the semi-trivial

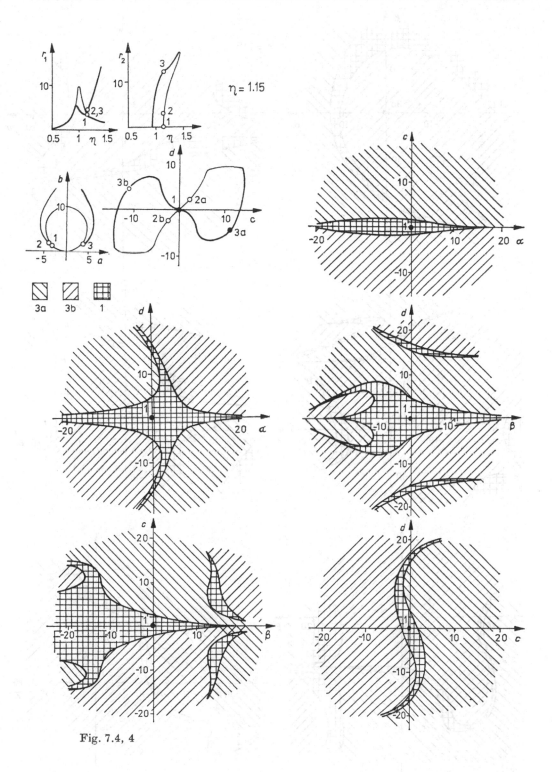

Fig. 7.4, 4

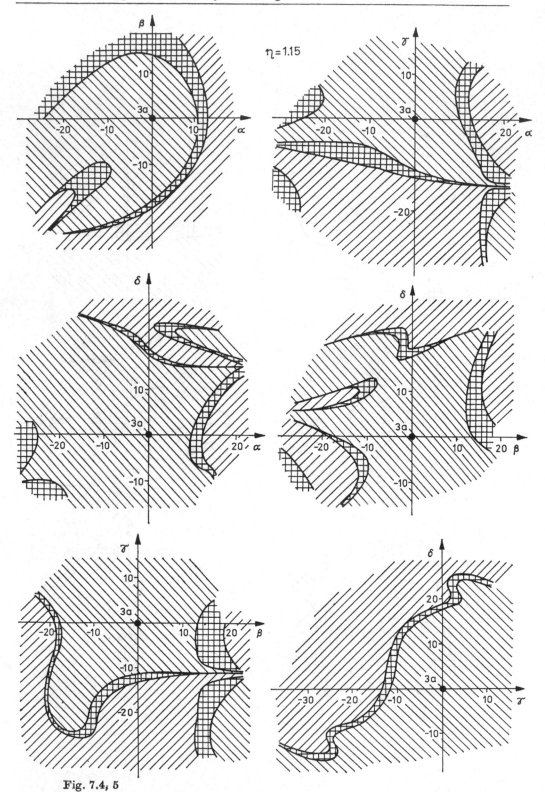

Fig. 7.4; 5

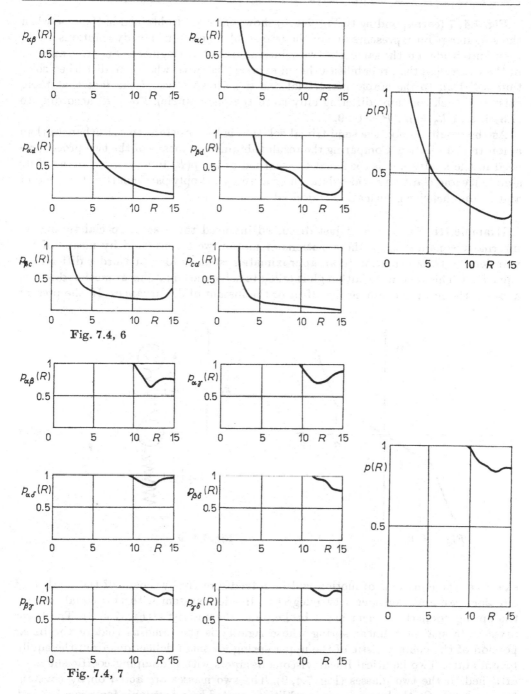

Fig. 7.4, 6

Fig. 7.4, 7

solution and the probabilities indicate that with which the disturbed solution will converge to the semi-trivial one. The auxiliary diagrams (drawn to a smaller scale) show the probabilities on the principal circles, $p_{\alpha\beta}(R)$, $p_{dc}(R)$, ... , used to obtain the course of the function $p(R)$. For a linear function of the density of distribution of the probability of occurrence of the disturbances on the various substitute spheres, defined by (7.2, 6) and $R_0 = 15$, the probability is $P = 0.64$.

Fig. 7.4, 7 (corresponding to Fig. 7.4, 5) shows the results obtained in the case when the singular point represents one of the two stable non-trivial steady solutions. Since their amplitudes are the same, the two solutions are not distinguished one from another in the diagrams; the probabilities shown express that with which the disturbed solution will lead to the stable non-trivial one (i.e. irrespective of the fact that there exist two such solutions differing only as to the phase angle). For $f(R)$ according to (7.3, 6) and $R_0 = 15$, $P = 0.99$.

As the results imply, the semi-trivial solution is less resistant to disturbances than a non-trivial solution. Comparing the results obtained by means of the two procedures used in the analysis, it is seen that the method of the $p(R)$ function, by virtue of its simplicity and clarity, provides data which are easy to apply particularly when prompt and comprehensive information is desired.

Example III. The example just discussed involved the case of special tuning (to internal resonance) when the equations of motion were converted by means of the van der Pol transformation to an approximate system of four first-order differential equations. This procedure, although it failed to solve the equations of motion directly, avoided the need to examine the effect of the instant of disturbances. In the present

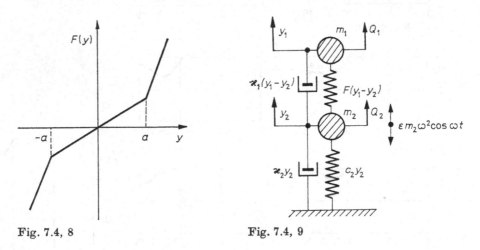

Fig. 7.4, 8   Fig. 7.4, 9

example, the equations of motion will be solved directly by means of the method of circular scanning. Consider a two-degree-of-freedom system with two equal masses; the spring connecting them has a broken-line characteristic (Fig. 7.4, 8). The lower mass rests against a linear spring whose rigidity is the same as that of the linear portion of the characteristic of the upper spring for small deflections from the equilibrium state. Two identical linear viscous dampers with a damping coefficient $\varkappa$ are attached to the two masses (Fig. 7.4, 9). The two masses are acted on by constant forces, Q_1 and Q_2, the lower one is in addition excited by a harmonic force $m\varepsilon\omega^2 \cos \omega t$. Denoting by y_1, y_2 the deflections of the upper and lower mass, respectively, the motion of the system is described by the differential equations

$$\left.\begin{array}{l} m\ddot{y}_1 + \varkappa(\dot{y}_1 - \dot{y}_2) + F(y_1 - y_2) = Q_1\,, \\ m\ddot{y}_2 - \varkappa(\dot{y}_1 - \dot{y}_2) - F(y_1 - y_2) + \varkappa\dot{y}_2 + cy_2 = Q_2 + m\varepsilon\omega^2 \cos \omega t \end{array}\right\} \quad (7.4, 5)$$

where

$$F(y_1 - y_2) = c(y_1 - y_2) \quad \text{for} \quad |y_1 - y_2| \leqq a, \qquad (a > 0),$$
$$= c(y_1 - y_2) + c_1(y_1 - y_2 - a) \quad \text{for} \quad y_1 - y_2 > a,$$
$$= c(y_1 - y_2) + c_1(y_1 - y_2 + a) \quad \text{for} \quad y_1 - y_2 < -a.$$

Introducing the notation

$$\frac{c}{m} = \omega_0^2, \qquad \frac{\varkappa}{m\omega_0} = D, \qquad \frac{Q_1}{c} = q_1, \qquad \frac{Q_2}{c} = q_2, \qquad \frac{\omega}{\omega_0} = \eta, \qquad (7.4, 6)$$

the relative coordinates

$$\frac{y_1}{\varepsilon} = x_1, \qquad \frac{y_2}{\varepsilon} = x_2 \qquad\qquad\qquad (7.4, 7)$$

and the transformation

$$\omega_0 t = \tau \qquad\qquad\qquad\qquad\qquad (7.4, 8)$$

equations (7.4, 5), after rearrangement, take the form

$$\left. \begin{aligned} x_1'' + D(x_1' - x_2') + (x_1 - x_2)[1 + \varphi(x_1 - x_2)] &= q_1, \\ x_2'' - D(x_1' - x_2') + Dx_2' - (x_1 - x_2)[1 + \varphi(x_1 - x_2)] + x_2 &= q_2 + \eta^2 \cos \eta\tau \end{aligned} \right\}$$
$$(7.4, 9)$$

where

$$\varphi(x_1 - x_2) = 0 \quad \text{for} \quad |x_1 - x_2| \leqq \frac{a}{\varepsilon} \qquad (a > 0)$$

$$= \frac{c_1}{c}\left(1 - \frac{a}{\varepsilon}\frac{1}{|x_1 - x_2|}\right) \text{ for } |x_1 - x_2| > \frac{a}{\varepsilon}.$$

Figs. 7.4, 10 to 7.4, 12 show the amplitude-frequency characteristics obtained by analogue computer solutions for the relative deflections x_1, x_2 and their difference

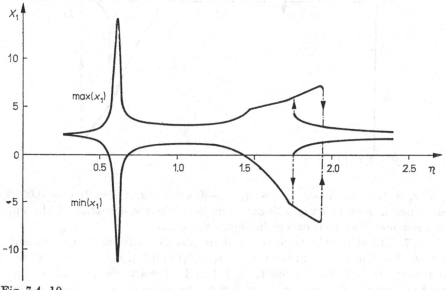

Fig. 7.4, 10

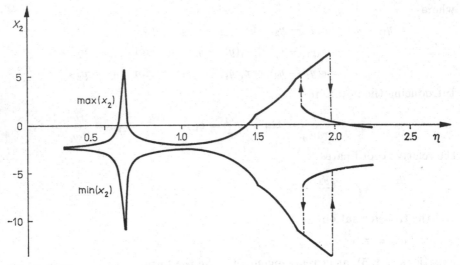

Fig. 7.4, 11

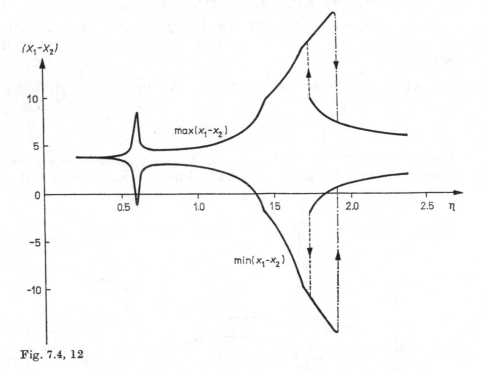

Fig. 7.4, 12

$x_1 - x_2$ in the case when $q_1 = 4$, $q_2 = -6$, $a/\varepsilon = 10$, $c_1/c = 2$, $D = 0.05$. The lower resonance is seen to have a linear character. The non-linearity of the upper spring does not manifest itself except in higher resonance.

Fig. 7.4, 13 shows the trajectories of the steady solutions (both resonant and non-resonant) in the partial phase planes $(x_1, x_1'/\eta)$ and $(x_2, x_2'/\eta)$ for a relative excitation frequency $= 1.86$. The points I, II, III and IV mark the position of the reference point at time $\tau = 0$, $\frac{1}{2}\pi/\eta$, π/η and $\frac{3}{2}\pi/\eta$. The coordinates of these points define the

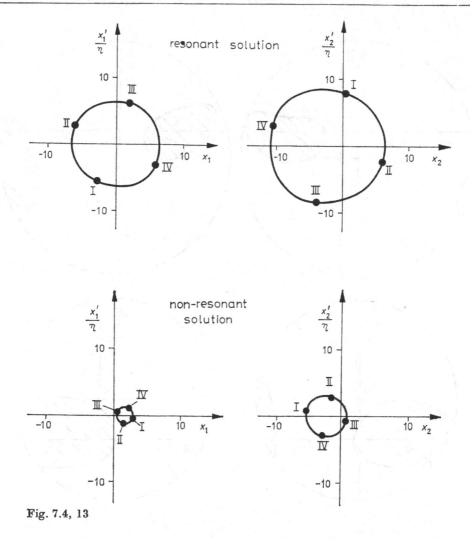

Fig. 7.4, 13

initial conditions for which the steady solutions at the specified times are obtained directly, without transient motions. They determine the position of the singular points in the stroboscopic phase plane. The method of circular scanning makes it possible to obtain not only the function $p(R)$ on the various principal circles of the spheres of disturbances but also, by means of a second plotter, the domains of attraction in the various plane sections of the space of the initial conditions. As in Example II, two initial conditions (corresponding to two coordinates of the singular point) are kept constant, and two are varied in accordance with the circular scanning. These variable coordinates of the singular point describe the centre of the circle of circular scanning. Since six combinations of the principal circles are possible for four initial conditions (hence the six diagrams of the function $p(R)$), there exist six combinations of the plane sections of the separatrix surface. The solution of the example was carried out for $\eta = 1.86$ and four initial times, i.e. $= 0, \frac{1}{2} \pi/\eta, \pi/\eta, \frac{3}{2} \pi/\eta$. Figs. 7.4, 14 to 7.4, 17 refer to the resonant solution and show the domains of attraction at these initial times for disturbances in the solution's neighbourhood. The regions of the initial conditions leading to the non-resonant solution are shown in hatching. Fig.

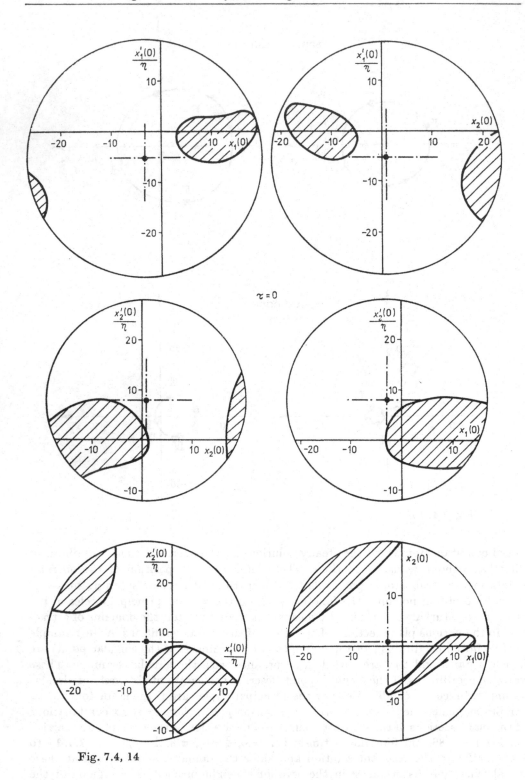

Fig. 7.4, 14

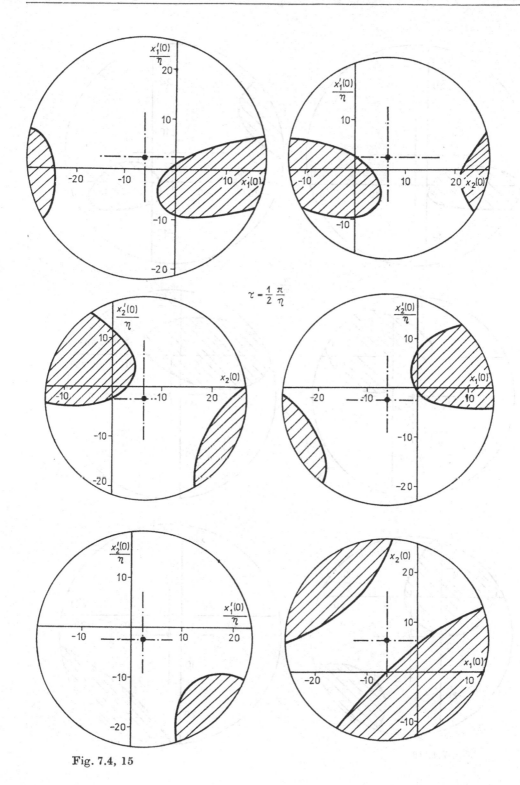

$$\tau = \frac{1}{2}\frac{\pi}{\eta}$$

Fig. 7.4, 15

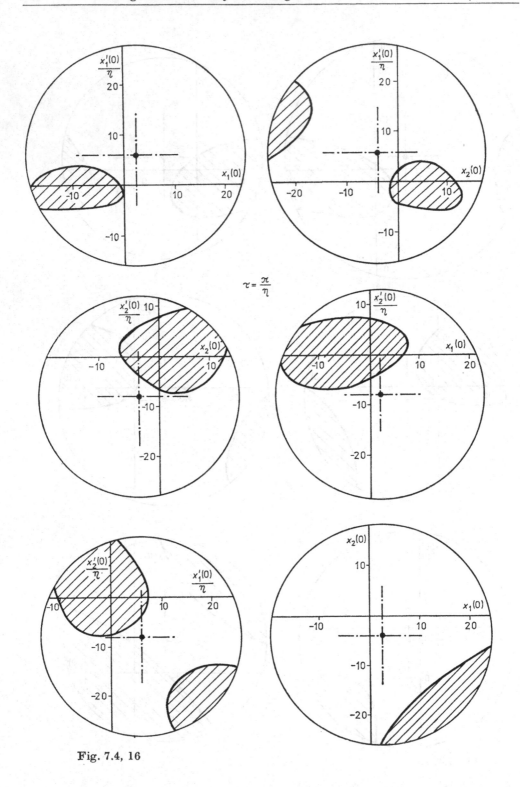

Fig. 7.4, 16

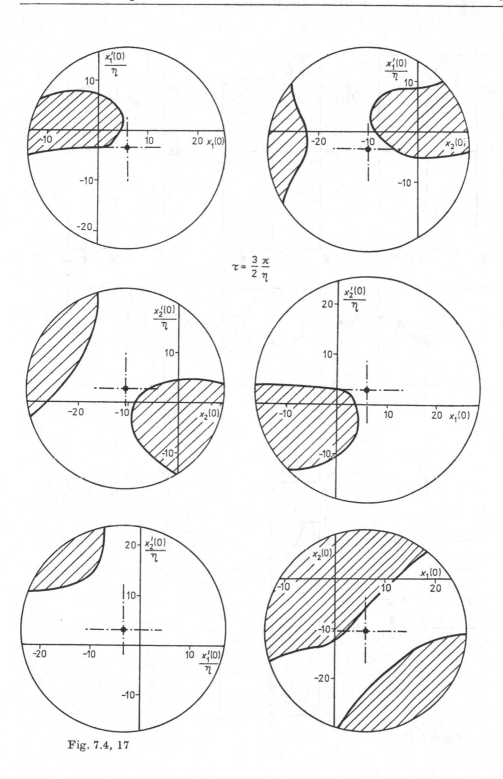

$$\tau = \frac{3}{2}\frac{\varkappa}{\eta}$$

Fig. 7.4, 17

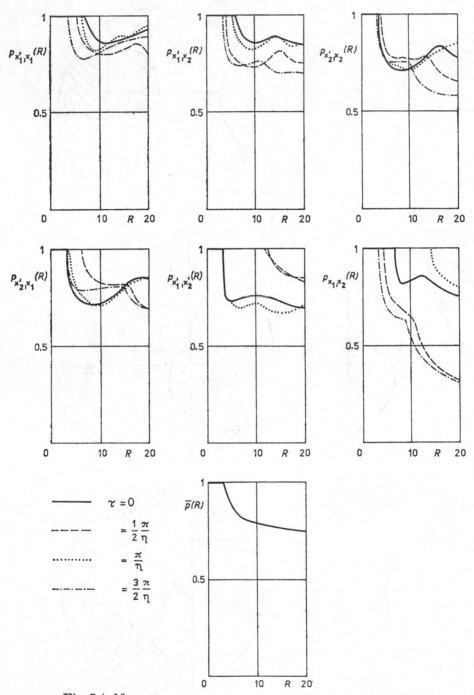

Fig. 7.4, 18

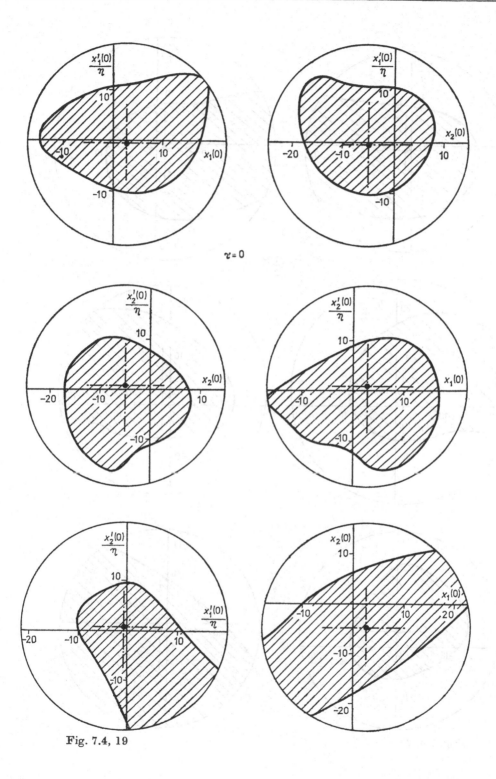

Fig. 7.4, 19

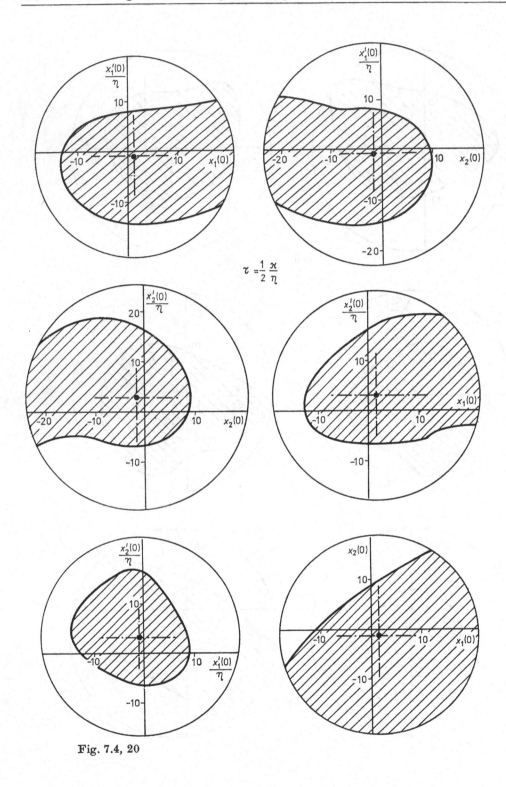

$$\tau = \frac{1}{2}\frac{\varkappa}{\eta}$$

Fig. 7.4, 20

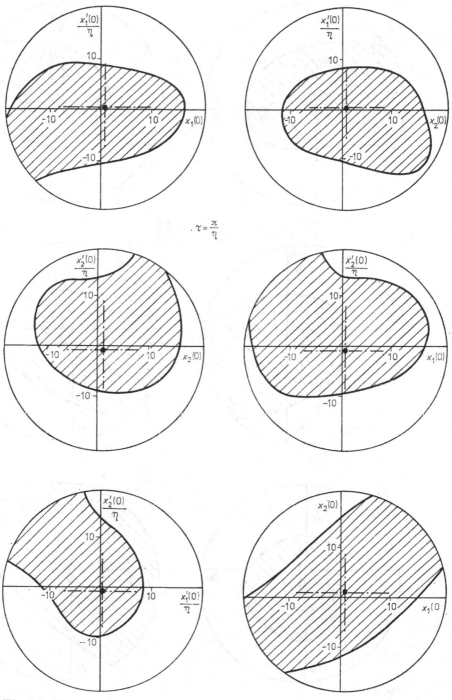

$$\tau = \frac{\pi}{\eta}$$

Fig. 7.4, 21

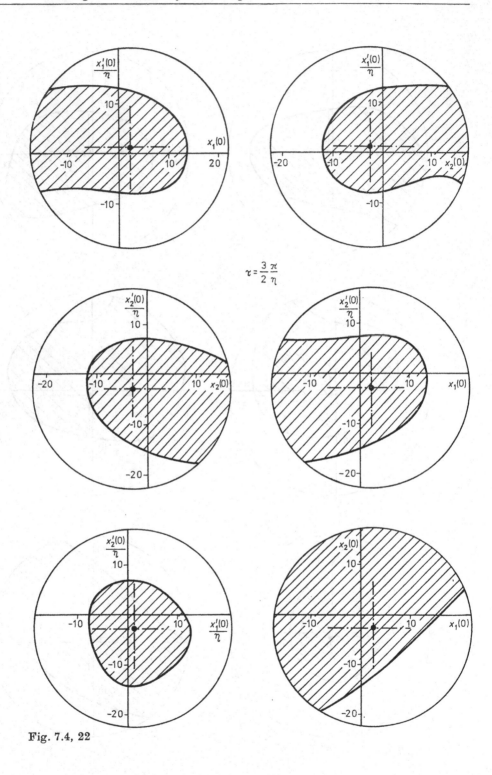

Fig. 7.4, 22

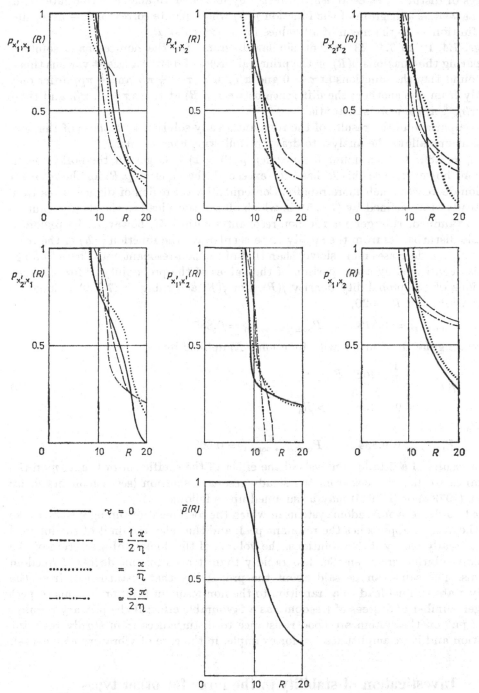

Fig. 7.4, 23

7.4, 18 shows the corresponding functions $p(R)$ on the various principal circles of the spheres of disturbances established directly by means of an analogue computer solution, as well as a diagram of the function $\overline{p}(R)$ whose points correspond — according to definition — to the mean of all values for a particular R.

Figs. 7.4, 19 to 7.4, 23 present similar diagrams for the non-resonant solution. Comparing the functions $p(R)$ on the principal circles of disturbances at various times, it is found that the functions at $\tau = 0$ and π/η, or at $\tau = \frac{1}{2}\,\pi/\eta$ and $\frac{3}{2}\,\pi/\eta$ differ only slightly from one another; the difference between $p(R)$ at times $\tau = 0$, π/η and those at $\frac{1}{2}\,\pi/\eta$, $\frac{3}{2}\,\pi/\eta$ is more substantial.

A comparison of the results of the resonant steady solution with those of the non-resonant one allows the analyst to draw the following conclusions:

At $\eta = 1.86$, R' (maximum R for which $\overline{p}(R) = 1$) is larger for the non-resonant solution; however, once this R' has been exceeded, the function $\overline{p}(R)$ for the resonant solution decreases much more rapidly. Consequently, the region of disturbances from the steady state defined by (7.4, 5), in which all solutions lead to the steady solution being examined, is larger for the non-resonant solution. If, however, the region of possible disturbances from the steady state is enlarged, the function $\overline{p}(R)$ of the resonant solution decreases more slowly than that of the non-resonant one. These findings are also confirmed by a comparison of the values of the probability P for specified functions of the probability density $f(R)$. For $f(R)$ according to (7.3, 6) (triangular distribution) and $R_0 = 20$,

$$P_{\text{res} \to \text{res}} = 0.878\,, \qquad P_{\text{nonres} \to \text{nonres}} = 0.885\,.$$

For rectangular distribution with function $f(R)$ defined by

$$f(R) = \begin{cases} \dfrac{1}{R_0} & \text{for} \quad R \leqq R_0 \\[2mm] 0 & \text{for} \quad R > R_0 \end{cases} \tag{7.4, 10}$$

and $R_0 = 20$,

$$P_{\text{res} \to \text{res}} = 0.834\,, \qquad P_{\text{nonres} \to \text{nonres}} = 0.668\,.$$

The results of a detailed analysis of the effect of the coefficient of tuning η on the resistance to disturbances from the steady resonant solution (see a monograph by Tondl (1973a) or (1979c)) may be summed up as follows:

For two-degrees-of-freedom systems in which the coefficient of tuning η is varied so that the system approaches the resonant peak and thus also the limit of the interval of two steady locally stable solutions, the volume of the domain of attraction of the resonant solution grows smaller less rapidly than it does for one-degree-of-freedom systems. The same can be said about the probability that disturbances from the steady state will not lead to a transition to the non-resonant solution. In this respect a larger number of degrees of freedom has a favourable effect if the primary requirements put on the system are good resistance to disturbances from steady resonant vibration and large amplitudes (as, for example, in the case of vibratory conveyors).

7.5. Investigation of stability in the large for other types of disturbances

The disturbances whose effect will now be examined are not determined by the initial conditions; they are of the type described in Section 7.1.

Depending on whether the steady state in question is non-oscillatory (equilibrium position) or oscillatory (periodic vibration) the methods of examination can be divided into two groups.

In the first case the system is subjected to a disturbance of a certain type; the initial conditions at time $t = 0$ (at the beginning of the disturbance application) correspond to the steady state. If a transient solution converges to this steady state, the parameters of the disturbance lie in the domain of attraction of the steady state being examined and can be varied, for example, by means of the method of TV scanning. Let the disturbance be characterized by two parameters, for example A (the pulse amplitude) and T_0 (the time of the disturbance duration) and let these parameters vary within the intervals

$$A_1 \leqq A \leqq A_2, \qquad T_{01} \leqq T_0 \leqq T_{02}.$$

One parameter, for example A, is kept constant, the other is varied in steps of ΔT_0 through the whole interval (T_{01}, T_{02}). Next, A is varied in steps of ΔA and the process is repeated. A transient solution in fast repetition is obtained for each value of the parameters and, by means of a suitable criterion, an estimate is made whether or not the transient process converges to the steady solution. The logical circuit which applies the criterion controls the dip and lift of the graph plotter stylus; its vertical and horizontal displacements are regulated by the circuits which effect the gradual changes of the disturbance parameters. The result of the procedure is the hatched area of the domain of attraction of the steady state in the (A, T_0) plane.

In the second case (the steady state being examined is oscillatory — periodic vibration) the procedure has several variants whose choice is controlled by the instant of the disturbance application. If the domain of attraction is to be examined for a definite time instant, the analysis proceeds as in the former case. At the instant the disturbance is applied the system is subjected to initial conditions which realize the steady state without a transient solution.

Consider now the case when the beginning of the disturbance application is stochastic rather than deterministic. Let the frequency of the steady oscillatory solution be ω (the vibration period $T = 2\pi/\omega$). Assume that the probability of occurrence of the instant of the disturbance application is the same throughout the whole vibration period T of the steady solution ($jT \leqq t \leqq (j + 1) T$ $(j = 1, 2, ...)$. Divide the period T into N equal time intervals and apply N times the disturbance characterized by the parameters A, T_0 at the beginning of each interval. The probability that the disturbance applied at any time will lead to the steady state is defined by the relation

$$p(A, T_0) = \frac{k}{N} \tag{7.5, 1}$$

where k is the number of transient solutions which converge to the steady state. In practice the subsequent procedure depends on the type of the stationary vibration, that is, whether it is excited externally or parametrically, or whether it is self-excited.

In the first case the disturbances are applied successively with a delay $(r/N) T$ $(r = 1, 2, ... , N)$. The logical circuit which applies the criterion of convergence of the transient solution controls the motion of the graph plotter stylus in one direction (for example, vertical). If a transient solution leads to the steady state, the stylus moves $1/N$ of the total pre-set range. After examination of N transient processes, the stylus marks out a segment k/N long of the total range corresponding to the probability 1. Following completion of N such solutions, one of the parameters, for example A, is changed by ΔA and the process is repeated until the whole interval of A is traversed. If the other parameter is also varied in steps, a set of $p(A)$ diagrams is obtained for various values of T_0. On the basis of these diagrams the analyst sets up the axonometric representation of the function $p(A, T_0)$.

In the second case — investigation of stability of stationary self-excited vibration — the procedure is more complicated. An analogue solution can proceed as follows: One part of the computer (system I) performs a slow solution of the equations of motion for parameters corresponding to the state without a disturbance. The other part of the analogue (system II) is programmed to solve the equations of motion in fast repetition. This part — complete with the logic circuits — carries out the actual solution of the problem; system I serves only as a source of the necessary initial conditions which are specified so as to lead to the solution whose stability is being examined without transient processes. Let T_1 be the period of the steady solution for system I. Divide this period into n intervals and use the deflections and velocities obtained after a T_1/n time interval for the initial conditions of the fast repetition solution in system II. Since the solution in system II proceeds at a very fast rate,

$$T_1 = N T_2 \tag{7.5, 2}$$

where T_2 is the period of the steady solution in system II, and N is a large integer, $N \gg n$; N/n is a large integer. The instant at which the initial conditions from system I are applied to system II represents the time at which system II is subjected to a disturbance of T_0 duration. The time in the course of which system II carries out the solution, is $(1/n)/T_1$; it is divided into three intervals, viz.

$$\frac{T_1}{n} = T_0 + T' + T'' \tag{7.5, 3}$$

T_0 is the time of the disturbance duration, T' is the time sufficient for stabilizing the transient response, and T'' the time during which a suitable criterion (for example, integral) is applied to the logic circuits. The time interval T' must be a multiple of T_2 such that during its course the transient vibration becomes stabilized. In most cases this is achieved with a factor of 15 to 30. This explains why special stress is laid on the requirement that N/n should be a large integer.

The logic and control circuits of the part of the analogue which handles system II are the same as those used in previous examples.

As in Section 7.3, given the function of the probability density of the disturbances, $f(A, T_0)$ (for disturbances of the type of a force pulse — Fig. 7.1, 1 or a step change of a parameter lasting a definite time — Fig. 7.1, 2 b) or $f(Q_I, Q_{II})$ (for disturbances of the type of a step change of a parameter — Fig. 7.1, 2 a), one can determine the probability that random disturbances of a certain type will lead to a given solution. Functions $f(A, T_0)$ or $f(Q_I, Q_{II})$ satisfy the relations

$$\int_{-\infty}^{\infty} \int_{-\infty}^{\infty} f(A, T_0) \, \mathrm{d}A \, \mathrm{d}T_0 = 1 , \qquad \int_{-\infty}^{\infty} \int_{-\infty}^{\infty} f(Q_I, Q_{II}) \, \mathrm{d}Q_I \, \mathrm{d}Q_{II}; \tag{7.5, 4}$$

they can be defined in a restricted interval, that is, be non-zero in finite intervals of the values of A, T_0 or Q_I, Q_{II}. The probability P is then defined by the equation

$$P = \int_{-\infty}^{\infty} \int_{-\infty}^{\infty} p(A, T_0) \, f(A, T_0) \, \mathrm{d}A \, \mathrm{d}T_0 \tag{7.5, 5a}$$

or by the equation

$$P = \int\limits_{-\infty}^{\infty} \int\limits_{-\infty}^{\infty} p(Q_I, Q_{II})\, f(Q_I, Q_{II})\, dQ_I\, dQ_{II}\,. \tag{7.5, 5b}$$

The reader interested in additional details of the method outlined above is referred to a special monograph by TONDL (1973a).

7.6.   Other applications of the results

Determination of the probability P (for disturbances in the initial conditions or for other types of disturbances) for a specified density of distribution by means of the function $f(R)$ in relation to one of the parameters of the system makes it possible to establish the dependence of P on that parameter. The function $P(\eta)$ (η is the variable parameter, for example, the excitation frequency) can then be used in solution of other problems, for example, that of comparing the resistance of various steady solutions in the whole interval of the excitation force, in which more than one steady locally stable solutions obtain.

By way of example, consider that two such solutions, one resonant (I), the other non-resonant (II) exist in the interval (η_1, η_2). Denoting by

$$P_I(\eta) \quad \text{where} \quad \eta_1 < \eta < \eta_2$$

the probability with which a disturbed solution at the specified function of the density of distribution of the disturbances from the steady solution I will again lead to that solution, and by

$$P_{II}(\eta) \quad \text{where} \quad \eta_1 < \eta < \eta_2$$

the probability with which a disturbed solution at the same function of the density of distribution of the disturbances from the steady solution II will again lead to that solution, the analyst can use the values

$$\tilde{P}_I = \frac{1}{\eta_2 - \eta_1} \int\limits_{\eta_1}^{\eta_2} P_I(\eta)\, d\eta\,, \qquad \tilde{P}_{II} = \frac{1}{\eta_2 - \eta_1} \int\limits_{\eta_1}^{\eta_2} P_{II}(\eta)\, d\eta \tag{7.6, 1}$$

as a measure of comparison of the global resistance of the two steady solutions to disturbances from the steady state. In practical applications, the integrals of (7.6, 1) can be replaced by summations.

Expressions (7.6, 1) can also be used in evaluation of the resistance to disturbances of various systems. It is, of course, necessary to arrange the differential equations of motion of the systems being compared to the same, preferably the dimensionless form and to operate with the same function of the density of distribution $f(R)$ or $f(A, T_0)$ or possibly $f(Q_I, Q_{II})$ and the same type of disturbances when determining $P(\eta)$.

Optimization of a parameter of a system, for example, tuning, also belongs to the class of problems being discussed. The system is required to possess, in the highest possible degree, the desired property (for example, a very large amplitude of vibration of a particular mass) in combination with maximum $P(\eta)$, maximum probability that no random disturbance will cause transition to another steady state. If η is the parameter to be optimized, the analyst should know the dependence $P(\eta)$ as well as the dependence $A(\eta)$ which expresses the relation between the desired property of the system and the parameter η. The optimum value of η can be obtained from the conditions that the function

$$\Phi(\eta) = [P(\eta)]^k [A(\eta)]^n \qquad (7.6, 2)$$

should be at its maximum. Exponents k, n express the weights of the respective functions; by their means the analyst can lay the stress on the requirement of max $P(\eta)$ or max $A(\eta)$. If $A(\eta)$ is to be very low, the exponent n must be negative.

7.7.   Examples

Example I. Consider a one-degree-of-freedom system whose parameter undergoes a sudden change (Fig. 7.1, 2a). The examination is to show whether or not this effect results in a qualitative change of the steady solution (for example, the change of the resonant to the non-resonant solution). The mass m rests against a spring with a broken-line characteristic of the restoring force $F(y)$ and is excited by a harmonic force with an amplitude Q. The mass is also acted on by a constant force Q_0; the damping is linear viscous, with the coefficient of proportionality $\varkappa$.

The motion of the system is described by the differential equation

$$m\ddot{y} + \varkappa\dot{y} + F(y) = Q_0 + Q \cos \omega t \qquad (7.7, 1)$$

where

$$F(y) = \begin{cases} cy & \text{for } |y| \leqq a \ (a > 0) \\ cy + c_1(y - a) & \text{for } y > a \\ cy + c_1(y + a) & \text{for } y < -a . \end{cases}$$

Introducing the notation

$$\frac{c}{m} = \Omega^2 , \qquad \frac{\varkappa}{m\Omega} = D , \qquad \frac{Q_0}{Q} = q , \qquad y_0 = \frac{Q}{c} , \qquad \frac{y}{y_0} = x \qquad (7.7, 2)$$

and the transformation $\Omega t = \tau$, equation (7.7, 1) can be rearranged to the dimensionless form

$$x'' + Dx' + x[1 + \varphi(x)] = q + \cos \eta\tau \qquad (7.7, 3)$$

where

$$\eta = \frac{\omega}{\Omega} , \qquad \varphi(x) = \begin{cases} 0 & \text{for } |x| \leqq \dfrac{a}{y_0} , \\ \dfrac{c_1}{c}\left(1 - \dfrac{a}{y_0}\dfrac{1}{|x|}\right) & \text{for } |x| > \dfrac{a}{y_0} . \end{cases}$$

The solution was obtained for the following values of the parameters:

$$D = 0.05 , \qquad \frac{c_1}{c} = 2 , \qquad \frac{a}{y_0} = 10 , \qquad \eta = 1.15 . \qquad (7.7, 4)$$

The prestress was varied (by a sudden change) within the range

$$-2 \leqq q_{\mathrm{I}} \leqq 6 , \qquad -2 \leqq q_{\mathrm{II}} \leqq 6 \qquad (7.7, 5)$$

By way of preliminary information, Figs. 7.7, 1 to 7.7, 4 show the amplitude-frequency characteristics for various values of prestress q. It is readily seen that

$$[\max (x)]_q = -[\min (x)]_{-q} \tag{7.7, 6}$$

The interval of η in which two stable steady solutions (one resonant, the other non-resonant) exist is found to grow narrower with increasing q. In the range of q considered here the value $\eta = 1.15$ always lies in the region of the two-valued stable steady solution.

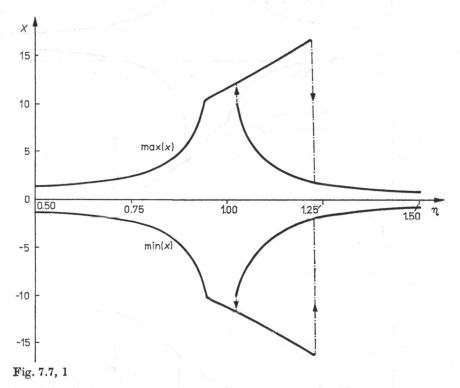

Fig. 7.7, 1

Additional information may be obtained by consulting Figs. 7.7, 5 and 7.7, 6, which show the diagrams of the domains of attraction for various values of q and $\tau = 0$. The region of the initial conditions which lead to the resonant solution can be seen to grow larger with increasing (in negative as well as positive sense) prestress.

The next figures show the diagrams of the function $p(q_I, q_{II})$ indicating the probability that the resonant (Fig. 7.7, 7) and the non-resonant (Fig. 7.7, 8) steady vibration will lead again to resonant (non-resonant) steady solution in consequence of a step change of the prestress (from q_I to q_{II}) at any time. Since $F(-y) = -F(y)$, i.e. the nonlinear characteristic is symmetric,

$$p(-q_I, -q_{II}) = p(q_I, q_{II}) . \tag{7.7, 7}$$

Using the simplest possible function of the probability density of occurrence

$$f(q_I, q_{II}) = \begin{cases} \dfrac{1}{(q_2 - q_1)^2} & \text{for} \quad q_I, q_{II} \text{ inside the interval } (q_1, q_2) , \\ 0 & \text{for} \quad q_I, q_{II} \text{ outside the interval } (q_1, q_2) , \end{cases} \tag{7.7, 8}$$

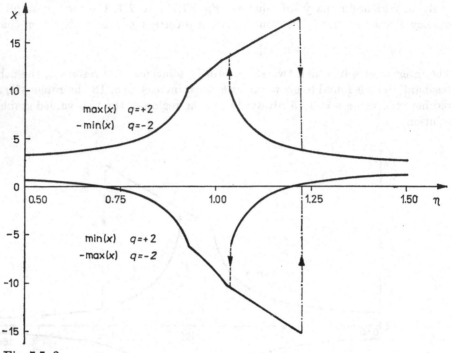

Fig. 7.7, 2

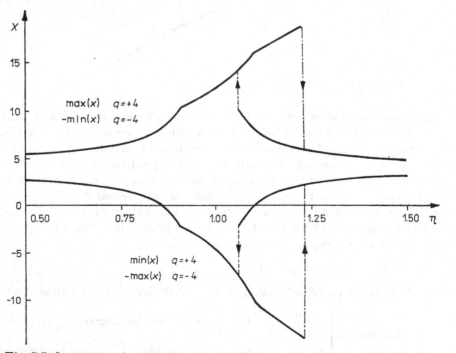

Fig. 7.7, 3

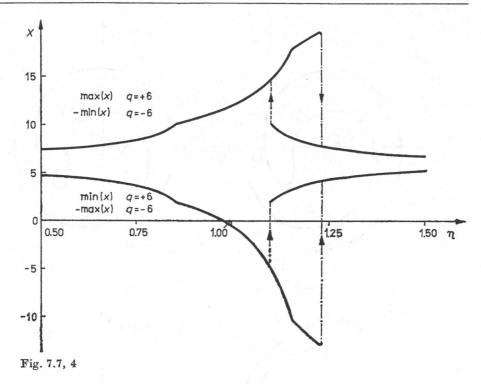

Fig. 7.7, 4

that is, assigning the same weight to any step change for $q_1 \leqq q_I \leqq q_2$, $q_1 \leqq q_{II} \leqq q_2$, we obtain the following values for the probability P:

$$P_{\mathrm{res} \to \mathrm{res}} = 0.858 \,, \qquad P_{\mathrm{nonres} \to \mathrm{nonres}} = 0.965 \,.$$

The conclusion which can be drawn on the basis of these results is as follows:

In systems described by (7.7, 3) and having parameters as indicated above, the probability of step changes of prestress which satisfy the assumptions made in Section 7.1, of changing resonant vibration to non-resonant and vice versa, is very low. Consequently, such systems are highly resistant to step changes of prestress. In the event that a qualitative change of steady vibration does occur, it is very difficult to return the system to the original state just by application of additional changes of prestress.

Example II. This example shows the effect of a change of a parameter applied for a definite time interval only (Fig. 7.1, 2b). The system to be analyzed is a very simple model of that in which relative dry friction is the cause of self-excitation. A rigid body of weight m rests on a continuous conveyor belt moving at velocity v (Fig. 7.7, 9). The body is bound by a linear spring of rigidity c and its motion is controlled by means of an absolute damper with damping of the dry friction character (idealized by Coulomb friction). Since the dry-friction absolute damper has a stabilizing effect, a stable equilibrium position always exists. (Rather than by a single value of the deflection coordinate, this position is described by an interval of values — for details see a monograph by TONDL (1970b)). However, steady locally stable self-excited vibration is apt to exist at a certain velocity v of the belt motion and relative dry friction between the body of mass m and the belt; in the phase plane it is

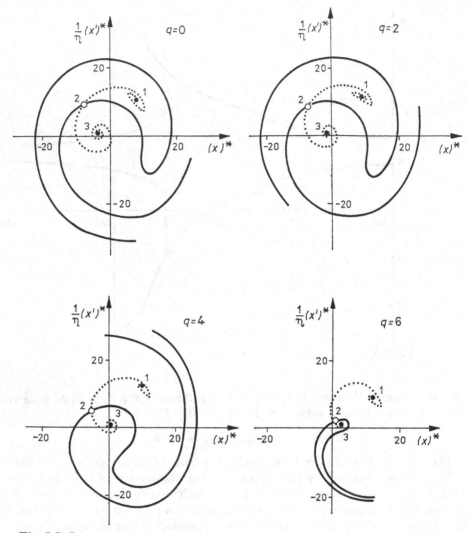

Fig. 7.7, 5

represented by a stable limit cycle. In the case being considered there exists, in addition, an unstable limit cycle which surrounds the steady equilibrium position and forms the separatrix; the separatrix marks out the region of the initial conditions which lead to the equilibrium position from that of the initial conditions which lead to steady self-excited vibration.

Analyze now the effect of a sudden disturbance characterized by a step change of a parameter, which is in action for a certain time. Examine the effect of a sudden change of the magnitude of the dry friction force at the point of contact of the body resting on the moving belt. This change can be caused, for example, by minute particles of a material which landed on the belt, was dragged by the belt's motion between the contact surfaces of the body mass m, remained there for some time and was removed from the contact surfaces by the belt's motion.

To simplify the analysis assume that only the magnitude of the friction force and not the character of the dry friction undergoes a change in this process. The change

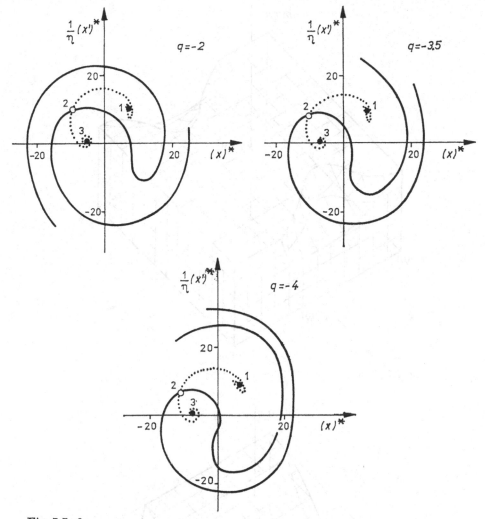

Fig. 7.7, 6

can be expressed by means of a coefficient, a multiple of the original value. The equation of motion (not considering the step change of the friction force) is

$$m\ddot{x} - mgf\varphi(v - \dot{x}) + \vartheta \, \mathrm{sgn} \, \dot{x} = 0 \tag{7.7, 9}$$

where g is the acceleration of gravity, f the coefficient of dry friction between body and belt, $\varphi(v - \dot{x})$ the function of the dependence of the relative friction, and ϑ the coefficient of Coulomb's friction of the absolute damper.

Introducing the notation

$$V = \frac{v}{\Omega}, \quad \Omega = \sqrt{\frac{c}{m}}, \quad \delta = \frac{\vartheta}{c}$$

and the time transformation $\Omega t = \tau$, we obtain the following equation

$$x'' + x - F(V - x') + \delta \, \mathrm{sgn} \, x' = 0 \tag{7.7, 10}$$

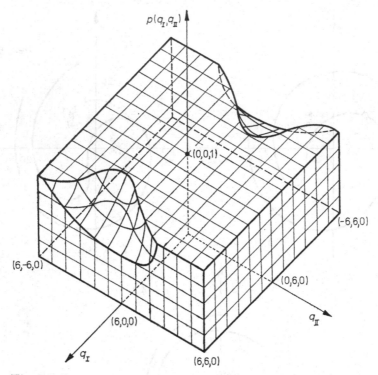

Fig. 7.7, 7

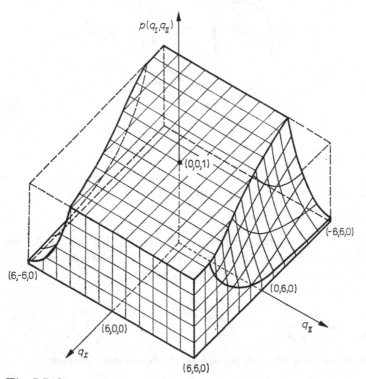

Fig. 7.7, 8

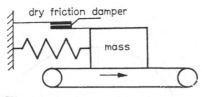

Fig. 7.7, 9

The function $F(V - x')$ is assumed to have the simple form (the decreasing slope of the dependence is important for self-excitation effects)

$$F(V - x') = f_0 \operatorname{sgn}(V - x') - f_1(V - x') \qquad (7.7, 11)$$

where $f_0 = 1$, $f_1 = 0.1$ for the case being considered.

Taking the step change of the friction force into consideration, the equation of motion becomes

$$x'' + x - [1 + K(\tau)] F(V - x') + \delta \operatorname{sgn} x' = 0 \qquad (7.7, 12)$$

where

$$K(\tau) = \begin{cases} 0 & \text{for} \quad \tau < \tau_1, \quad \tau > \tau_2 \\ K = \text{const} & \text{for} \quad \tau_1 \leqq \tau \leqq \tau_2. \end{cases}$$

Assume that K can take values only in the interval $-1 \leqq K \leqq 1$ and that $F(V) \gg \delta$; consequently, for $K = 0$, vibration represented by a stable cycle in the phase plane exists in addition to the stable equilibrium position.

Examine now the effect of a step change of the relative dry friction on the equilibrium position. For purposes of information, Fig. 7.7, 10 shows the records of the limit cycles in the phase plane for various values of δ and $V = 4$. The larger (stable) limit cycle is seen to surround the unstable one.

Fig. 7.7, 11 shows the boundaries of the domain of attraction for the non-oscillatory solution (equilibrium position) in the case of $V = 4$ and $\delta = 0.02$ and 0.03 in a diagram having coordinates K, T_0/T where $T = 2$ is the period of self-excited vibration. The boundary $K(T_0/T)$ can be seen to have local minima in the neighbourhood of $T_0/T = n + 1/2$ and maxima in the neighbourhood of $T_0/T = n$ $(n = 0, 1, 2, ...)$. The resistance to disturbances of the type specified increases very rapidly with increasing δ. In the often-quoted monograph by TONDL (1973a), which discusses this and other cases in greater detail, the diagrams shown in Fig. 7.7, 11 are additionally processed by specifying the density of the probability of occurrence of disturbances in the form of rectangular areas; points lying inside these areas represent the values of the disturbance parameters having the same probability of occurrence.

Assume now that prior to a change of a parameter, the system vibrated in steady self-excited vibration. The results obtained in terms of the probability function $p(K, T_0/T)$ are shown in Fig. 7.7, 12 $(\delta = 0.2)$ and Fig. 7.7, 13 $(\delta = 0.3)$ $(V = 4$ in both cases). Unlike the non-oscillatory solution, the resistance of the steady self-excited vibration to disturbances decreases with increasing δ. For small values of δ, the probability function $p(K, T_0/T)$ was equal to one in the whole range of K, T_0/T. As the results suggest, a disturbance caused by a sudden increase of the force of relative dry friction lasting a certain time cannot result in transition to the steady equilibrium position. Such transition can be achieved for larger values of δ by application of a disturbance caused by a sudden decrease of the force of relative dry friction lasting

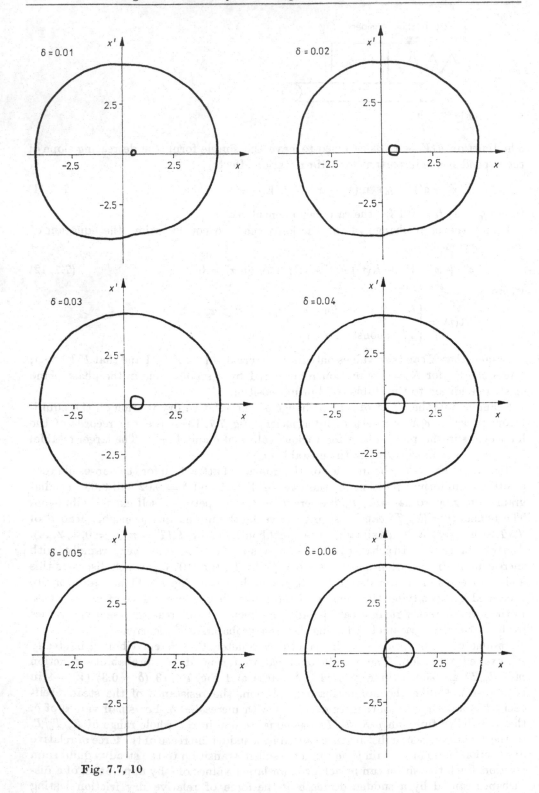

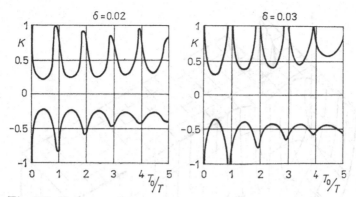

Fig. 7.7, 11

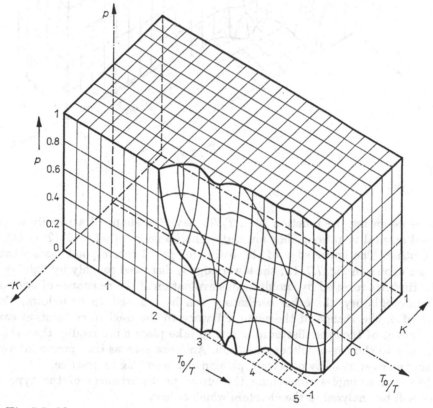

Fig. 7.7, 12

a certain time. For larger values of δ and longer time of duration (larger T_0/T) a comparatively small decrease of the relative dry friction can result in transition to the steady equilibrium position. The effect of the duration of disturbance, T_0, is also of interest. Compared with the case when the system is in equilibrium position prior to the application of the disturbance, it is less marked and decreases comparatively rapidly with increasing T/T_0; consequently, for larger values of δ, the function $p(K, T_0/T)$ after three to five periods ($T_0/T = 3$ to 5) becomes largely independent of T_0/T and takes the value of unity or zero. For $p = 1$ and $p = 0$ and increasing T_0/T, the edges of the

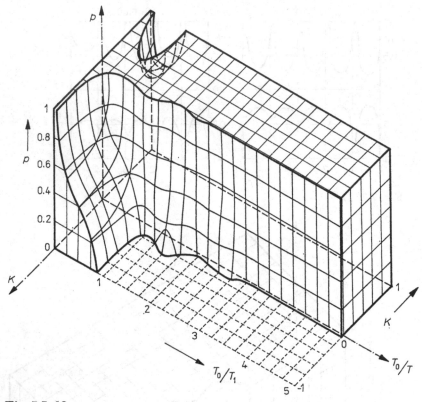

Fig. 7.7, 13

body representing the function $p(K, T_0/T)$ have the shape of a rapidly attenuating damped natural vibration. They are shifted half a period, that is, $T_0/T = 1/2$, against one another. The upper edge of the body (for $p = 1$) corresponds somewhat to the diagram shown in Fig. 7.7, 11; the lower edge is displaced roughly by $T_0/T = 1/2$.

The findings revealed by simultaneous evaluation of the resistance of two solutions, one non-oscillatory, the other oscillatory can be summed up as follows: For small values of δ, the change of the non-oscillatory to the oscillatory solution caused by disturbances of the type discussed is apt to take place more readily than that of the oscillatory to the non-oscillatory solution. Analyses such as that presented above can explain the irreversibility of certain phenomena occurring in practice.

Additional examples concerning the effect of disturbances of the type of force pulses will be analyzed in the chapters which follow.

8. Analysis of some excited systems

8.1. Duffing system with a softening characteristic

Consider a Duffing system which differs from the conventional model (Section 4.4) by a negative coefficient of the cubic term. An analysis of this system is interesting from physical as well as methodological aspects. It will be shown later in the section that some methods of analysis fail to provide a complete answer to the problems of stability (both in the small and in the large) of the steady solutions. This fact was pointed out in two papers by TONDL (1976 c).

The system being examined is described by the equation (written in dimensionless form and including the shift of the time origin)

$$y'' + \varkappa y' + y - \gamma y^3 = \cos(\eta \tau + \varphi) \tag{8.1, 1}$$

where $\varkappa$, γ are positive constants, and φ is the phase shift angle. If the steady solution is sought in the form

$$y = A \cos \eta \tau \tag{8.1, 2}$$

use of the harmonic balance method leads to the following equations for determining A and φ:

$$\left.\begin{array}{c} A(1 - \tfrac{3}{4}\gamma A^2 - \eta^2) = \cos \varphi , \\ A\varkappa\eta = \sin \varphi . \end{array}\right\} \tag{8.1, 3}$$

The dependence $A(\eta)$ is obtained by help of the inverse function $\eta(A)$ which is readily determined from the equation

$$\eta^4 - 2\left(1 - \frac{3}{4}\gamma A^2 - \frac{1}{2}\varkappa^2\right)\eta^2 + \left(1 - \frac{3}{4}\gamma A^2\right)^2 - \frac{1}{A^2} = 0 . \tag{8.1, 4}$$

The dependence $\varphi(\eta)$ is described by the equation

$$\tan \varphi = \varkappa\eta(1 - \tfrac{3}{4}\gamma A^2 - \eta^2)^{-1} . \tag{8.1, 5}$$

The backbone curve is defined by the equation

$$A_{\mathrm{S}} = \frac{2}{\sqrt{3\gamma}}(1 - \eta^2)^{1/2} ; \tag{8.1, 6}$$

this equation implies the necessity of satisfying the inequality $\eta < 1$. The limit envelope has the form of the rectangular hyperbola

$$A_{\mathrm{L}} = \frac{1}{\varkappa\eta}. \tag{8.1, 7}$$

The following three cases can occur:

(a) The limit envelope and the backbone curve intersect at two points.
(b) The limit envelope touches the backbone curve.
(c) The limit envelope and the backbone curve do not intersect.

The case (b) constitutes the boundary between the cases (a) and (c). The curve $A(\eta)$ of case (a) consists of two branches. The lower branch has the usual form of resonance curve of a system with a softening characteristic for which it is typical that A at any point is smaller than a particular value. The lower branch is chatacterized by that A at any point is always greater than this particular value (Fig. 8.1, 1).

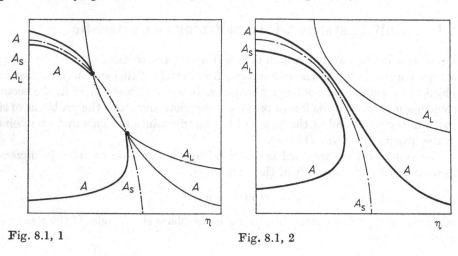

Fig. 8.1, 1   Fig. 8.1, 2

The curve $A(\eta)$ of case (c) also consists of two branches, one lying to the right, the other to the left of the backbone curve (Fig. 8.1, 2).

The same curves are obtained when using the van der Pol or the Krylov and Bogoljubov method. Stability of the solution is established by application of the rule of vertical tangents (see Chapter 2). This procedure, however, fails to provide a comprehensive picture of the solution stability.

If $y = A \cos \eta\tau + x$, where x is a variation of the variable y, is substituted for y in (8.1, 1) the equation in variations takes the form

$$x'' + \varkappa x' + x - \tfrac{3}{2}\,\gamma A^2\,(1 + \cos 2\eta\tau)\,x = 0 \,. \tag{8.1, 8}$$

This is the Mathieu equation. The solution on the boundary of the region of first-order instability may be approximated by

$$x = u \cos \eta\tau + v \sin \eta\tau \,. \tag{8.1, 9}$$

Substituting (8.1, 9) in (8.1, 8) and comparing the coefficients of $\cos \eta\tau$ and $\sin \eta\tau$ leads to a system of homogeneous equations in u and v. As the condition of non-triviality of the solution implies,

$$A = \frac{2}{3\sqrt{\gamma}}\,\{2(1 - \eta^2) \mp [(1 - \eta^2)^2 - 3\varkappa^2\eta^2]^{1/2}\}^{1/2} \,. \tag{8.1, 10}$$

This establishes the boundary of the region of first-order instability of the steady solution (8.1, 2), which corresponds to the approximate boundary of the region of first-order instability of (8.1, 8). It should, of course, be noted that other instability regions are also likely to play a role. In the case being examined a major influence is exercised by the region of zero-order instability (the solution on the boundary of which is the Mathieu equation of zero order) which comes into consideration only for negative values of the constant term of the coefficient of x in (8.1, 8). The boundary can be approximately determined from the condition

$$1 - \tfrac{3}{2}\gamma A^2 \geqq 0 \,. \tag{8.1, 11}$$

In the chart of the regions of instability of the Mathieu equation this approximation corresponds to the replacement of the boundary of the region of zero-order instability by a vertical straight line passing through the origin. The same condition is obtained when checking the stability of the solution of (8.1, 8) for the average values of the coefficients. The characteristic equation thus arrived at takes the form

$$\lambda^2 + \varkappa\lambda + 1 - \tfrac{3}{2}\gamma A^2 = 0 \,.$$

As subsequent examples will show, the boundary implied by (8.1, 10) yields the same result as the rule of vertical tangents. However, the boundary implied by the condition (8.1, 11) cannot be obtained when using the van der Pol or the Krylov and Bogoljubov method. Figs. 8.1, 3 to 8.1, 5 show A as a function of η for $\gamma = 0.01$ and $\varkappa = 0.2$ (Fig. 8.1, 3), $\varkappa = 0.175$ (Fig. 8.1, 4) and $\varkappa = 0.15$ (Fig. 8.1, 5). The stable solutions are drawn in heavy solid lines, the unstable ones in dashed lines, and the various regions of instability (obtained using (8.1, 10) and (8.1, 11) are shown crosshatched or dotted. Fig. 8.1, 3 also shows the backbone curve (the dot-and-dash line) and the limit envelope (the light solid line). The arrows in Figs. 8.1, 4 and 8.1, 5 indicate the transient (jump) phenomenon arising as the excitation frequency is slowly increased or decreased. As Fig. 8.1, 5 reveals, if the excitation frequency is decreased

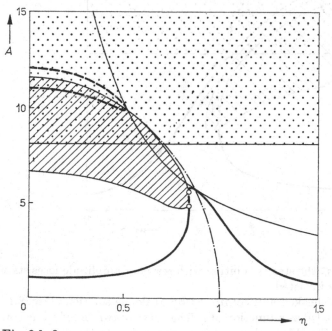

Fig. 8.1, 3

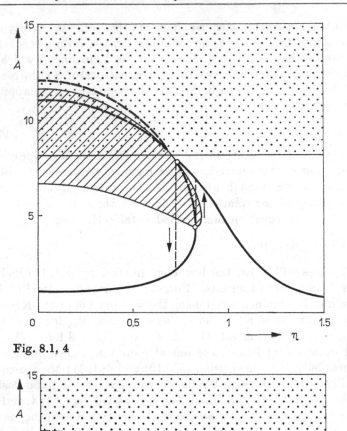

Fig. 8.1, 4

Fig. 8.1, 5

at a definite value of η, the steady solution with resonant amplitude loses stability and divergent vibration is initiated.

These results were checked and confirmed by analogue solutions. Fig. 8.1, 6 shows the extreme deflection $[y]$ as a function of η. The curve drawn in light line corresponds to the case shown in Fig. 8.1, 4, that shown in heavy line, to the case in Fig. 8.1, 5.

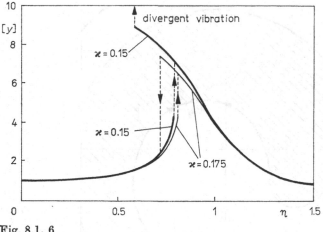

Fig. 8.1, 6

In the investigation of stability in the large which follows the approach using the van der Pol (or the Krylov and Bogoljubov) method will first be shown to yield incorrect results in the case being examined. Since they are identical with (4.4, 2) except for the minus sign of the coefficient γ, the corresponding transformed differential equations will not be repeated here. The results obtained in investigations of the domains of attraction fot $\gamma = 0.01$ and $\varkappa = 0.175$ are shown in Fig. 8.1, 7 ($\eta = 0.5$) and Fig. 8.1, 8 ($\eta = 0.8$). In the first case, the resonant solution corresponds to a point of the upper branch, in the second, to a point of the lower branch. As a stability analysis (see condition (8.1, 11)) and an analogue solution reveal, the resonant solution of the first case is unstable rather than stable as implied by Fig. 8.1, 7.

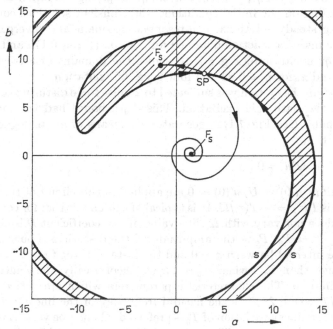

Fig. 8.1, 7

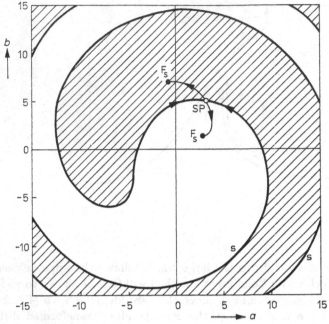

Fig. 8.1, 8

The domains of attraction expressed in terms of the coordinates of the initial conditions were obtained by a direct analogue solution of (8.1, 1) in the course of which the initial conditions $y(0)$, $y'(0)$ were varied in the interval $(-15, 15)$ by means of the TV scanning method. In some cases the equation was solved for the excitation $\cos \eta\tau$ as well as for the alternatives $-\sin \eta\tau$, $-\cos \eta\tau$ and $\sin \eta\tau$, applying the initial conditions at time $\tau = 0$. Depending on the initial conditions, the transient solutions could converge to the following steady solutions: non-resonant, resonant and divergent. Fig. 8.1, 9 shows the domains of attraction obtained for $\gamma = 0.01$, $\varkappa = 0.175$ and various values of η (the $\cos \eta\tau$ excitation). Fig. 8.1, 10 shows the domains of attraction for identical values of γ and $\varkappa$, for $\eta = 0.82$ and four types of excitation.

In another examination the system was subjected to a pulse force disturbance applied in combination with the harmonic excitation. This disturbance had the form of a decaying oscillatory pulse (see Fig. 7.1, 1) generated by means of an analogue circuit described by the equation

$$u'' + 2D\Omega u' + \Omega^2 u = 0 \,. \tag{8.1, 12}$$

If the initial conditions $u(0) = U$, $u'(0) =: 0$ are applied to this circuit at time $\tau = 0$, the disturbance pulse is $P(\tau) = -u'(\tau)/\Omega$. It is typical of this circuit that for constant U the value of $P(\tau)_{max}$ does not vary with Ω. The value of the coefficient D is chosen so as to make $P_{max} = \frac{1}{2} U = P_0$ (P_0 is the amplitude of the disturbance pulse). Owing to the effect of D, the interval between $\tau = 0$ and the instant of the first zero crossing of $P(\tau)$ is slightly longer than $T_0/2$ where $T_0 = 2\pi/\Omega$. Theoretically, the length of this pulse duration is unlimited. The characteristic parameters which are varied, are P_0 ($P_0 = P_{max}$) and T_0/T where $2\pi/T$ is the natural frequency of the linearized system without damping. Since a linear variation of T_0 is preferred, Ω must be varied according to a rectangular hyperbola relation. The value of T_0/T was varied in the interval

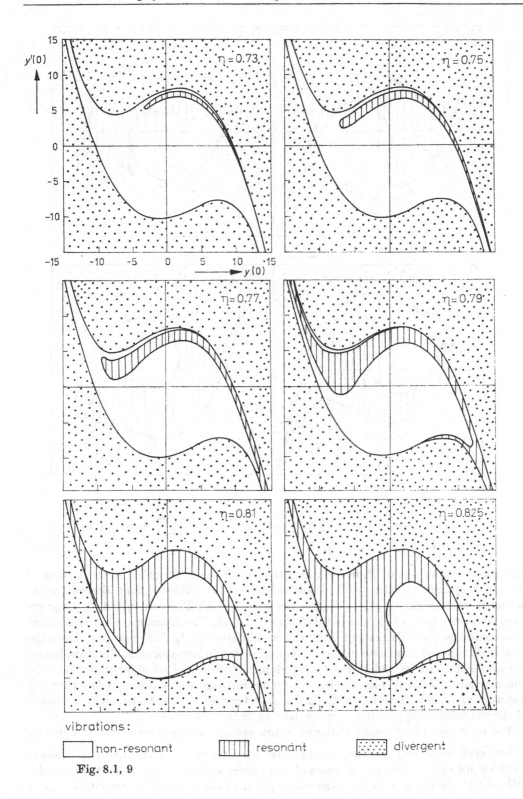

vibrations:

| | non-resonant | | | resonant | | | divergent |

Fig. 8.1, 9

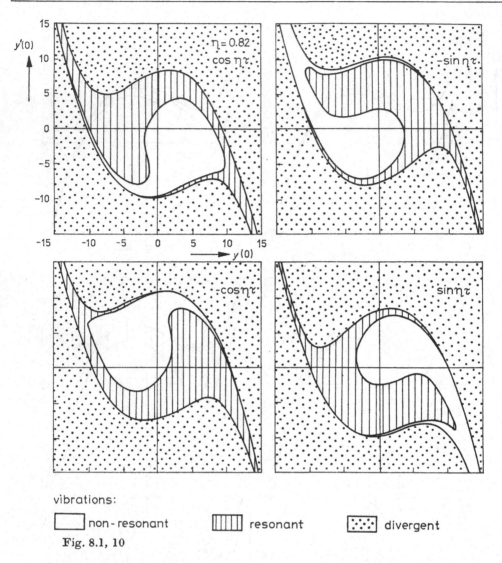

vibrations:

[ ] non-resonant   [||||||] resonant   [:·:·:] divergent

Fig. 8.1, 10

(0.4, 8), and the value of P_0 in the interval (0,5). To show the effect of the instant of the pulse application, the problem was solved for the following four alternatives of harmonic excitation: $\cos \eta t$, $-\sin \eta t$, $-\cos \eta t$ and $\sin \eta t$. The corresponding diagrams were obtained for $\gamma = 0.01$, $\varkappa = 0.175$ and $\eta = 0.75$. The domains of attraction are represented in plane diagrams having the coordinates P_0, T_0/T. Fig. 8.1, 11 shows the results obtained for the case of the system vibrating in non-resonant vibration before the application of the pulse. Fig. 8.1, 12 depicts the case of resonant vibration. The non-resonant vibration is seen to be much more resistant to disturbing pulses than the resonant vibration. The effect of the phase shift between the excitation and the instant of the disturbance application is more distinct in the latter case.

The most important results obtained in this section can be summarized as follows:

For systems having a softening characteristic of the restoring force the classical methods for establishing the domains of attraction, which use approximate first-order differential equations based on the assumption of slowly varying amplitudes (or a

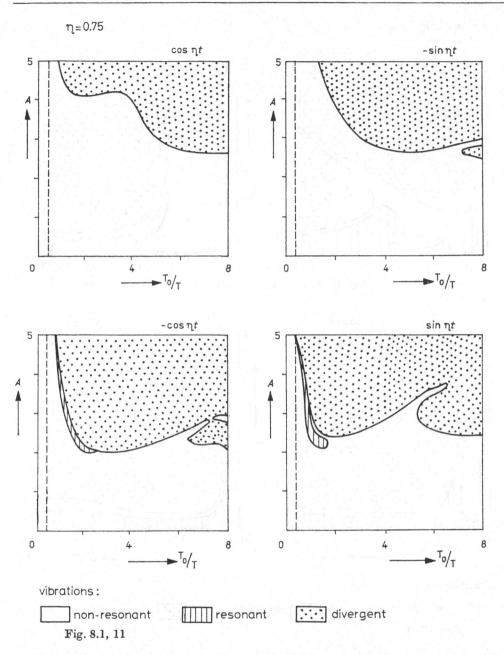

$\eta = 0.75$

vibrations:

▢ non-resonant   ▥ resonant   ⣿ divergent

Fig. 8.1, 11

slowly varying amplitude and phase), do not yield results that are quantitatively or even qualitatively correct. Neither do they disclose the existence of divergent vibration.

The domains of attraction expressed in terms of the coordinates of the initial conditions differ substantially from those obtained in investigations relating to the resistance of a particular steady locally stable solution to disturbances of a specified type.

Three domains of attraction, that is, that of resonant, of non-resonant and of divergent vibration, can be obtained when a decaying oscillatory pulse is applied to the Duffing system having a softening characteristic. Under its effect the steady resonant

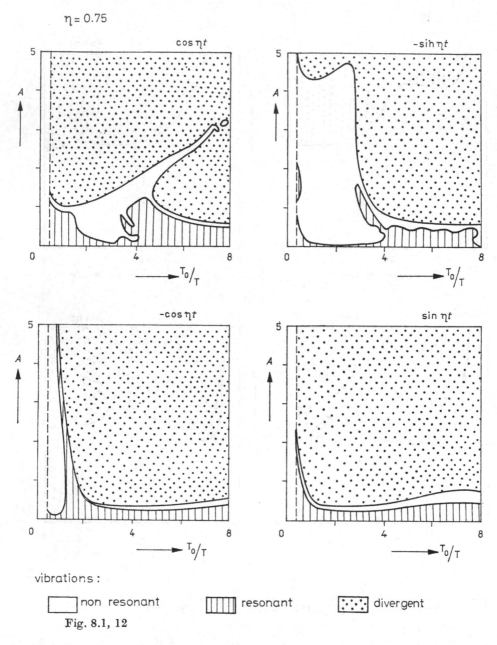

$\eta = 0.75$

vibrations:

⬜ non resonant   ▥ resonant   ⦂⦂⦂ divergent

Fig. 8.1, 12

vibration is apt to change more readily to the divergent than to the non-resonant vibration.

8.2.   Some special cases of kinematic (inertial) excitation

This section is devoted to an analysis of systems I and II considered in Section 1.2 (Fig. 1.2, 6). Equation (1.2, 5) which describes the motion of system I is not only non-homogeneous but — as in the case of parametric excitation — also has periodically variable coefficients. It is woth noting that parametric excitation can be linear as well as non-linear. Introducing the dimensionless deflection $y = x/a$ and carrying out the time transformation

$$\Omega t = \tau \tag{8.2, 1}$$

where $\Omega^2 = 2k/m$, (1.2, 5) takes the dimensionless form

$$y'' + 2\delta y' + y + \varepsilon_1 y^3 + \varepsilon_2 (y - \cos \eta\tau)^3 = \tfrac{1}{2} \cos \eta\tau - \delta\eta \sin \eta\delta \tag{8.2, 2}$$

where

$$\delta = \frac{\varkappa}{m\Omega}, \qquad \eta = \frac{\omega}{\Omega}, \qquad \varepsilon_1 = \frac{\gamma_1 a^2}{m\Omega^2}, \qquad \varepsilon_2 = \frac{\gamma_2 a^2}{m\Omega^2}.$$

The following alternatives are studied:

(a) $\varepsilon_1 = \varepsilon_2 = \varepsilon$;

(b) $\varepsilon_1 = -\varepsilon_2 = -\varepsilon$; $\left.\vphantom{\begin{matrix}1\\1\\1\end{matrix}}\right\}$ combinations of a softening and a hardening spring

(c) $\varepsilon_1 = -\tfrac{1}{2}\varepsilon_2 = -\varepsilon$.

The fundamental analysis is made by means of the harmonic balance method; to facilitate the formal calculation, the time shift Φ/η is introduced where Φ is the phase shift between response and excitation.

For alternative (a), (8.2, 2) becomes

$$y'' + 2\delta y' + y + \varepsilon\{y^3 + [y - \cos(\eta\tau + \Phi)]^3\}$$
$$= \tfrac{1}{2} \cos(\eta\tau + \Phi) - \delta\eta \sin(\eta\tau + \Phi); \tag{8.2, 3}$$

the stationary solution of (8.2, 3) can be approximated by

$$y = A \cos \eta\tau . \tag{8.2, 4}$$

Application of the harmonic balance method leads to the following equations for obtaining A and Φ:

$$A[1 - \eta^2 + \tfrac{3}{2}\varepsilon(1 + A^2) + \tfrac{3}{4}\varepsilon(\cos 2\Phi - 3A \cos \Phi)]$$
$$= (\tfrac{1}{2} + \tfrac{3}{4}\varepsilon) \cos \Phi - \delta\eta \sin \Phi , \tag{8.2, 5}$$

$$A[2\delta\eta - \tfrac{3}{4}\varepsilon(A \sin\Phi - \sin 2\Phi)] = (\tfrac{1}{2} + \tfrac{3}{4}\varepsilon) \sin \Phi + \delta\eta \cos \Phi . \tag{8.2, 6}$$

In the calculation, (8.2, 5) and (8.2, 6) are used for determining the function $\eta(\Phi)$ for various values of A; the points of intersection of the curves of the two sets yield the values of η and Φ corresponding to the gradually varied A. Equations (8.2, 5) and (8.2, 6) rearranged for the purpose of this calculation have the form

$$\eta^2 - \frac{\delta \sin \Phi}{A}\eta + \frac{\left(\dfrac{1}{2} + \dfrac{3}{4}\varepsilon\right) \cos \Phi}{A} - 1 - \frac{3}{4}\varepsilon\,[2(1 + A^2) + \cos 2\Phi$$
$$- 3A \cos \Phi] = 0 , \tag{8.2, 7}$$

$$\eta = \frac{\sin \Phi}{\delta(2A - \cos \Phi)}\left[\frac{1}{2} + \frac{3}{4}\varepsilon(1 + A^2 - 2A \cos \Phi)\right]. \tag{8.2, 8}$$

Fig. 8.2, 1 shows the calculated relation between A and η for $\delta = 0.05$ and several values of ε; Fig. 8.2, 2 shows the corresponding functions $\Phi(\eta)$. It can be seen that the results obtained for alternative (a) are not qualitatively different from those established for the model shown in Fig. 1.2, 5.

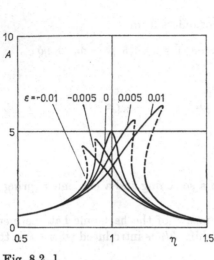

Fig. 8.2, 1   Fig. 8.2, 2

Different results, however, are obtained for alternative (b) (i.e. for $\varepsilon_1 = -\varepsilon_2 = -\varepsilon$). For this alternative the rearranged equations of motion takes the form

$$y'' + 2\delta y' + \varepsilon\{3y^2 \cos (\eta\tau + \Phi) + \tfrac{3}{2} y[1 + \cos 2(\eta\tau + \Phi)]\}$$
$$= (\tfrac{1}{2} + \tfrac{3}{4} \varepsilon) \cos (\eta\tau + \Phi) - \delta\eta \sin (\eta\tau + \Phi) + \tfrac{1}{4} \varepsilon \cos 3(\eta\tau + \Phi). \quad (8.2, 9)$$

Proceeding as before, the equations for obtaining A and Φ become

$$\eta^2 - \frac{\delta \sin \Phi}{A} \eta + \frac{\left(\dfrac{1}{2} + \dfrac{3}{4} \varepsilon\right) \cos \Phi}{A} - 1 - \frac{3}{4} \varepsilon(2 + \cos 2\Phi - 3A \cos \Phi) = 0,$$
$$(8.2, 10)$$

$$\eta = \frac{\sin \Phi}{\delta(2A - \cos \Phi)} \left[\frac{1}{2} + \frac{3}{4}\varepsilon(1 + A^2 - 2A \cos \Phi)\right]. \quad (8.2, 11)$$

The curves $A(\eta)$ and $\Phi(\eta)$ for $\delta = 0.05$ are shown in Figs. 8.2, 3 and 8.2, 4. The unstable solutions are drawn in dashed lines assuming the applicability of the rule of vertical tangents.

To illustrate the effect of ε, Fig. 8.2, 5 shows the axonometric view of the response curve $A(\eta)$ (without indicating the stability of the solution). The assumption concerning the solution stability will be checked for correctness in a later analysis.

For the case of $\varepsilon > 0$, that is, excitation via a hardening spring: they differ qualitatively from both those of (a). As will be shown later, there exists no stationary solution for larger values of $\varepsilon > 0$ in a certain range of $\eta - 1$, and the transient solutions represent divergent vibration for any initial conditions. For smaller values of $\varepsilon > 0$, there exist two domains of the initial conditions: one for which all solutions converge to a stable stationary solution with a finite vibration amplitude, and another (lying outside the former) which leads to divergent vibration.

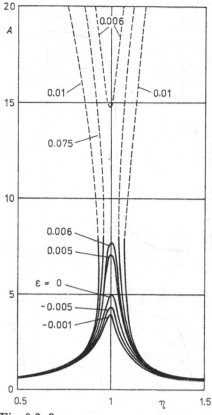

Fig. 8.2, 3

These results were confirmed both by an analysis using the van der Pol method, and analogue solutions. The solution to (8.2, 2) is now sought in the form

$$y = u \cos \eta \tau + v \sin \eta \tau \tag{8.2, 12}$$

where u and v are slowly varying functions of time for a transient solution. Application of the well-known procedure leads to the following equations

$$
\left.
\begin{aligned}
u' &= \left(\frac{1}{2\eta}\right)\left\{-\delta\eta(2u - 1) + \left[1 - \eta^2 - \frac{3}{4}\varepsilon(2u - 1)\right]v\right\}, \\
v' &= \left(\frac{1}{2\eta}\right)\left\{-2\delta\eta v + \frac{1}{2} - (1 - \eta^2)\,u + \frac{3}{4}\varepsilon[u(u - 1) + 1 + v^2]\right\}.
\end{aligned}
\right\} \tag{8.2, 13}
$$

The singular points of the system represent stationary solutions which are identical with those obtained by means of the harmonic balance method recalling that the relations

$$u = A \cos \Phi\,, \qquad v = A \sin \Phi \tag{8.2, 14}$$

apply. The roots of the equation

$$
\begin{vmatrix}
\dfrac{\partial U}{\partial u} - \lambda & \dfrac{\partial U}{\partial v} \\[2ex]
\dfrac{\partial V}{\partial u} & \dfrac{\partial V}{\partial v} - \lambda
\end{vmatrix} = 0
\tag{8.2, 15}
$$

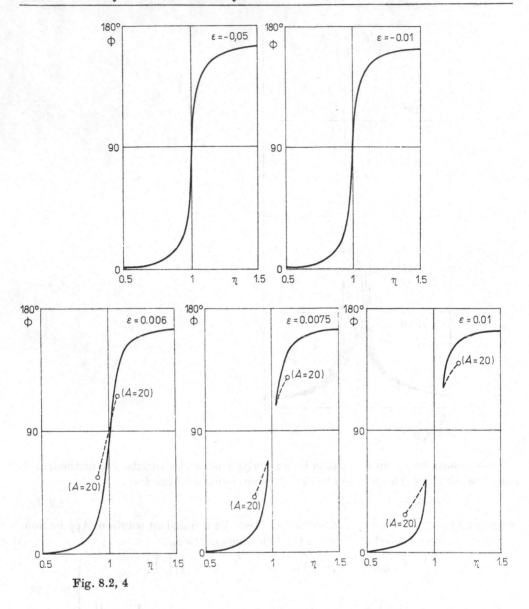

Fig. 8.2, 4

where $U(u, v)$ and $V(u, v)$ are the right-hand sides of (8.2, 13) for the arguments of the coordinates of the singular points, define the stability and type of a singular point. Substitution in (8.2, 15) results in

$$[2\eta(\delta + \lambda)]^2 - (\tfrac{3}{2}\,\varepsilon v)^2 + [1 - \eta^2 - \tfrac{3}{4}\,\varepsilon(2u - 1)]^2 = 0 \,.$$

As this equation implies, satisfying the inequality

$$[1 - \eta^2 - \tfrac{3}{4}\,\varepsilon(2u - 1)]^2 - (\tfrac{3}{2}\,\varepsilon v)^2 + (2\delta\eta)^2 > 0 \qquad (8.2, 16)$$

is the condititition of stability for a singular point (assuming that $\delta > 0$).

If the condition

$$[1 - \eta^2 - \tfrac{3}{4}\,\varepsilon(2u - 1)]^2 < (\tfrac{3}{2}\,\varepsilon v)^2 \qquad (8.2, 17)$$

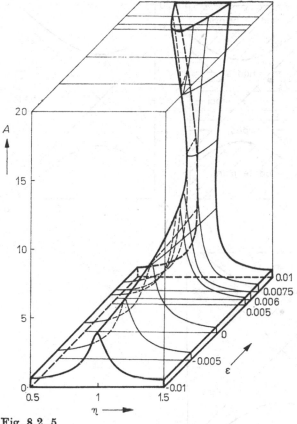

Fig. 8.2, 5

is satisfied, the singular point is a saddle or a node; if it is not satisfied, the singular point is a focus. Since, in the case of equality, condition (8.2, 16) is a boundary between saddles and nodes, the singular point — if conditions (8.2, 16) and (8.2, 17) are satisfied — is a stable node. If condition (8.2, 17) is not satisfied, the singular point is a stable focus because inequality (8.2, 16) is always satisfied. Fig. 8.2, 6 shows the boundaries of the domain of stability, the types and positions of the singular points in the (u, v)-plane for the case corresponding to Figs. 8.2,3 and 8.2, 4 for $\varepsilon = 0.0075$; (a) represents the case for $\eta = 0.95$, (b) that for $\eta = 1.05$, and (c) that for $\eta = 1$, $\varepsilon = 0.006$.

A solution of (8.2, 13) yields the phase portraits which provide information about the solution stability for any initial conditions, and, if several singular points exist, also the domains of attraction — in the case being discussed, the domain of attraction of the stable singular point and the domain of divergent vibration. Some examples of the phase portraits are shown in Fig. 8.2, 7: (a) to (c) are drawn for $\varepsilon = 0.0075$ and $\eta = 0.95$, 1 and 1.05; (d) for $\varepsilon = 0.006$ and $\eta = 1$ ($\delta = 0.05$ in all cases).

The separatrix (drawn in heavy full line) is formed by two trajectories entering the saddle point (SP) and divides the domain of attraction of the stable focus (F_S) or the stable node (N_S) from the domain of divergent vibration. Since only divergent vibration exists for the whole range of values of u and v corresponding to the case shown in Fig. 8.2, 7c, the earlier conclusions are fully confirmed.

Fig. 8.2, 8 shows the maximum deflection ($[y]$) as a function of frequency η for the case of $\delta = 0.05$ and $\varepsilon = 0.0075$ (in the vicinity of $\eta = 1$, the vibration becomes diver-

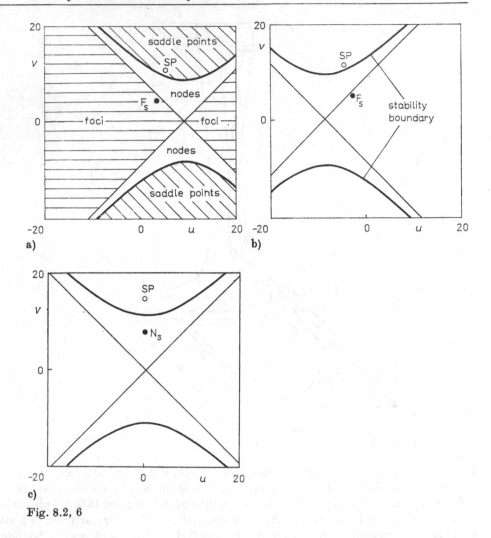

Fig. 8.2, 6

gent) and of $\varepsilon = 0.006$ obtained in an analogue solution of (8.2, 3). Fig. 8.2, 9 shows vibration records for the case of $\varepsilon = 0.0075$ and two different values of η. As the first record reveals, at $\eta = 1$ the vibration becomes divergent even for zero initial conditions. The other two records correspond to initial conditions close to the separatrix.

For alternative (c) ($\varepsilon_1 = -\varepsilon_2/2 = -\varepsilon$) the $A(\eta)$ curves corresponding to $\delta = 0.05$ and several values of ε are shown in Fig. 8.2, 10. As ε is increased, the maximum values of A grow larger at a fast rate, and after a certain positive value of ε has been exceeded, the resonance peak is no longer well defined. In fact, as A grows larger, the two branches of the $A(\eta)$ curve move ever further apart. Starting from a certain value of η, two stationary stable solutions, one resonant, the other non-resonant, exist for every $\eta > 1$.

Applying the procedure and notation adopted for system I, the equation of motion (1.2, 6) which describes system II can be converted (for the alternative $\varepsilon_1 = -\varepsilon_2 = -\varepsilon$) to the following dimensionless form:

$$y'' + 2\delta y' + y + \tfrac{3}{4}\,\varepsilon[(4y^2 + 1)\sin\varphi\,\sin\eta\tau - 2y\sin 2\varphi\,\sin 2\eta\tau]$$

$$= \cos\varphi\,(\cos\eta\tau - 2\delta\eta\,\sin\eta\tau) - \tfrac{1}{4}\,\varepsilon\,\sin 3\varphi\,\sin 3\eta\tau. \qquad (8.2, 18)$$

The case when $\varphi = 0$ (the motions of the suspension points of the two springs are in phase) will not be discussed here because it can readily be dealt with by introducing the relative deflection or, alternatively, by solving (8.2, 18) which, for $\varphi = 0$, turns out to be a simple, non-linear non-homogeneous differential equation. If the motions of the suspension points are in opposition ($\varphi = \pi/2$, $\sin \varphi = 1$, $\cos \varphi = 0$), there exist only a very small external excitation with frequencies η and 3η and a linear and a non-linear parametric excitation with frequency η; of these, the latter has the greatest effect.

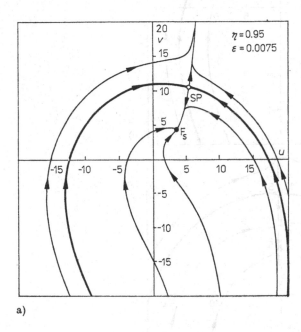

a)

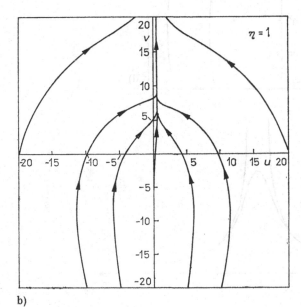

b)

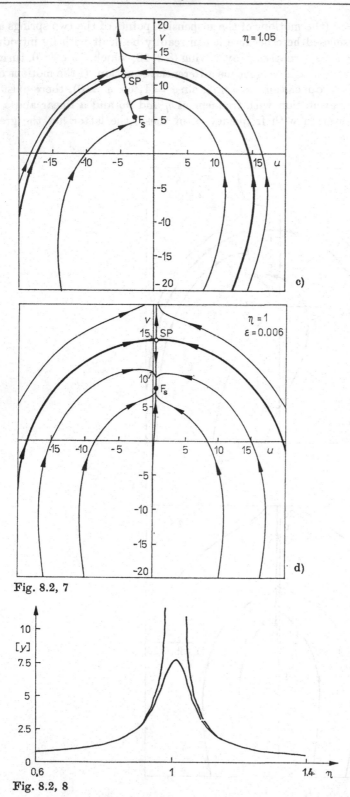

Fig. 8.2, 7

Fig. 8.2, 8

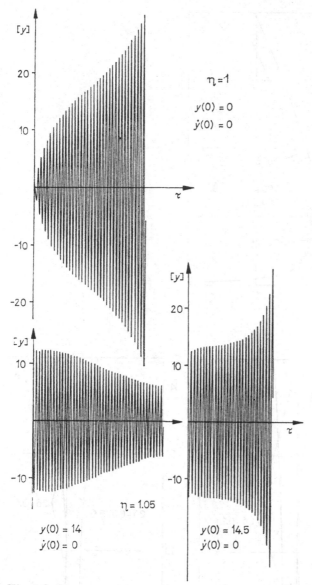

Fig. 8.2, 9

Applying the procedure adopted for system I, the equations for obtaining A and Φ turn out to be

$$\eta^2 - \frac{2\delta C \sin \Phi}{2A}\eta + \frac{C \cos \Phi}{A} - 1 - \frac{3}{4}\varepsilon \sin \Phi \left[S\left(3A + \frac{1}{A}\right) - 4SC \cos \Phi \right]$$
$$= 0 , \qquad\qquad\qquad\qquad\qquad\qquad\qquad\qquad (8.2, 19)$$

$$\eta = \{C \sin \Phi + \tfrac{3}{4}\varepsilon[S(A^2 + 1)\cos \Phi - 2SCA \cos 2\Phi]\}\,[2\delta(A - C \cos \Phi)]^{-1} \qquad (8.2, 20)$$

where $S = \sin \varphi$, $C = \cos \varphi$.

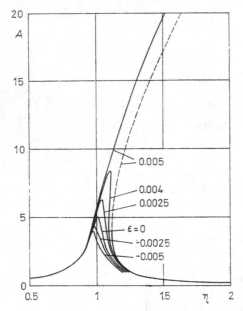

Fig. 8.2, 10

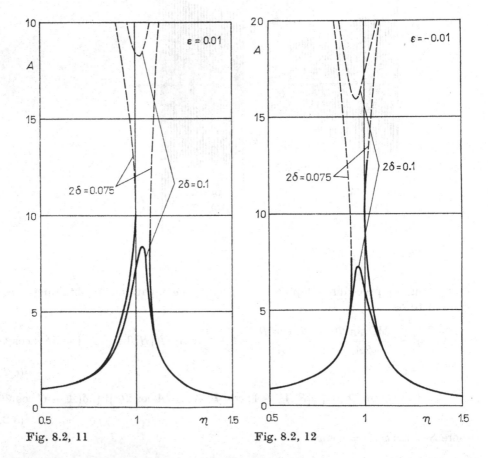

Fig. 8.2, 11   Fig. 8.2, 12

The results obtained for the alternative $\varphi = \pi/4$ are presented in Figs. 8.2, 11 and 8.2, 12. The first figure shows the $A(\eta)$ curves for $\varepsilon = 0.01$, the second for $\varepsilon = -0.01$ and two values of 2δ. The dependences $\Phi(\eta)$ corresponding to these resonance curves are shown in Figs. 8.2, 13 and 8.2, 14. The results obtained for the alternative $\varphi = \pi/2$, two different values of 2δ and $\varepsilon = \pm 0.01$ are presented in Figs. 8.2, 15 and 8.2, 16 (the $A(\eta)$ curves are the same and the $\Phi(\eta)$ are merely shifted for the two values of φ). The $A(\eta)$ curve has two branches. One branch corresponds to the stable solution, with an amplitude so small as practically to coincide with the η axis in the diagram. The other branch, which corresponds to the unstable solution (this is always a saddle point for any η in the (u, v) plane), is a curve whose minimum value of A is obtained for η slightly less than 1.

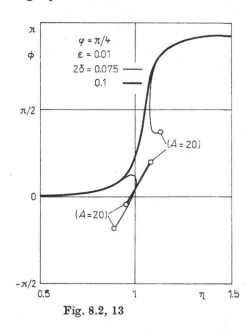

Fig. 8.2, 13

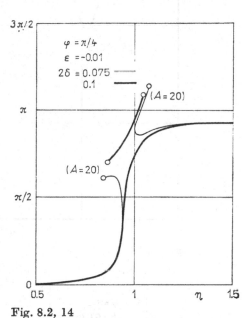

Fig. 8.2, 14

As in the case of system I, the solution of (8.2, 18) can be sought in the form (8.2, 12); for $\varphi = \pi/2$ this leads to the equations

$$u' = \left(\frac{1}{2\eta}\right)\left[-2\delta\eta u + (1 - \eta^2)\,v + \frac{3}{4}\,\varepsilon(1 + u^2 + 3v^2)\right],$$

$$v' = \left(\frac{1}{2\eta}\right)\left[-2\delta\eta v - (1 - \eta^2)\,u - \frac{3}{4}\,\varepsilon u v\right]. \qquad \left.\right\} \qquad (8.2, 21)$$

It can be shown that for $\eta \sim 1$, there always exist two singular points, one being a stable focus or node (for $\eta = 1$), the other a saddle. Fig. 8.2, 17 shows the phase portraits obtained for $\varepsilon = 0.01$, $2\delta = 0.075$ and three different values of η (0.9, 1, 1.1). The separatrix divides the (u, v) phase plane into two domains of attraction — the domain of the stable singular point and the domain of divergent vibration. The vibration records obtained by solving (8.2, 18) for $\varphi = \pi/2$, $\varepsilon = 0.01$ and $2\delta = 0.075$ (Fig. 8.2, 18) confirm the results arrived at earlier. (Fig. 8.2, 18a represents the case of $\eta = 0.9$, Fig. 8.2, 18b that of $\eta = 1$ and different initial conditions $y(0)$ (indicated in the diagrams; $\dot{y}(0) = 0$ in all cases).

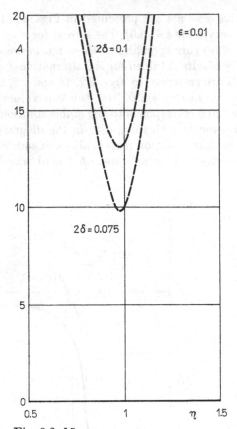

Fig. 8.2, 15

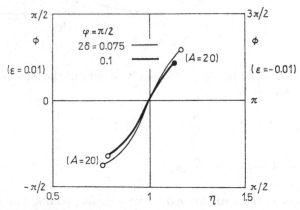

Fig. 8.2, 16

To give a general idea of the effect of the phase shift φ, Fig. 8.2, 19 shows the $A(\eta)$ curves for $\varepsilon = 0.01$, $2\delta = 0.075$ and three different values of φ. As mentioned earlier, the case of $\varphi = 0$ is a commonly occurring external excitation corresponding to alternative (a) of system I; the case of $\varphi = \pi/4$ corresponds to alternative (b) of system I. For $\varphi = \pi/2$ when the external excitation is very small and a trivial solution is practically the only stable steady solution, divergent vibration is also possible for $\eta \sim 1$.

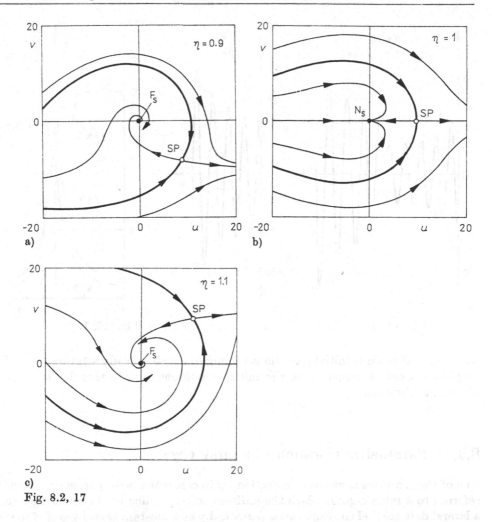

Fig. 8.2, 17

This means that for a definite value of η $(\eta \sim 1)$ there exist two domains of initial conditions — one leading to small, the other to divergent vibration.

The main results obtained in this section can be summed up as follows:

In special cases the response to kinematic (inertial) excitation differs substantially, both quantitatively and qualitatively, from the response to an excitation force acting on the mass or from that to a conventional kinematic excitation. Special cases of this sort are encountered, for example, in systems whose mass is mounted between two non-linear springs, one softening, the other hardening. When the kinematic excitation derives from the motion of the end of one of the springs, the special case occurs when the spring involved is the hardening one. When the suspension points of both springs perform harmonic motions, the special case arises as the two motions become shifted in phase. The difference in the response grows particularly striking in systems having a combination of non-linear springs for which a similar system under an external excitation acting directly on its mass would appear linear because the non-linearities of the two springs compensate one another. It should be stressed that the qualitative difference may in fact become so great that no stable steady solution exists in a particular interval of the excitation frequency and all unsteady solutions are divergent.

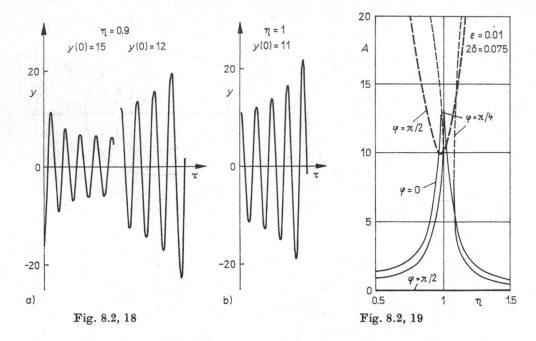

Fig. 8.2, 18   Fig. 8.2, 19

Although a domain of initial conditions leading to a stable steady solution with a finite amplitude exists in some cases, the initial conditions outside this domain result in divergent vibration.

8.3. Parametric vibration of a mine cage

One of the examples mentioned in Section 1.2 in connection with parametric excitation referred to a mine cage in which the stiffness corresponding to the restoring force for a lateral deflection of the cage varies periodically at a constant travel speed of the cage. The cage is carried along guide bars by means of cage guides spaced distance l apart; in the case being considered, $l_0 < l$ where l_0 is the spacing of the bar supports as well as the length of the guide bar. As Fig. 8.3, 1 (drawn horizontally for ease of clarity) shows, the restoring force acting on the cage guide as the guide bar deforms is defined by both the stiffness of the supports and the stiffness of the bar and depends on the position of the cage guide, i.e. on the distance between the cage guide and the guide

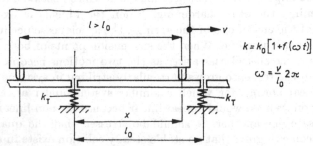

$$k = k_0 \left[1 + f(\omega t) \right]$$

$$\omega = \frac{v}{l_0} 2\pi$$

Fig. 8.3, 1

support. Assuming a constant speed of travel of the cage, v, and linear springs, the stiffness at the point at which the cage guide rests against the guide bar is defined by

$$k = k_0[1 + f(\omega t)] \tag{8.3, 1}$$

where $\omega = \dfrac{v}{l_0} 2\pi$, k_0 is the mean stiffness and $f(\omega t)$ is a periodic function with period $2\pi/\omega$. Fig. 8.3, 2 shows the stiffness ratio k/k_S (k_S is the stiffness midway of the span) as a function of the cage guide position. The curves are drawn for several values of the k_T/k_{S0} ratio; k_T is the stiffness of the guide bar support and k_{S0} the stiffness of the guide bar on absolutely stiff supports midway of the span. The optimum value of the k_T/k_{S0} ratio is 0.5. This value, however, does not correspond to the k_T/k_{S0} ratio of actual structures.

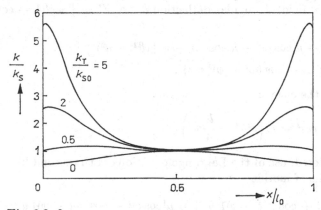

Fig. 8.3, 2

To simplify subsequent considerations, replace $f(\omega t)$ by a harmonic function. The stiffness of the restoring force which is acting on the first cage guide then becomes

$$k_1 = k_0(1 + \mu \cos \omega t) \tag{8.3, 2}$$

and that of the restoring force acting on the next cage guide

$$k_2 = k_0[1 + \mu \cos(\omega t - \psi)] \tag{8.3, 3}$$

where

$$\psi = 2\pi \frac{l - l_0}{l_0}.$$

In the derivation of the equations of motion of the cage it is assumed that the cage travels at a uniform speed v in guides without clearance and that the centroid of the cage lies at its centre.

The cage has a mass m and a moment of inertia about the axis passing through the centroid normal to the axis of travel $I = mr^2$. The motion of the cage can be described either by a lateral deflection y and an angular deflection ϑ (Fig. 8.3, 3) or by deflections y_1 and y_2 of points 1 and 2. Assuming ϑ to be small, these coordinates are related as follows:

$$\left. \begin{array}{l} y_1 = \tfrac{1}{2} l \sin \vartheta + y \doteq \tfrac{1}{2} l\vartheta + y = \tfrac{1}{2} l(\vartheta + u), \\[2mm] y_2 = \tfrac{1}{2} l \sin \vartheta - y \doteq \tfrac{1}{2} l\vartheta - y = \tfrac{1}{2} l(\vartheta - u) \end{array} \right\} \tag{8.3, 4}$$

where $u = 2y/l$.

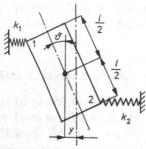

Fig. 8.3, 3

The potential energy (ignoring the part of the potential energy which is imparted to the cage in the gravity field at angular deflection ϑ, i.e. $V' = \frac{1}{2} mgl(1 - \cos \vartheta)$) is defined by

$$2V = \frac{1}{4} l^2 k_0 \{[2 + \mu \cos \omega t + \mu \cos(\omega t - \psi)] (\vartheta^2 + u^2)$$
$$+ 2\mu[\cos \omega t - \cos (\omega t - \psi)] \vartheta u\} . \tag{8.3, 5}$$

The kinetic energy of the cage is

$$2T = r^2 m \dot{\vartheta}^2 + m \dot{y}^2 = m \left(r^2 \dot{\vartheta}^2 + \frac{l^2}{4} \dot{u}^2 \right) . \tag{8.3, 6}$$

If damping is not considered, use of the Lagrange equations of the second kind results in the following equations of motion

$$\left. \begin{aligned} &\ddot{\vartheta} + \Omega_1^2 \{ [1 + \tfrac{1}{2} \mu[\cos \omega t + \cos (\omega t - \psi)] \, \vartheta + \tfrac{1}{2} \mu[\cos \omega t - \cos (\omega t - \psi)] \, u \} = 0 , \\ &\ddot{u} + \Omega_2^2 \{ [1 + \tfrac{1}{2} \mu[\cos \omega t + \cos (\omega t - \psi)]] \, u + \tfrac{1}{2} \mu[\cos \omega t - \cos (\omega t - \psi)] \, \vartheta \} = 0 \end{aligned} \right\}$$
$$\tag{8.3, 7}$$

where $\Omega_1 = (l/r)\sqrt{k_0/2m}$, $\Omega_2 = \sqrt{2k_0/m}$ are the mean natural frequencies of torsional and lateral vibrations, respectively.

The instability intervals of the first kind and first order are described (in the first approximation) by the relations

$$2\Omega_s - \tfrac{1}{2} \mu a_{ss} < \omega < 2\Omega_s + \tfrac{1}{2} \mu a_{ss} \qquad (s = 1, 2); \tag{8.3, 8}$$

the instability interval of the second kind and first order is defined by the relation

$$|\Omega_1 \pm \Omega_2| - \tfrac{1}{2} \mu \sqrt{\pm a_{12} a_{21}} < \omega < |\Omega_1 \pm \Omega_2| + \tfrac{1}{2} \mu \sqrt{\pm a_{12} a_{21}} \tag{8.3, 9}$$

where

$$a_{11} = a_{22} = \tfrac{1}{2} \sqrt{2(1 + \cos \psi)} , \qquad a_{12} = a_{21} = \tfrac{1}{2} \sqrt{2(1 - \cos \psi)} .$$

Since $a_{12} a_{21} = a_{12}^2 = a_{21}^2 > 0$, only the plus sign has a meaning in inequalities (8.3, 9).

For $\psi = 0 + 2n$, i.e. $l/l_0 = n$ ($n = 1, 2, ...$), $a_{11} = a_{22} = 1$, $a_{12} = a_{21} = 0$, the interval of the second kind does not exist.

For $\psi = (1 + 2n) \pi$, i.e. $l/l_0 = \frac{1}{2} + n$ ($n = 1, 2, ...$), $a_{11} = a_{22} = 0$, $a_{12} = a_{21} = 1$, that is, the intervals of the first kind disappear.

On the basis of these findings one can draw the important conclusion that only one kind of instability interval can be quenched by altering the pitch of the cage guides.

The width of all instability intervals can be affected by changing the value of the coefficient μ, that is, by reducing the variability of the stiffness of the restoring forces. This can be done by mounting the cage guides elastically on, for example, flexible rubber elements having a stiffness k_2. For the arrangement shown schematically in Fig. 8.3, 4 the total stiffness K is then described by

$$\frac{1}{K} = \frac{1}{k_2} + \frac{1}{k_0(1 + \mu \cos \omega t)} = \frac{k_2 + k_0(1 + \mu \cos \omega t)}{k_0 k_2 (1 + \mu \cos \omega t)}.$$

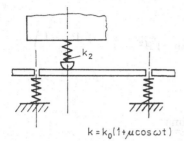

$$k = k_0(1 + \mu \cos \omega t)$$

Fig. 8.3, 4

Assuming $\mu < 1$, one obtains from the above equation the approximate relation

$$K = \frac{k_2 k_0 (1 + \mu \cos \omega t)}{k_2 + k_0(1 + \mu \cos \omega t)} \doteq \frac{k_2 k_0}{k_2 + k_0} \left(1 + \frac{k_0}{k_2 + k_0} \mu \cos \omega t \right) \qquad (8.3, 10)$$

which implies that the effect of the stiffness variability is the smaller, the lesser is the ratio $k_2/(k_2 + k_0)$. In other words, the use of soft-mounted cage guides is advantageous.

It has been assumed so far that the cage moves in its guides without clearance. To estimate, at least qualitatively, the effect of the cage-guides clearance, consider a simplified case when $a_{12} = a_{21} = 0$ and system (8.3, 7) decomposes into two independent differential equations of the same type. This means that the lateral and the torsional vibrations are not bound one to another. Either equation can be rearranged (using the time transformation $\Omega_s t = \tau$ and the notation $\eta = \omega/\Omega_s$, $w = \vartheta/\vartheta_0$ or $w = u/u_0$ where ϑ_0 and u_0 denote the values of the coordinates for which the clearance is taken up) to the dimensionless form

$$w'' + f(w - 1)(1 + \mu \cos \eta \tau) = 0 \qquad (8.3, 11)$$

where

$$f(w - 1) = \begin{cases} 0 & \text{for} \quad |w| < 1 \\ w - 1 & \text{for} \quad w > 1 \\ w + 1 & \text{for} \quad w < -1 \end{cases}$$

Note that this is a case of non-linear parametric exittation. To obtain a limited steady solution for any η, assume the damping to be linear and progressive and write the differential equation of motion in the form

$$w'' + (b + \delta w^2) w' + f(w - 1)(1 + \mu \cos \eta \tau) = 0 \qquad (8.3, 12)$$

where b is the coefficient of linear damping, and δ that of the progressive one. Both b and δ are assumed to be small compared to unity. Approximate the solution at main parametric resonance by

$$w = A \cos \left(\tfrac{1}{2} \eta \tau + \varphi \right) \qquad (8.3, 13)$$

where A, φ are coefficients slowly varying in time for unsteady solutions. Accordingly, (8.3, 12) can be converted to a set of two first-order differential equations

$$
\left.
\begin{aligned}
A' &= \left(\frac{1}{\eta}\right)\left[-\frac{1}{2}\,\eta\left(b+\frac{1}{4}\,\delta A^2\right)+\frac{1}{2}\,\mu Q \sin 2\varphi\right]A\,, \\
\varphi' &= \left(\frac{1}{\eta}\right)\left[-\left(\frac{1}{2}\,\eta\right)^2+Q\left(1+\frac{1}{2}\,\mu\cos 2\varphi\right)\right]
\end{aligned}
\right\}
\tag{8.3, 14}
$$

where

$$
Q=
\begin{cases}
\dfrac{2}{\pi}\left[\cos^{-1}\dfrac{1}{A}-\dfrac{\sqrt{A^2-1}}{A^2}\right]=\dfrac{2}{\pi}\left(\tan^{-1}\sqrt{A^2-1}-\dfrac{\sqrt{A^2-1}}{A^2}\right) \\
\hspace{6cm}\text{for}\quad A>1\,, \\[4pt]
0 \quad\text{for}\quad A\leqq 1\,.
\end{cases}
$$

The steady solution is obtained from the equations

$$
\left.
\begin{aligned}
\tfrac{1}{2}\,\eta\left(b+\tfrac{1}{4}\,\delta A^2\right) &= \tfrac{1}{2}\,\mu Q(A)\sin 2\varphi\,, \\
(\tfrac{1}{2}\,\eta)^2-Q(A) &= \tfrac{1}{2}\,\mu\,Q(A)\cos 2\varphi\,.
\end{aligned}
\right\}
\tag{8.3, 15}
$$

Eliminating φ leads to the equation

$$
(\tfrac{1}{4}\,\eta^2)_{1/2}=Q-\tfrac{1}{2}\,(b+\tfrac{1}{4}\,\delta A^2)^2\mp[\tfrac{1}{4}\,\mu^2 Q^2-Q(b+\tfrac{1}{4}\,\delta A^2)^2+\tfrac{1}{4}\,(b+\tfrac{1}{4}\,\delta A^2)^4]^{1/2}
\tag{8.3, 16}
$$

from which $\eta(A)$ and thus also $A(\eta)$ can be determined. From the equation

$$
\tan 2\varphi=\frac{\left(b+\dfrac{1}{4}\,\delta A^2\right)\eta}{\left(\dfrac{1}{2}\,\eta\right)^2-Q}
$$

or the equation

$$
\tan\varphi=\frac{\dfrac{1}{2}\left(b+\dfrac{1}{4}\,\delta A^2\right)\eta}{\left(\dfrac{1}{2}\,\eta\right)^2-\left(1-\dfrac{1}{2}\,\mu\right)Q}
\tag{8.3, 17}
$$

one can establish the relationship $\varphi(\eta)$. The backbone curve is described by the equation

$$
\eta=2\sqrt{Q(A_s)}
\tag{8.3, 18}
$$

and the limit envelope by the equation

$$
\eta=\frac{\mu Q(A_{\mathrm{L}})}{b+\dfrac{1}{4}\,\delta A_{\mathrm{L}}^2}\,.
\tag{8.3, 19}
$$

As shown by TONDL (1976b), the system being analyzed belongs to the class of systems to which applies the rule of vertical tangents, that is, the points of the resonance curve at which the tangents are vertical form the boundaries between stable and unstable solutions. The author has also shown that the effect of non-linear parametric excitation causes subharmonic resonances of order $1/N$ ($N = 2, 3, ...$) to exist besides parametric resonances of order N.

Fig. 8.3, 5 shows the resonance curve $(A(\eta))$, the backbone curve $(A_S(\eta))$ and the limit envelope $(A_L(\eta))$ drawn for the case of $b = 0$, $\delta = 0.01$ and $\mu = 0.4$. In contrast with the case of linear parametric excitation when the equilibrium position is unstable in a certain interval of the excitation frequency, the equilibrium position is found to be

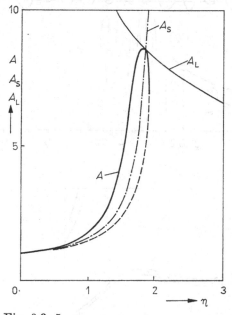

Fig. 8.3, 5

stable over the whole range of η; however, in the intervals in which parametric resonance occurs, the equilibrium position is only locally stable. The parametric resonance shown in the figure is larger than that which occurs in the case of linear parametric excitation and extends, theoretically, from zero value of the excitation frequency. Fig. 8.3 6 shows the results of analogue computations: the curve of the extreme deflection $[w]$ vs. the relative excitation frequency, and the vibration records obtained for $\eta = 1.8$ (A, B) and $\eta = 3.2$ (C) rae added to complete the information. It can clearly be seen that not only the main resonance of the first order (marked 1) but also the subharmonic resonances of order 1/2 and 1/3 (marked 1/2, 1/3) are present. The domains of attraction shown in Fig. 8.3, 7 (for $\eta = 1.4$) and in Fig. 8.3, 8 (for $\eta = 1.8$) were obtained by solving (8.3, 14).

As the theoretical analysis has revealed, only one type of parametric resonance can be quenched by altering the pitch of the cage guides. A more effective means is a flexible mounting of the guides having a slight prestress which eliminates the unfavourable effect of clearance, that is, the non-linear parametric excitation and in turn, the possible occurrence of subharmonic resonances. Such an arrangement makes it possible to raise the speed of cage travel.

The above study of the vibration pattern of a mine cage was made not just as an academic excursion into the field of systems with non-linear parametric excitation;

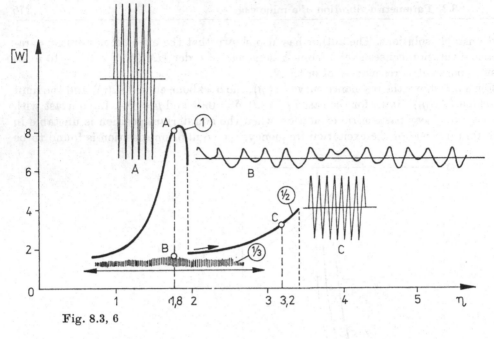

Fig. 8.3, 6

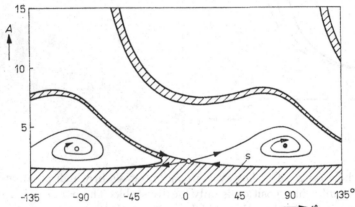

Fig. 8.3, 7

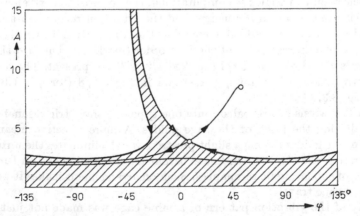

Fig. 8.3, 8

rather it was to assist in finding the cause of failure of an actual structure. The collapse was due to the effect of severe torsional vibrations of the cage which caused every second guide bar to break alternately on the two sides of the guide structure (Fig. 8.3, 9).

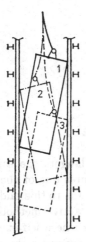

Fig. 8.3, 9

9. Quenching of self-excited vibration

9.1. Basic considerations and methods of solution

Since self-excited vibration impairs reliable operation and endangers the safety of miscellaneous machinery and structures, its suppression constitutes one of the major tasks of vibration engineering. An ideal and very expedient means to this end is the removal of the source of self-excitation. However, this so-called active method is not applicable in all cases. The systems in which it fails include those where self-excitation is an inherent characteristic of the technological process (for example, the cutting forces in machine tools) or is inherent in the function of the device (for example, the hydrodynamic forces in journal bearings). Sometimes, as in the case of self-excited oscillations — galloping — of high-voltage transmission lines, application of the active method is not feasible because of economic or operational reasons. In cases of these sorts resort must be made to passive methods, that is, to paralyzing the destabilizing effect of negative damping, which is obtained in the equations of motion when expressing the action of forces producing self-excitation, by an increase in the level of positive damping. This chapter deals exclusively with the passive methods of quenching self-excited vibration.

The practicability and efficacy of the various means used in connection with these methods will be examined using systems which belong to the class represented in its simplest form by the van der Pol oscillator. Only systems with a finite number of degrees of freedom will be considered.

Let the portion of the system being examined, whose masses are acted on by self-excitation, be called the basic self-excited subsystem, or briefly, the basic system. Consider the self-excitation to have the classical form of that of the Van der Pol oscillator that is, the mass is acted on by damping having a negative linear viscous component as well as a positive progressive component, which can be expressed as the product of the square of the corresponding deflection and its velocity. The study that follows concerns two means of suppressing self-excited vibration, i.e. absorbers and resilient foundations with damping or, more exactly, absorber and foundation subsystems.

Absorbers, a common and popular means of suppressing forced vibrations, are not frequently used in self-excited systems. An analysis of their action in those systems was published only recently. MANSOUR (1972) was the first author to explain the basic properties of absorbers as applied to a two-mass, two-degrees-of-freedom system (the subsystem of the Van der Pol oscillator attached to the subsystem of an absorber). As he pointed out, an absorber is not always capable of compensating the self-excitation effect represented by negative linear damping, and there exists an optimum value of the absorber's damping coefficient. These results were later extended by other authors, for example, TONDL (1975 a, b; 1976d; 1977), HAGEDORN (1978), ROWBOTTOM (1981).

The principle of mounting on resilient or sprung foundations has been analyzed for discrete systems whose various masses perform only a lateral motion, as well as for rotor systems. The literature relating to the first group of systems includes studies by TONDL (1975 a, b), (1976 d) and a comprehensive treatise including the action of absorbers published in a book by TONDL (1980 b). Of the fairly ample literature on the second group, mention should be made of the work of TONDL (1965, 1970 c, 1971 a, 1980 a), and KELZON and YAKOVLEV (1971 a, b).

Systems of the van der Pol type belong to the class of systems with soft self-excitation (provided that the action of forces (for example, friction) causing a qualitative change in a system with hard self-excitation is not taken into consideration). The possibility of the occurrence of self-excited vibration is ascertained by means of an analysis of the equilibrium position stability, that is, an analysis of the linearized equations of motion written in terms of the coordinates of the deflections from the equilibrium position.

The conventional method of investigating the equilibrium position stability, which involves an analysis of the roots of the characteristic equation using the Routh-Hurwitz criterion, has two disadvantages: it provides no more than an answer to the question of whether or not the system being examined is stable, and its calculations grow ever more time-consuming as the systems become more complicated. Consequently, it is not capable of indiating either the natural mode with respect to which the system is unstable or the frequency of the resulting self-excited vibration. Below is a description of methods which are free of these disadvantages if certain assumptions are satisfied. It must be assumed that none of the roots of the characteristic equation is real and positive (the characteristic equations of a vast majority of mechanical discrete systems with n degrees of freedom has n complex roots). The exposition also refers to a method of determining the dependence of the amplitudes of single-frequency vibration on a particular parameter of the system. Although the method is approximate, it is more exact than the conventional procedures because the solution is not approximated by the natural modes of the abbreviated (undamped linearized) system.

The two methods to be described are both based on the assumption concerning the roots of the characteristic equation. For values of the parameters lying on the boundary of the equilibrium position stability, one of these roots becomes imaginary.

The first method — named the boundary values method — is suitable for dealing with fairly simple systems (having only a few degrees of freedom). It involves finding a set of values of two parameters of the system (for example, the coefficient of negative damping β and that of positive linear damping $\varkappa$), lying on the stability boundary, where it holds for a root of the characteristic equation that $\lambda = i\Omega$ (Ω is real and represents the frequency of self-excited vibration initiated on this boundary. Substituting for λ in the characteristic equation must cause the real and the imaginary parts of the equation to become zero. The two equations thus obtained, viz.

$$F_1(\Omega; \varkappa, \beta) = 0 \, , \qquad F_2(\Omega; \varkappa, \beta) = 0 \qquad\qquad (9.1, 1)$$

are polynomials in Ω and their coefficients are functions of parameters $\beta, \varkappa$. The calculations can generally proceed as follows: the value of one of the parameters (for example $\varkappa$) is varied in steps and the corresponding values of Ω and β are obtained from the two equations (9.1, 1). The two-mass systems to be analyzed by means of this method are characterized by the fact that one of (9.1,1) can be written in the form

$$F_1(\Omega; K) = 0 \qquad\qquad (9.1, 2)$$

where

$$K = \varkappa\beta \tag{9.1, 3}$$

and the other in the form

$$F_2(\Omega; K, \varkappa) = 0 . \tag{9.1, 4}$$

For the two-mass systems being solved here, (9.1, 2) is quadratic in Ω. This fact permits ready determination of the $\Omega - K$ dependence as well as the corresponding values of $\varkappa$ ((9.1, 4)) and β ((9.1, 3)). Using a calculator with a graph plotter the curve $\beta(\varkappa)$ is drawn directly by increasing the parameter K in steps and calculating the corresponding Ω, β, $\varkappa$ in each step. Since two values of Ω^2 exist for two-mass systems, we thus obtain a set of two curves, each representing the boundary of the stability region for one natural frequency. The boundary of the region of parameters, which is common to both stability regions, is identical with that obtained by using the Routh-Hurwitz criterion. It represents the region of the equilibrium position stability with respect to both natural frequencies.

The second method has several features reminiscent of the procedure of determining the amplitude of single-frequency self-excited vibration as a function of a particular parameter of the system, for example, the tuning coefficient Q. Consider a system with N degrees of freedom whose motion is described in terms of the coordinates of the deflections from the equilibrium position x_k $(k = 1, 2, ... , N)$. The system is capable of vibrating in n $(n \leq N)$ modes of single-frequency vibration of frequencies Ω_s $(s = 1, 2, ... , n)$. The single-frequency solution can be approximated by the form

$$\left.\begin{array}{l} x_1 = A \cos \Omega t , \\[2mm] x_j = A_j \cos \Omega t + B_j \sin \Omega t , \qquad (j = 2, 3, ... , N) . \end{array}\right\} \tag{9.1, 5}$$

By establishing the dependence of A on the system parameter Q whose optimum value is to be found, one obtains the curves $A(Q)$ corresponding to the various frequencies Ω_s. These curves can be either simply continuous or consist of branches. As shown in Fig. 9.1, 1, they can feature either:

(a) an interval of Q for which no real solution $A \neq 0$ exists, or
(b) separate values, rather than an interval, of Q for which $A = 0$, or, finally,
(c) a whole range of Q for which $A \neq 0$.

An investigation of the boundaries of the equilibrium position stability is aimed at determining the limit values $Q = Q^+$ for which $\lim\limits_{Q \to Q^+} A = 0$.

If the solution of (9.1, 5) is expressed in terms of the relative quantities

$$y_k = \frac{x_k}{A} \qquad (k = 1, 2, ... , N)$$

the solution of the parameters on the stability boundary may take the form

$$\left.\begin{array}{l} y_1 = \cos \Omega t , \\[2mm] y_j = a_j \cos \Omega t + b_j \sin \Omega t \qquad (j = 2, 3, ... , N) \end{array}\right\} \tag{9.1, 6}$$

where $a_j = A_j/A$, $b_j = B_j/A$ are coefficients which are generally different from zero for $A \to 0$. Based on the assumptions put forward above, substitution of (9.1, 6) in the linearized equations of motion leads to $2N$ non-homogeneous algebraic equations con-

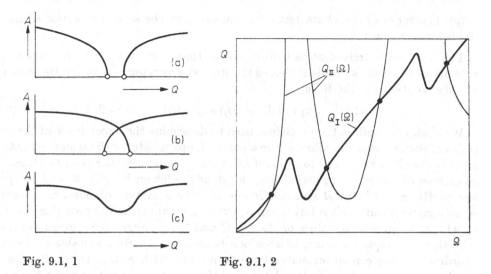

Fig. 9.1, 1   Fig. 9.1, 2

taining the unknowns a_j, b_j, Ω and Q. These equations are linear with respect to the coefficients a_j and b_j. In most cases, parameter Q is not present in all the equations. Accordingly, the equations can be rearranged, for example, so as to contain the absorber tuning coefficient Q (its square at most) in only two of them. Coefficients a_j, b_j as functions of Ω which are then readily obtained from the $2(N-1)$ equations are substituted in the remaining two equations containing Q, which enable Q to be determined as a function of $\Omega(Q_{\mathrm{I}}(\Omega), Q_{\mathrm{II}}(\Omega))$. The points of intersection of these two curves give the values of Q and Ω corresponding to the stability boundary (Fig. 9.1, 2).

By ascertaining the boundary values of Q for various values of another parameter, say $\varkappa$, we obtain curves $\varkappa(Q)$, the so-called boundary curves in the $(\varkappa, Q)$ plane. These curves connect the values of parameters Q, $\varkappa$ at which the amplitudes of self-excited vibration converge to zero, that is, only a part of the boundary curves forms the boundary of the region of stability of the equilibrium position. This is the procedure used for determining the points corresponding to cases (a) and (b) shown in Fig. 9.1, 1. A thorough analysis of the boundary curves is, therefore, necessary. Further details concerning this subject may be found in a book by TONDL (1980b).

The method of boundary curves can be extended to systems governed by differential equations of dependent variables z_k ($k = 1, 2, \ldots, n$) which are two-dimensional vectors of the deflections from the equilibrium position whenever the following assumptions are satisfied:

(a) The differential equations are homogeneous; this means, for example, in the case of rotor systems, that the rotors are fully balanced.

(b) All forces (elastic, damping, etc.) are expressed in terms of central symmetric fields — the absolute value of a force vector undergoes no change if the absolute values of the deflection and velocity vectors are held constant. Thus, for example, the vector of the restoring force is defined by the relation (denoting by z the deflection vector)

$$\boldsymbol{P} = -[f_1(|\boldsymbol{z}|) + \mathrm{i}f_2(|\boldsymbol{z}|)]\,\boldsymbol{z}\;.$$

In case of rotor systems, this means that the axis of rotation is vertical (or the effect of gravitation can be neglected) and all the elements such as bearings, glands, etc. are symmetrical about the axis of rotation.

(c) The roots of the characteristic equation have the same properties as those of the former systems.

The solution is arrived at as before, except that in rotor systems it is the angular velocity of the rotor which is chosen as the step-wise varying parameter. On the boundary the solution has the form

$$z_1 = \exp(i\Omega t), \quad z_j = (A_j + iB_j)\exp(i\Omega t), \quad (j = 2, 3, \dots, n). \quad (9.1, 7)$$

We shall now outline the procedure used to determine the dependence of the amplitudes of single-frequency vibration on a particular parameter of the system (denoted Q). In this case it is necessary to proceed from the complete (rather than the linearized) equations of motion. Approximating the steady solution by (9.1, 5) and comparing the coefficients of $\cos\Omega t$ and $\sin\Omega t$ one obtains a system of non-linear algebraic equations whose difficult solution can be avoided by applying the procedure by means of which the unknown values of A_j, B_j, Ω and Q are sought for a given value of X (the algebraic equations thus become non-homogeneous). By a suitable choice of the coordinate x_1 one can either make the system linear with respect to A_j, B_j or, as will be shown by way of examples, determine these quantities successively from linear equations.

For systems described by vector equations the steady solution can be approximated by

$$z_1 = A\exp(i\Omega\tau), \quad z_j = (A_j + iB_j)\exp(i\Omega\tau), \quad (j = 2, 3, \dots, n). \quad (9.1, 8)$$

9.2.   Two-mass systems with two degrees of freedom

The analysis presented below is concerned with three basic two-mass, two-degrees-of-freedom systems whose basic self-excited subsystem (basic system for short) is a Van der Pol oscillator described by the van der Pol equation. To curtail the exposition, the differential equations are presented already rearranged by means of the introduction of dimensionless coefficients of the linear terms. The time transformation is in all cases related to the natural frequency of the abbreviated (undamped) basic system.

The following schematic notation is used to simplify the description:

(a) Scheme in Fig. 9.2, 1 a denotes positive linear damping.
(b) Scheme in Fig. 9.2, 1 b denotes a negative coefficient of the linear term of damping having a positive progressive component-self-excitation of the Van der Pol type.
(c) Scheme in Fig. 9.2, 1 c denotes the positive Coulomb dry friction.

The lateral deflections of masses m_k are denoted by x_k and the stiffnesses of springs by c_k.

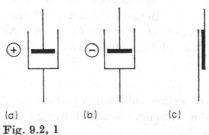

(a)   (b)   (c)
Fig. 9.2, 1

System I. The basic system having mass m_2 and a spring of stiffness c_2 is attached to an absorber system (m_1, c_1)-Fig. 9.2, 2.

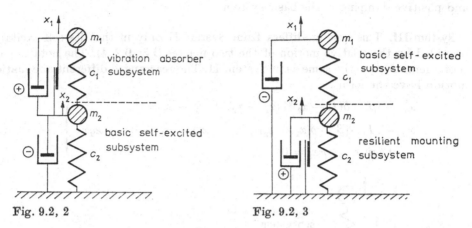

Fig. 9.2, 2   Fig. 9.2, 3

The system is described by the equations

$$x_1'' + \varkappa Q(x_1' - x_2') + \vartheta \, \mathrm{sgn}\,(x_1' - x_2') + Q^2(x_1 - x_2) = 0 \,,$$
$$x_2'' - (\beta - \delta x_2^2)\, x_2' - \mu[\varkappa Q(x_1' - x_2') + \vartheta \, \mathrm{sgn}\,(x_1' - x_2')$$
$$+ Q^2(x_1 - x_2)] + x_2 = 0 \qquad (9.2,\,1)$$

where $\mu = m_1/m_2$ is the ratio between the absorber mass and the basic system mass, $Q = \sqrt{\dfrac{c_1/m_1}{c_2/m_2}}$ is the coefficient of absorber tuning, β is the coefficient of negative linear damping, δ is the coefficient of the component of progressive positive damping, $\varkappa$ is the coefficient of positive linear damping of the absorber, and ϑ is the coefficient of dry friction. By introducing the relative deflection

$$x = x_1 - x_2$$

equations (9.2, 1) may be given the form

$$x'' + x_2'' + \varkappa Q x' + \vartheta \, \mathrm{sgn}\, x' + Q^2 x = 0 \,,$$
$$x_2'' - (\beta - \delta x_2^2)\, x_2' - \mu(\varkappa Q x' + \vartheta \, \mathrm{sgn}\, x' + Q^2 x) + x_2 = 0 \qquad (9.2,\,1\text{a})$$

or the form

$$x'' + x_2'' + \varkappa Q x' + \vartheta \, \mathrm{sgn}\, x' + Q^2 x = 0 \,,$$
$$(1 + \mu)\, x_2'' + \mu x'' - (\beta - \delta x_2^2)\, x_2' + x_2 = 0 \,. \qquad (9.2,\,1\text{b})$$

System II. The basic system (m_1, c_1) is mounted on a foundation subsystem (m_2, c_2) — Fig. 9.2, 3. The corresponding differential equations of motion can be rearranged to give the form

$$x_1'' - (\beta - \delta x_1^2)\, x_1' + x_1 - x_2 = 0 \,,$$
$$x_2'' - M(x_1 - x_2) + \varkappa q x_2' + \vartheta \, \mathrm{sgn}\, x_2' + q^2 x_2 = 0 \qquad (9.2,\,2)$$

where $M = m_1/m_2$ is the ratio between the masses of the basic system and the foundation subsystem, $q = \sqrt{\dfrac{c_2/m_2}{c_1/m_1}}$ is the coefficient of foundation tuning, $\varkappa$ is the coefficient of linear positive damping of the foundation mass motion, ϑ is the coefficient of dry friction of the foundation mass motion, and β and δ are the coefficients of negative and positive damping of the basic system.

System III. This system differs from System II only in that its self-excitation is provided by the relative motion of the two masses (Fig. 9.2, 4). The notation of the coefficients M, q is the same as in System II. The rearranged differential equations of motion have the form

$$\left.\begin{aligned}
&x_1'' - [\beta - \delta(x_1 - x_2)^2]\,(x_1' - x_2') + x_1 - x_2 = 0\,, \\
&x_2'' - M\{-[\beta - \delta(x_1 - x_2)^2]\,(x_1' - x_2') + x_1 - x_2\} + \varkappa q x_2' \\
&\quad + \vartheta\,\mathrm{sgn}\,x_2' + q^2 x_2 = 0
\end{aligned}\right\} \tag{9.2, 3}$$

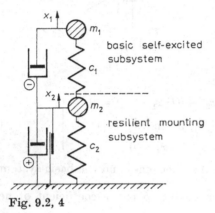

basic self-excited subsystem

resilient mounting subsystem

Fig. 9.2, 4

or, on substituting $x = x_1 - x_2$, the form

$$\left.\begin{aligned}
&x'' + x_2'' - (\beta - \delta x^2)\,x' + x = 0\,, \\
&(1 + M)\,x_2'' + M x'' + \varkappa q x_2' + \vartheta\,\mathrm{sgn}\,x_2' + q^2 x_2 = 0\,.
\end{aligned}\right\} \tag{9.2, 3 a}$$

Since dry friction is ignored ($\vartheta = 0$) in the investigation of the equilibrium position stability, all three systems belong to the class with soft self-excitation.

Consider first System I. The limit values of β, $\varkappa$ on the stability boundary can be established by means of the first method of Section 9.1, that is by substituting the solution $x_k = X_k \exp{(\mathrm{i}\Omega\tau)}$ in the linearized differential equations of motion. Carrying out this substitution in (9.2, 1 a) we obtain the characteristic equation which — recalling the considerations of the foregoing section — we can write as two equations, that is,

$$\left.\begin{aligned}
&\Omega^4 - [1 + (1 + \mu)\,Q^2 - \varkappa\beta Q]\,\Omega^2 + Q^2 = 0\,, \\
&[\varkappa(1 + \mu)\,Q - \beta]\,\Omega^2 - \varkappa Q + \beta Q^2 = 0\,.
\end{aligned}\right\} \tag{9.2, 4}$$

Introducing the parameter

$$K = \varkappa\beta \tag{9.2, 5}$$

leads to the equation

$$(\Omega^2)_{1,2} = \tfrac{1}{2}\,[1 + (1 + \mu)\,Q^2 - KQ] \pm \{\tfrac{1}{4}\,([1 + (1 + \mu)\,Q^2 - KQ]^2 - Q^2\}^{1/2} \tag{9.2, 6}$$

as well as to the equation

$$\varkappa = \{K(\Omega^2 - Q^2)/Q[\Omega^2(1 + \mu) - 1]\}^{1/2} . \tag{9.2, 7}$$

The calculation of the limit values of parameters $\varkappa$ and β can be conveniently carried out as follows:

(a) Frequency Ω corresponding to given values of $\varkappa$, β is obtained from (9.2, 6) for various values of parameter K (which is varied step-wise on the computer).

(b) The values of Ω are substituted in (9.2, 7), which yields the corresponding values of $\varkappa$.

(c) β is obtained from (9.2, 5).

Since (9.2, 6) has two roots, two values of Ω are obtained for each K. The resulting two curves $\beta(\varkappa)$ represent the boundary of the equilibrium position stability corresponding, respectively, to the lower and the higher natural frequency of the system. Both curves start from the origin of the coordinates of the $(\varkappa, \beta)$ plane. The area close to the $\varkappa$ axis is the region of stability corresponding to the lower or the higher natural frequency. Only those values of $\varkappa$, β which lie in both regions form the region of stability of the equilibrium position.

Curves $\beta(\varkappa)$ were obtained for various values of the tuning coefficient Q and the ratio of masses μ. They are presented in comprehensive diagrams, one of which is shown as an illustrative example. Curves $\beta(\varkappa)$ shown in Fig. 9.2, 5a correspond to the lower frequency Ω, those in Fig. 9.2, 5b to the higher Ω (both were obtained for $\mu = 0.1$). Fig. 9.2, 6 shows the various regions of the equilibrium position stability

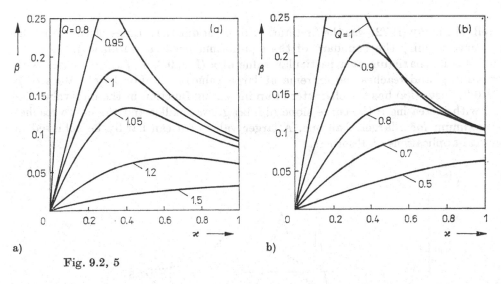

a)

Fig. 9.2, 5

b)

for $Q = 1$; the blank area represents the region of the equilibrium position stability. For small values of $\varkappa$ the boundary curve $\beta(\varkappa)$ is close to a straight line passing through the origin. With increasing $\varkappa$ the slope of the curve decreases, and the curve reaches its maximum. For a further increase of $\varkappa$, the curve becomes a decreasing function. The dash-line curve in Fig. 9.2, 6 represents the boundary of the equilibrium position stability obtained by approximation effected by means of the diagonal terms of the quasinormal system (see, for example, a monograph by TONDL (1970b)). For larger values of $\varkappa$, the two curves are seen to differ not only quantitatively but qualitatively

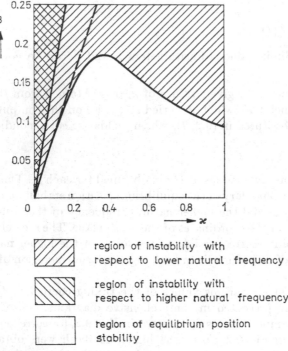

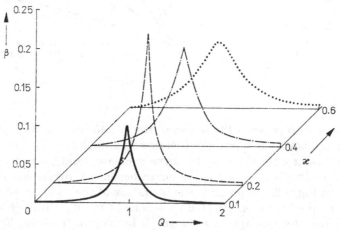

Fig. 9.2, 6

as well. MANSOUR (1972) was the first author to point out the strange character of the $\beta(\varkappa)$ curve forming the boundary of the equilibrium position stability (for $Q = 1$), which has its maximum at a particular value of $\varkappa$ (Fig. 9.2, 6). $\beta_{\max}$ increases with increasing μ and reaches the extreme at large values of $\varkappa$. For large values of μ ($\mu > 0.3$) curve $\beta(\varkappa)$ has the character of an increasing function in the interval $0 < \varkappa < 1$; with increasing $\varkappa$ the curve slope $(\mathrm{d}\beta/\mathrm{d}\varkappa)$ grows smaller and the curve reaches its maximum for $\varkappa$ larger than one. A larger value of μ can hardly be achieved in practical applications of absorbers.

Fig. 9.2, 7

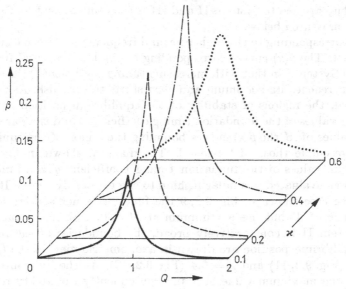

Fig. 9.2, 8

Figs. 9.2, 7 and 9.2, 8 show the results of the examination drawn in comprehensive diagrams. The diagrams of the boundary values of the equilibrium position stability drawn in coordinates of the coefficient β and the tuning coefficient Q for various values of the damping coefficient $\varkappa$ (the first diagram corresponds to $\mu = 0.05$, the second to $\mu = 0.1$) reveal the following: The optimum value of the tuning coefficient is less than 1, that is the optimum natural frequency of the absorber subsystem alone is slightly lower than the natural frequency of the basic system and the difference grows larger with increasing mass ratio μ. There exists an optimum value of the damping coefficient $\varkappa$. The maximum of the boundary value β increases with increasing μ.

As can be seen in the diagram of Fig. 9.2, 5, the stability region corresponding to the lower natural frequency grows larger with diminishing Q; the opposite applies to the stability region corresponding to the higher natural frequency. The most favourable case as regards the region of stability of the equilibrium position occurs when the regions corresponding to the two ranges of natural frequencies overlap. Since the character of the boundary curves is then the same, overlapping in general occurs whenever the slopes of the two curves coincide. Fig. 9.2, 9 shows the curves of the optimum values of Q as a function of the mass ratio μ obtained on the basis of this criterion (heavy line; the light line refers to linear material damping — see TONDL (1980b)).

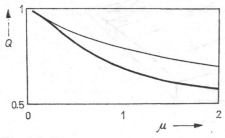

Fig. 9.2, 9

Since the procedure applied to Systems II and III is the same as that for System I, only its results are presented below.

The $\beta(\varkappa)$ curves corresponding to the higher natural frequency have same character as those of System I. The $\beta(\varkappa)$ curves corresponding to the lower natural frequency differ from those of System I in that with increasing tuning coefficient Q the stability region grows larger, reaches its maximum and then starts to diminish. As the mass ratio M is increased, the regions of stability of the equilibrium position reach this maximum at higher values of the foundation tuning coefficient q and their area grows smaller. For low values of M the $\beta(\varkappa)$ curves bounding the region of the equilibrium position stability resemble those of System I. The diagram shown in Fig. 9.2, 10 plotting the optimum values of the fundation tuning coefficient q as a function of the mass ratio M was evaluated in similar fashion to the case of System I. It can be seen that unlike the case shown in Fig. 9.2, 9 the function is not a decreasing one over the whole range of M but has a minimum at $M = 0.7$. Since in practice the range of M of System II is comparatively broad, the boundaries of the region of stability of the equilibrium position are drawn in axonometric view in the (β, q, M) space for $\varkappa = 0.2$ (Fig. 9.2, 11) and $\varkappa = 0.4$ (Fig. 9.2, 12). As the diagrams reveal, with increasing M the maximum value of β and consequently the ability to ensure compensation of the self-excitation effect, decrease even for optimum tuning. On the

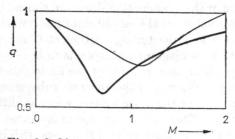

Fig. 9.2, 10

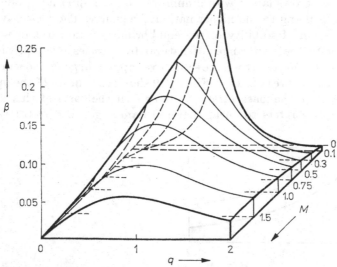

Fig. 9.2, 11

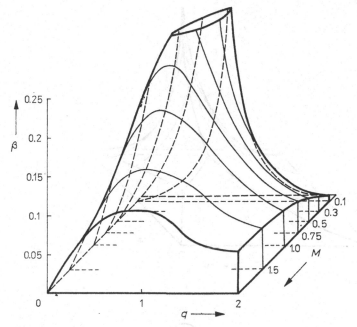

Fig. 9.2, 12

other hand, as M is increased, the system becomes less sensitive to accurate tuning of the foundation.

For System III, the $\beta(\varkappa)$ curves corresponding to both frequencies have the same features as those for System I, the only difference being the fact that $q > 1$ for the optimum value of the tuning coefficient and that the optimum value of q increases with increasing M (Fig. 9.2, 13). The regions of stability of the equilibrium position are shown in axonometric view in Fig. 9.2, 14 ($\varkappa = 0.2$) and Fig. 9.2, 15) ($\varkappa = 0.4$). Their character is similar to that of the corresponding regions of System II.

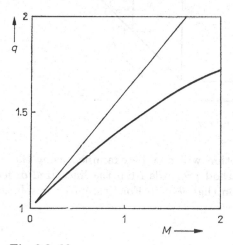

Fig. 9.2, 13

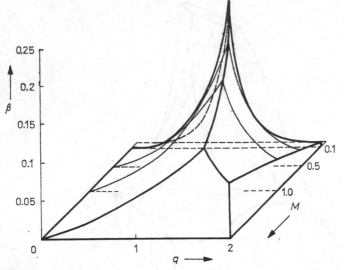

Fig. 9.2, 14

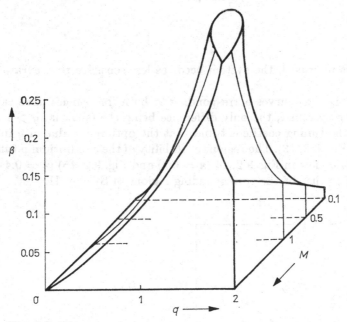

Fig. 9.2, 15

The stability of the equilibrium position will now be examined using the method of the boundary value curves. The method proceeds from the linearized differential equations of motion and the assumption that $\vartheta = 0$. For System I the solution

$$\left.\begin{array}{l} x_2 = \cos \Omega\tau , \\ x = a \cos \Omega\tau + b \sin \Omega\tau \end{array}\right\}$$

(9.2, 8)

is substituted in (9.2, 1 b) and the following equations are obtained to determine a, b, Ω, Q:

$$\left.\begin{aligned} a &= \frac{[1 - (1 + \mu)\,\Omega^2]}{\mu\Omega^2}\,, \\[2ex] b &= \frac{\beta}{\mu\Omega}\,, \\[2ex] Q_{\mathrm{I}} &= \Omega\left(1 + \frac{a}{a^2 + b^2}\right)^{1/2}, \\[2ex] Q_{\mathrm{II}} &= \frac{\Omega}{\varkappa}\,\frac{b}{a^2 + b^2}\,. \end{aligned}\right\} \qquad (9.2, 9)$$

For System II and System III, the problem is solved in a similar manner.

The results obtained by means of this procedure are presented in Fig. 9.2, 16 which shows an example of the boundary value curve for $\beta = 0.2$ and $\mu = 0.1$. The equilibrium position is stable only for the values lying inside the loop. The coordinates of the centre of the loop can be taken for optimum values of the parameters $\varkappa$ and Q. Fig. 9.2, 17 shows the boundary value curves for $\mu = 0.05$ and three values of β. As suggested by the diagram, no value of $\varkappa$ can stabilize the equilibrium position for $\beta = 0.3$. Sets of the boundary value curves were arranged into comprehensive diagrams, showing, for example, the effect of β when $\mu = 0.1$ (Fig. 9.2, 18) and the effect of μ when $\beta = 0.1$ (Fig. 9.2, 19). An increasing β or a decreasing μ causes the region of stability of the equilibrium position to diminish substantially.

The results obtained for System II are shown in Fig. 9.2, 20 to Fig. 9.2, 22, the first two figures indicating the boundary value curves for $M = 0.3$ and $M = 1$ ($\beta = 0.1$ in both cases), the third containing the comprehensive diagrams ((a) — front view, (b) — rear view; the cut-out portion of the cube represents the region of stability of the equilibrium position). For small values of M stabilization can be achieved for low values of $\varkappa$ in a narrow interval of q; for large values of M, $\varkappa$ must be increased but the system is less sensitive to accurate tuning.

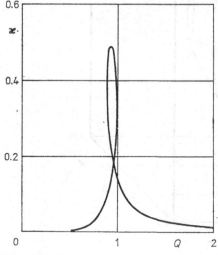

Fig. 9.2, 16

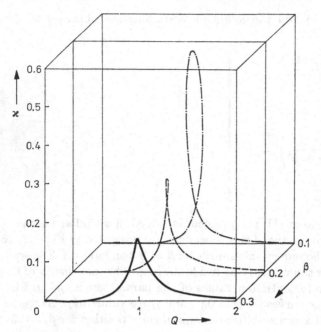

Fig. 9.2, 17

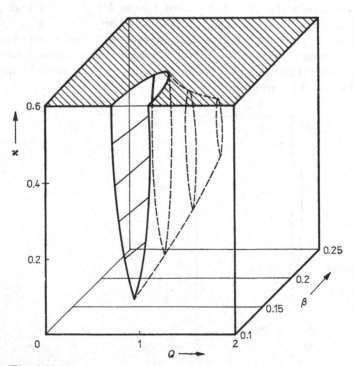

Fig. 9.2, 18

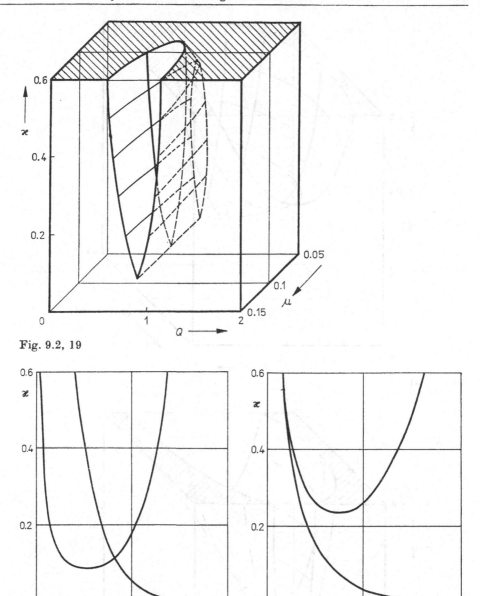

Fig. 9.2, 19

Fig. 9.2, 20   Fig. 9.2, 21

The results obtained for System III are shown in Figs. 9.2, 23 and 9.2, 24; the first figure presents the boundary value curves for $M = 0.3$ and $\beta = 0.1$, the second (axonometric view) reveals the effect of the mass ratio M (for $\beta = 0.1$).

The procedure for calculating the amplitudes of single-frequency self-excited vibration as a function of a system's parameter will now be explained using System I as an example. Substituting the approximate solution

$$
\left.
\begin{aligned}
x &= a \cos \Omega\tau + b \sin \Omega\tau, \\
x_2 &= X_2 \cos \Omega\tau
\end{aligned}
\right\}
$$

(9.2, 10)

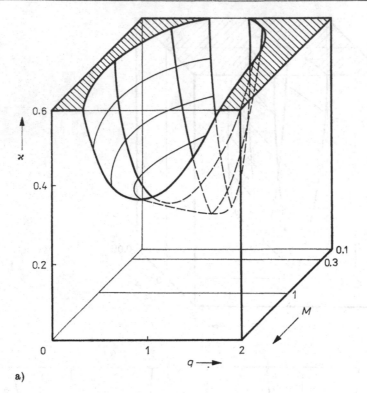

a)

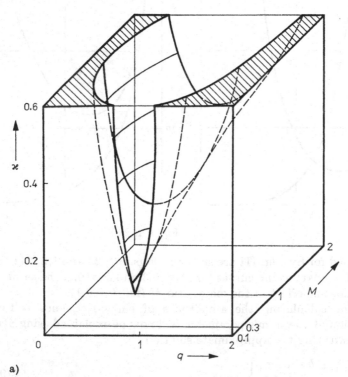

a)

Fig. 9.2, 22

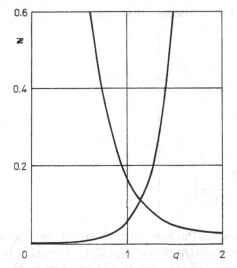

Fig. 9.2, 23

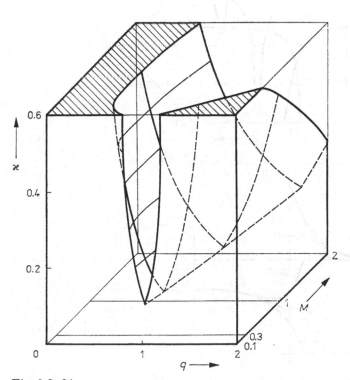

Fig. 9.2, 24

in (9.2, 1 b) we obtain, for a given X_2, the following equations

$$a = \frac{X_2[1 - (1 + \mu)\,\Omega^2]}{\mu\Omega^2}\,,$$

$$b = X_2\frac{\left(\beta - \frac{1}{4}\,\delta X_2^2\right)}{\mu\Omega}\,,$$

$$Q_{\mathrm{I}} = \Omega\left(1 + \frac{aX_2}{a^2 + b^2}\right)^{1/2}\,,$$

$$Q_{\mathrm{II}} = \frac{\left(\Omega^2\dfrac{bX_2}{a^2 + b^2} - \dfrac{4}{\pi}\,\dfrac{\vartheta}{(a^2 + b^2)^{1/2}}\right)}{\varkappa\Omega}\,.$$

(9.2, 11)

The calculation carried out on a Hewlett-Packard 9830 A calculator proceeds as follows: Using (9.2, 11) one obtains functions $Q_{\mathrm{I}}(\Omega)$ and $Q_{\mathrm{II}}(\Omega)$ (Ω is varied in steps) for several values of X_2, and from the points of intersection of these curves, the corresponding values of Q. In this way the relationships $X_2(Q)$ and $\Omega(Q)$ are established.

Some results obtained by means of this procedure for System I and System II are shown below. Fig. 9.2, 25 shows functions $X_2(Q)$ and $\Omega(Q)$ for various values of $\varkappa$

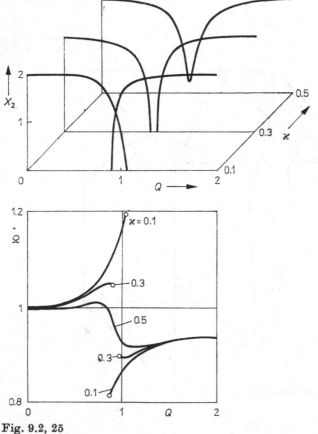

Fig. 9.2, 25

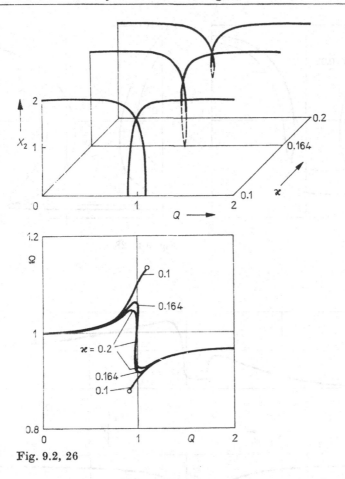

Fig. 9.2, 26

and $\beta = \delta = 0.2$, $\mu = 0.1$. Fig. 9.2, 26 shows the two curves for $\delta = 0.2$, $\beta = 0.3$ and $\mu = 0.05$. In Fig. 9.2, 25, the boundary value curve $\varkappa(Q)$ has a loop form, in Fig. 9.2, 26, a peak form. In the latter case, the absorber is not capable of quenching self-excited vibration for any tuning and damping coefficient.

The effect of dry friction is evaluated using the results obtained for System II. In all cases discussed, $\beta = \delta = \varkappa = 0.2$. Fig. 9.2, 27 shows the curve $X(q)$ for the mass ratio $M = 0.3$, Fig. 9.2, 28 for $M = 1$; the coefficient of dry friction $\vartheta = 0.05$ and $\vartheta = 0.1$ in both figures. As the diagrams suggest, no stable equilibrium position exists under the effect of dry friction; however, there exists an interval of the tuning coefficient in which only small-amplitude vibrations are present.

Many analogue solutions were carried out to check and complement the theoretical analyses. One of them, for example, was concerned with the dependence of extreme deflections (denoted by $[x_1]$, $[x_2]$, etc., for the purpose of differentiation) on the tuning coefficient Q (System I) or q (System II and System III). In all the diagrams, $\varkappa = \beta = \delta = 0.2$. Fig. 9.2, 29 shows $[x_1]$, $[x_2]$ drawn as functions of Q (System I) for $\mu = 0.1$ and $\vartheta = 0.2$. Dry friction can be seen to cause the interval of stable equilibrium position to be replaced by an interval of Q slightly broader than that corresponding to the case of $\vartheta = 0$, in which small-amplitude vibration occurs. Fig. 9.2, 30 drawn for $\mu = 0.1$ shows the effect of a varying dry friction coefficient. As ϑ is increased, not only the interval but also the amplitude of the small-amplitude vibra-

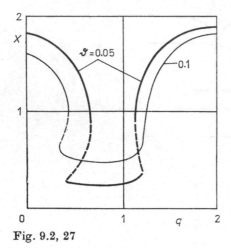

Fig. 9.2, 27

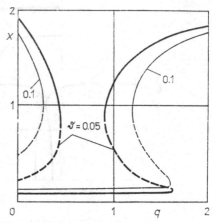

Fig. 9.2, 28

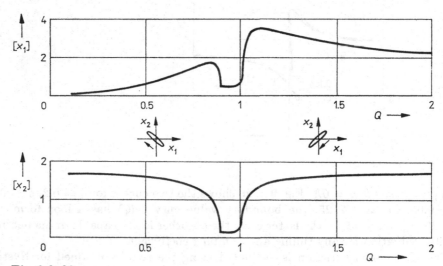

Fig. 9.2, 29

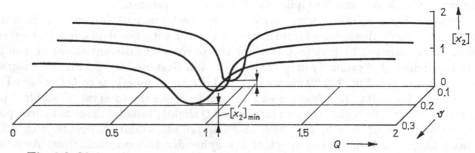

Fig. 9.2, 30

tion grows larger. Fig. 9.2, 31 shows the effect of the ratio between the absorber and the basic system masses, μ, in the case of $\vartheta = 0.1$. An increasing μ causes the interval of occurrence of small-amplitude vibration to broaden. The small-amplitude vibration does not always represent the only stable steady solution; in some cases there exist intervals of Q with two stable steady solutions, one represented by small-amplitude, the other by large-amplitude vibration.

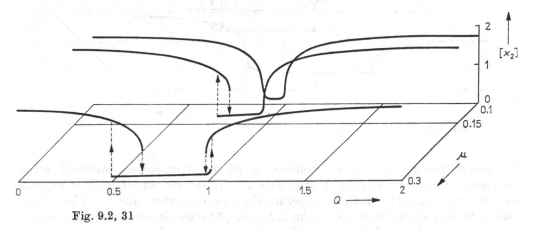

Fig. 9.2, 31

An example of the results obtained for System II for $M = 0.3$ is shown in Fig. 9.2, 32. As seen in the diagram, dry friction has a highly favourable effect, even in the case when for $\vartheta = 0$ no interval of q exists in which the equilibrium position is stable. A similar result is obtained for System III having the same value of M (Fig. 9.2, 33).

The next example relates to an investigation of stability in the large of System II. Fig. 9.2, 32 showed the case when two locally stable steady solutions existed for certain intervals of the tuning coefficient q. The investigation that follows concerns the resistance of the small-amplitude vibration to disturbances in the form of a sinusoidal pulse (Chapter 7). Let the disturbance function (a sinusoidal pulse) be described by the equation

$$P(\tau) = \begin{cases} A \sin \dfrac{2\pi}{\tau_0} \tau & \text{for} \quad \tau \leqq \dfrac{T_0}{2}, \\[2mm] 0 & \text{for} \quad \tau > \dfrac{T_0}{2}. \end{cases} \tag{9.2, 12}$$

Fig. 9.2, 32

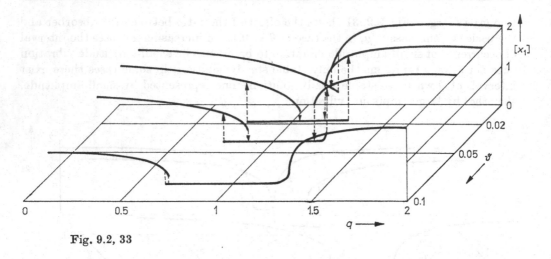

Fig. 9.2, 33

The values $\beta = \delta = \varkappa = 0.2$ are common to all the alternatives considered in the discussion. Since the vibration, particularly that in the coordinate x_2, is small, this steady state can, approximately, be regarded as non-oscillatory and the effect of the instant of the pulse action can be ignored. Fig. 9.2, 34 shows an example of the domain of attraction in the $(A, T_0/T_1)$ plane ($T_1 = 2\pi$ is the period of the natural vibration of the basic self-excited subsystem) for $M = 0.5$, $q = 0.5$ and $\vartheta = 0.015$, obtained by means of the automatic process on analogue computer using the method described in Chapter 7. The values of the disturbance parameters lying in the cross-hatched area lead to large-amplitude vibration. Diagrams of this kind drawn for various values of ϑ can be arranged in an axonometric pattern. Those corresponding to the case of $M = 0.3$ are shown in Fig. 9.2, 35 ($q = 0.5$), Fig. 9.2, 36 ($q = 0.75$) and Fig. 9.2, 37 ($q = 1$). All diagrams clearly show the influence of the tuning coefficient q on the parameters A, T_0/T_1. The closer the value of q is to unity, the more resistant is the steady state being examined to disturbing pulses. For a certain value of ϑ, the boundary function $A(T_0/T_1)$ has two distinct local minima.

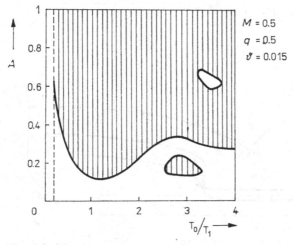

Fig. 9.2, 34

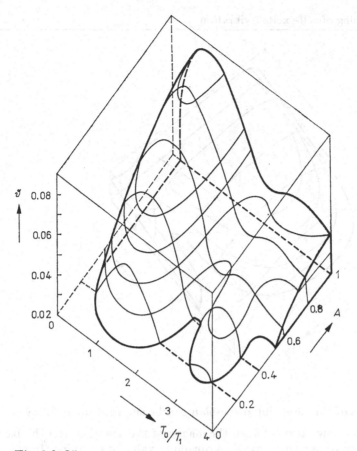

Fig. 9.2, 35

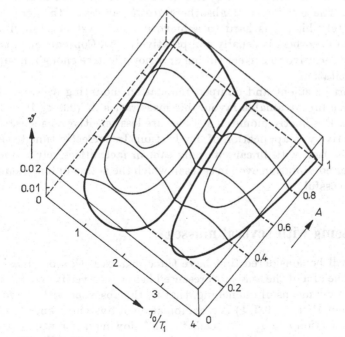

Fig. 9.2, 36

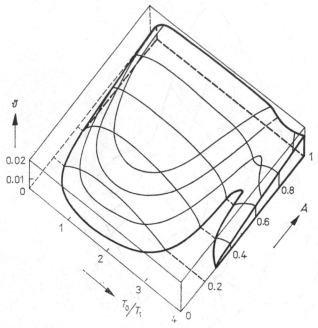

Fig. 9.2, 37

The main results of the foregoing discussion can be summed up as follows:

For system with a low ratio between the masses of the absorber and the basic system, μ, absorbers were shown to have an optimum value of damping. In some cases of intensive self-excitation an absorber may not be able to ensure stabilization of the equilibrium position. The efficiency of absorbers grows rapidly with increasing μ. However, an adequately high μ is hard to achieve in most systems, for in actual structures the mass of absorbers is usually comparatively low. Consequently, the absorber is either highly sensitive to correct tuning or is not effective enough in suppressing intensive self-excitation.

In cases of this sort, resilient and damped foundation mounting (System II and System III) having readily attainable favourable mass ratios M $(0.3 < M < 1)$ can be used to advantage. Resiliently mounted systems are less sensitive to accurate tuning and respond effectively to application of dry friction. In correctly tuned systems, dry friction in combination with linear damping can, in fact, effectively reduce the amplitudes of self-excited vibration even in cases in which the action of linear damping alone has been unsuccessful.

9.3.   Chain systems with several masses

The systems which will be considered first have three masses with a one-mass basic self-excited system. The aim of the analysis presented below is to verify the possibility of combining two of their means of quenching, that is, the absorber with the foundation subsystem. System IV (Fig. 9.3, 1) is a combination of Systems I and II. Introducing the relative deflection $v = x_1 - x_2$ leads to the following differential equations of motion (after rearrangement to the dimensionless form in the coefficients of the

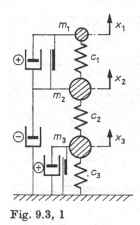

Fig. 9.3, 1

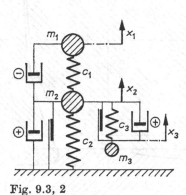

Fig. 9.3, 2

linear terms):

$$v'' + x_2'' + \varkappa_1 Qv' + \vartheta_1 \operatorname{sgn} v' + Q^2 v = 0 \,, \\ (1 + \mu) x_2'' + \mu v'' - (\beta - \delta x_2^2) x_2' - x_2 - x_3 = 0 \,, \\ x_3'' + \varkappa_3 q x_3' - \vartheta_3 \operatorname{sgn} x_3' + q^2 x_3 - M(x_2 - x_3) = 0 \quad \Bigg\} \qquad (9.3, 1)$$

where

$$Q^2 = \frac{c_1/m_1}{c_2/m_2} \,; \qquad q^2 = \frac{c_3/m_3}{c_2/m_2} \,, \qquad \mu = \frac{m_1}{m_2} \,, \qquad M = \frac{m_2}{m_3} \,,$$

System V (Fig. 9.3, 2) is a combination of Systems I and III, with the absorber connected to the foundation mass. Writing

$$x_1 = x_2 = u \,, \qquad x_3 - x_2 = v \,, \qquad x_2 = x$$

we obtain the differential equations

$$u'' + x'' + u - (\beta - \delta u^2)\, u' = 0 \,, \\ x'' + M(u'' + x'') + \mu(v'' + x'') + q^2 x + \varkappa_2 q x' + \vartheta_2 \operatorname{sgn} x' = 0 \,, \\ v'' + x'' + Q^2 v + \varkappa_3 Q v' + \vartheta_3 \operatorname{sgn} v' = 0 \quad \Bigg\} \qquad (9.3, 2)$$

where

$$q^2 = \frac{c_2/m_2}{c_1/m_1} \,, \qquad Q^2 = \frac{c_3/m_3}{c_1/m_1} \,, \qquad M = \frac{m_1}{m_2} \,, \qquad \mu = \frac{m_3}{m_2} \,.$$

Since the procedure adopted for the solution was already outlined in the preceding section, only some results obtained are presented here (for further details refer to Tondl (1980b)). The examples of the boundary value curves for System IV drawn in the (Q, q) plane, that is, in the coordinates of the tuning coefficients of the absorber and the foundation, are shown first. All the alternatives have the following parameters in common: $\varkappa_1 = \beta = 0.2$, $\mu = 0.1$, $\vartheta_1 = \vartheta_3 = 0$. Fig. 9.3, 3a shows the boundary value curves for $\varkappa_3 = 0.2$, $M = 0.3$ (for better clarity, Fig. 9.3, 3b shows the curves of Ω corresponding to the boundaries). Fig. 9.3, 4 shows the boundary value curves for $\varkappa_3 = 0.15$, $M = 0.3$, and Fig. 9.3, 5 those for $\varkappa_3 = 0.2$, $M = 0.5$. The domains of the parameter values in which the equilibrium position is unstable with respect to the lower or the higher natural frequency of the system, are done in hatching. Two sepa-

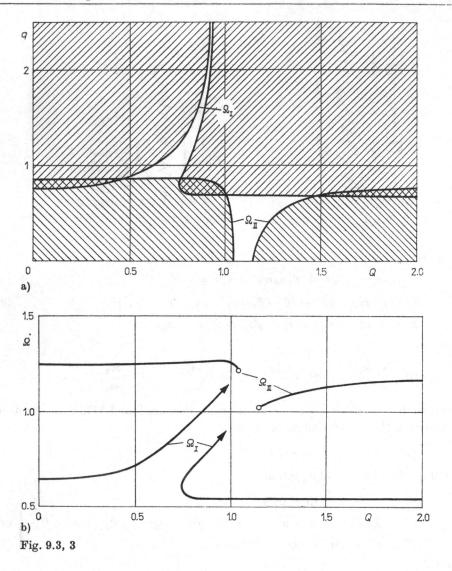

Fig. 9.3, 3

rate stability domains of the equilibrium position exist for the first example shown in Fig. 9.3, 3. For the second and third example, owing to the effect of either decreased damping or increased mass ratio M, the lower region disappears, although it is broader in the first example than the stability domain for larger values of q. It is therefore advantageous to provide a heavy enough damping of the foundation mass as well as a large enough foundation mass. Fig. 9.3, 6 is an axonometric view of the stability domain of the equilibrium position for the case of $\mu = 0.1$, $M = 0.3$, $\beta = \varkappa_3 = 0.2$ drawn in the coordinates Q, q (the tuning coefficients) and $\varkappa_1$ (the absorber damping coefficient).

In three-mass systems, the effect of dry friction is similar to that observed in two-mass systems. To demonstrate it by way of an example, Fig. 9.3, 7 shows a diagram plotting the amplitude of the foundation mass vibration X_3 as a function of the absorber tuning coefficient Q for $\varkappa_1 = \varkappa_2 = 0.2$, $\mu = 0.1$, $M = 0.3$, $q = 0.9$ and three values of the dry friction coefficients of the motion of the absorber and the foundation

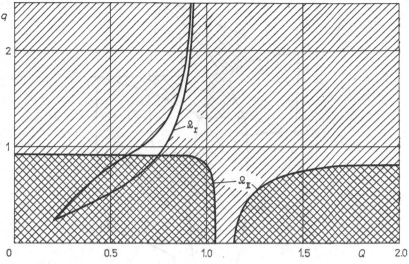

Fig. 9.3, 4

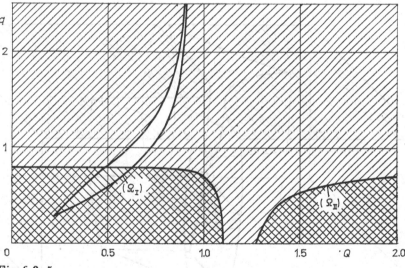

Fig. 9.3, 5

mass subsystems, ϑ_1 and ϑ_3. As the diagram suggests, the action of dry friction of the absorber relative motion alone ($\vartheta_1 = 0.05$, $\vartheta_3 = 0$) causes the interval of Q in which the equilibrium position is stable when dry friction is not present to be replaced by a somewhat broader interval of Q in which small-amplitude vibration occurs. The effect of dry friction of the foundation mass motion ($\vartheta_1 = 0$, $\vartheta_3 = 0.05$) is greater and as a result only small-amplitude vibration exists in the interval investigated, $0 < Q < 1.5$. These analytical results are confirmed in full by analogue computations. If the foundation mass is properly tuned, application of dry friction of the foundation mass motion can bring about a very efficient reduction of the self-excited vibration amplitude in a broad interval of values of the absorber tuning coefficient.

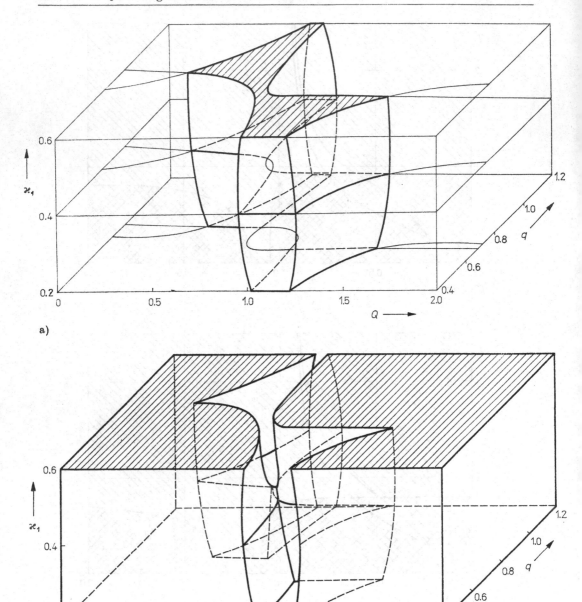

a)

b)

Fig. 9.3, 6

Examples of the results of an analysis of System V which follow are presented in diagrams all drawn for $\beta = \varkappa_2 = 0.2$ and $\mu = 0.1$. Figs. 9.3, 8 ($M = 0.3$) and 9.3, 9 ($M = 1$) show the regions of stability of the equilibrium position in the coordinates Q, q, $\varkappa_2$ (the absorber damping coefficient). Although the diagrams recording the effect of dry friction are on the whole similar to those obtained for System IV, they show an interesting anomalous effect of dry friction of the foundation mass motion. How-

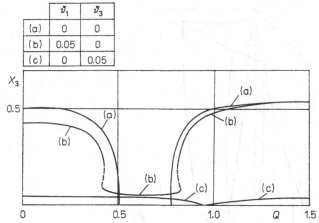

	ϑ_1	ϑ_3
(a)	0	0
(b)	0.05	0
(c)	0	0.05

Fig. 9.3, 7

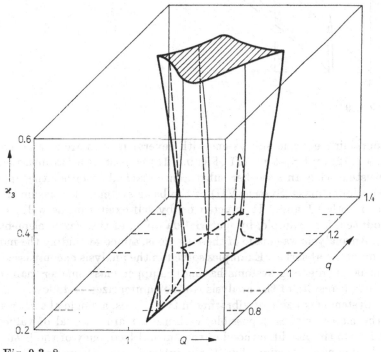

Fig. 9.3, 8

ever, this phenomenon, not noted in other cases, was present only for $M = 0.3$ and only within a certain interval of the foundation tuning coefficient q. Fig. 9.3, 10 shows the relation between the amplitude of the relative deflection $u = x_1 - x_2$ and the tuning coefficient Q for $q = 1.3$, $\mu = 0.1$, $M = 0.3$, $\beta = \delta = \varkappa_2 = \varkappa_3 = 0.2$ and for the friction coefficients of the foundation mass motion and absorber relative motion ϑ_2, ϑ_3 listed in the figure. For the alternatives (b) and (d) ($\vartheta_2 = 0.05$ and 0.06), the vibration amplitude reaches its maximum in the region in which the tuning is optimal for other cases. As Fig. 9.3, 11 representing the alternative having $q = 1.3$ suggests, the presence of this anomaly is confirmed by analogue computations. It is also evident, to a lesser extent, in cases when $q = 1$ and 1.5.

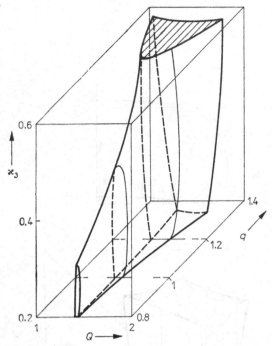

Fig. 9.3, 9

Systems containing a basic subsystem with several masses are treated next. System VI (Fig. 9.3, 12) and System VII (Fig. 9.3, 13) consist of a two-mass self-excited basic subsystem with an absorber subsystem attached to either the upper (System VI) or the lower mass (System VII) of the basic system. It is assumed that the basic system of both VI and VII is acted on by self-excitation as well as positive damping (produced, for example, by material damping of the elastic elements of the basic system). An analysis was made of these systems, aimed at finding the most satisfactory position of the absorber. Examples solved in the analysis encompassed a range of the most usual engineering systems having the upper mass smaller than the lower one. The important results of the analysis can be summarized as follows.

If the basic system is capable of vibrating in two modes, a single absorber attached to either of the masses makes it possible, within a certain interval of values of the tuning coefficient in the neighbourhood of the natural frequency of the basic system, to quench only one mode of vibration. If the intervals do not overlap (as is usually the case), application of a single absorber cannot ensure complete quenching of self-excited vibration (Fig. 9.3, 14). The first vibration mode (of the lower frequency) is effectively quenched by an absorber attached to the upper mass of the basic system (System VI), the second, by an absorber attached to the lower mass (System VII). The intensity of self-excitation in the first or second vibration mode is affected by the level of positive damping of the motion of the various masses of the basic system. Reduction of the positive damping of the lower mass brings about an increase in the effect of self-excitation, particularly in the second vibration mode.

System VIII (Fig. 9.3, 15) consists of a three-mass basic system (masses, connecting springs, relative positive damping and self-excitation are the same for all three masses: $m_2 = m_3 = m_4 = m$; $c_2 = c_3 = c_4 = c$; $\varkappa_2 = \varkappa_3 = \varkappa_4 = \varkappa$) and an absorber sub-

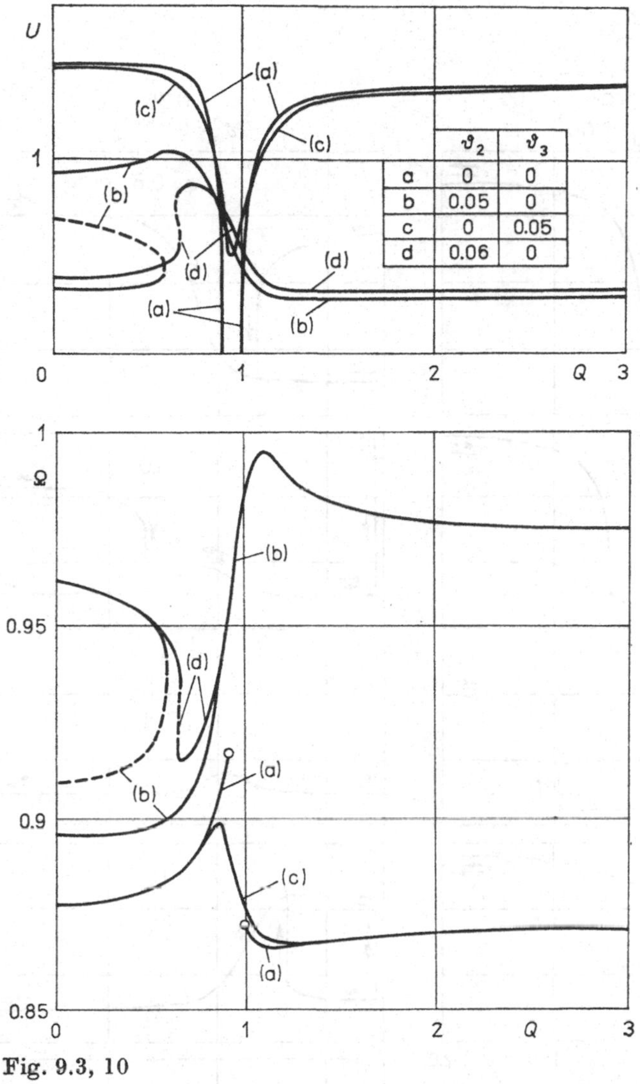

	ϑ_2	ϑ_3
a	0	0
b	0.05	0
c	0	0.05
d	0.06	0

Fig. 9.3, 10

system attached to the upper mass of the basic system. Fig. 9.3, 16 shows the boundary value curves in the $(\varkappa_1, Q)$ plane for $\mu = 0.05$ and several values of the coefficient of positive damping of the basic system $\varkappa$. Fig. 9.3, 17 shows the boundary value curves for $\varkappa = 0.1$ and several values of the absorber mass ratio $\mu = m_1/m$. $\beta = 0.2$ in both diagrams. As the diagrams imply, a region of $\varkappa_1$, Q in which the equilibrium position is stable, is obtained for fairly large values of μ. In case of high intensity self-excitation and light positive damping in the basic system, a low-mass absorber is not capable of stabilizing the equilibrium position.

As an analysis of System IX (Fig. 9.3, 18) reveals (Tondl (1980)) the equilibrium position of self-excited basic systems (System IX has a two-mass one) which can vibrate in two modes, can be stabilized by application of two sufficiently large and unequally tuned absorbers.

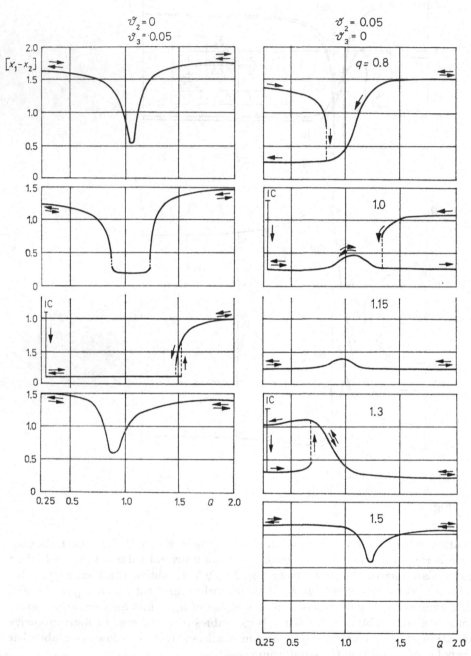

Fig. 9.3, 11 (IC denotes that different non-trivial initial conditions were applied)

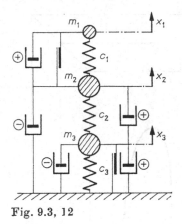

Fig. 9.3, 12

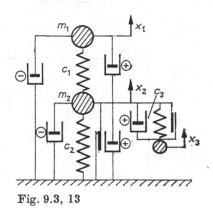

Fig. 9.3, 13

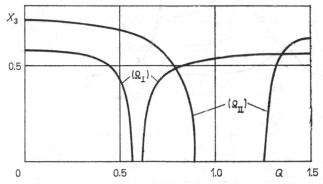

Fig. 9.3, 14

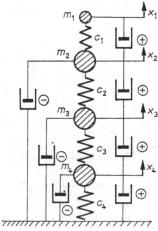

Fig. 9.3, 15

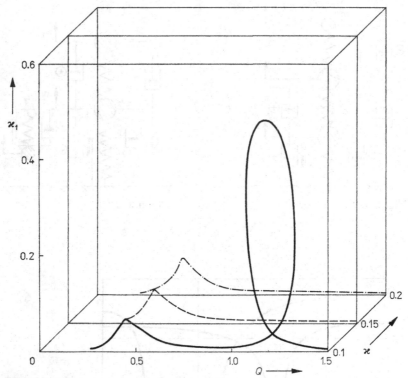

Fig. 9.3, 16

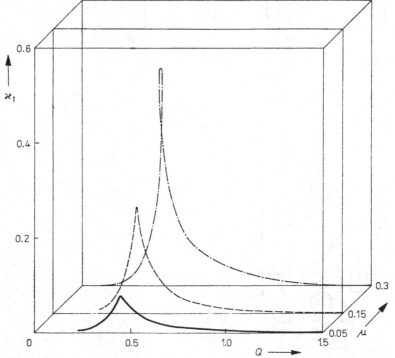

Fig. 9.3, 17

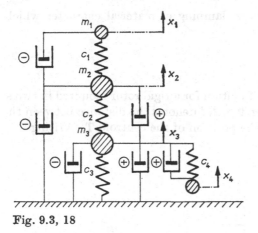

Fig. 9.3, 18

9.4.   Example of a rotor system

Consider a system consisting of a rigid, ideally balanced rotor supported in air pressurized bearings and rotating about a rigid but elastically mounted vertical axis (Fig. 9.4, 1 a) hinged in the plane of bearing 1. This arrangement corresponds, approximately, to the mounting of the axis in a rubber ring (shown schematically in Fig. 9.4, 1 b), i.e. to the case of rigidity in the radial direction being several times greater than that against the angular deflections of the axis. The model represents the spindle of a cotton-yarn spinning machine. Its motion is defined by three vectors of the plane deflections, z_1, z_2, z_0. Let m denote the rotor mass, I the equatorial moment of inertia about the axis perpendicular to the axis of rotation passing through the centroid T, and I_ω the polar moment of inertia about the axis of rotation. Let m_0 be the mass of the elastically mounted axis reduced to the plane of the second bearing. The rigidity of the axis mounting c_0 is reduced to the same plane. Mounting by

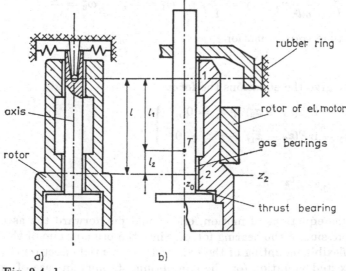

Fig. 9.4, 1

means of a rubber ring also represents damping of material character which can be expressed as

$$-c_0 k\,|z_0|\,\frac{\dot{z}_0}{|\dot{z}_0|}$$

To simplify, write the equations of motion for a rigid rotor mounted in two supports of rigidity c_1, c_2, respectively; in Fig. 9.4, 2, l denotes the distance between the planes of the supports, and l_1, l_2 describe the position of the centroid T. Writing

$$\alpha_1 = \frac{l_2}{l}, \qquad \alpha_2 = \frac{l_1}{l} \tag{9.4, 1}$$

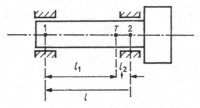

Fig. 9.4, 2

and denoting by ω the angular velocity of rotor rotation, and by z_1, z_2 the vectors of the deflections in the planes of mounting, one can describe the system shown in Fig. 9.4, 2 by the following equations of motion (see TONDL (1973c):

$$\left.\begin{array}{l} A\ddot{z}_1 + b\ddot{z}_2 - i\omega E(\dot{z}_1 - \dot{z}_2) + \omega_1^2 z_1 = 0\,, \\[2mm] b\ddot{z}_1 + B\ddot{z}_2 - i\omega E(\dot{z}_2 - \dot{z}_1) + \omega_2^2 z_2 = 0 \end{array}\right\} \tag{9.4, 2}$$

where

$$A = \alpha_1^2 + \frac{I}{ml^2}, \qquad b = \alpha_1\alpha_2 - \frac{I}{ml^2}, \qquad B = \alpha_2^2 + \frac{I}{ml^2},$$

$$E = \frac{I}{ml^2}\frac{I_\omega}{I} = \frac{I}{ml^2}\gamma\,, \qquad \gamma = \frac{I_\omega}{I}\,, \qquad \omega_1^2 = \frac{c_1}{m}\,, \qquad \omega_2^2 = \frac{c_2}{m}\,.$$

Introducing the time transformation

$$\omega_1 t = \tau \tag{9.4, 3}$$

makes it possible to give the equations the form

$$\left.\begin{array}{l} Az_1'' + bz_2'' - i\nu E(z_1' - z_2') + z_1 = 0\,, \\[2mm] bz_1'' + Bz_2'' - i\nu E(z_2' - z_1') + p^2 z_2 = 0 \end{array}\right\} \tag{9.4, 4}$$

where

$$\nu = \frac{\omega}{\omega_1}\,, \qquad p^2 = \frac{c_2}{c_1}\,.$$

Before writing the equations of motion, one should put forward the assumptions concerning the expression of the bearing forces. Since the primary aim of the analysis is to examine the flexible mounting of the axis and its essential effect on the limit of initiation of self-excited vibration (on the equilibrium stability in case of an ideally balanced rotor), a very simple expression of the bearing forces will be used. Assume the bearings to be identical, both of the hybrid type, and the resulting vector of the bearing forces to be expressed in terms of two independent components, one representing the aerostatic, the other the aerodynamic contribution. In linear form, the aerostatic component can be described by the product of rigidity c and the vector of the deflection (this component has the character of a central force). The second component represents the aerodynamic force which arises owing to the effect of shaft rotation and viscosity of the gas. For an incompressible medium the aerodynamic component in its simplest form (see TONDL (1974a)) is described by (for the deflection vector z)

$$h(2\dot{z} - i\omega z) \, . \tag{9.4, 5}$$

For a compressible medium and small amplitudes of the whirling motion, MARSH (1965) deduced the following definition of the vector of the aerodynamic forces

$$(P_e - iP_\varphi) \exp (i\varphi) \, . \tag{9.4, 6}$$

where φ is the position angle; the radial component P_e and the tangential component P_φ are defined by

$$P_e = \pi p_a L r_1 \varepsilon \frac{\Lambda^2 (1 - 2\dot{\varphi}/\omega)^2}{1 + \Lambda^2 (1 - 2\dot{\varphi}/\omega)^2} \, , \tag{9.4, 7}$$

$$P_\varphi = \pi p_a L r_1 \varepsilon \frac{\Lambda (1 - 2\dot{\varphi}/\omega)}{1 + \Lambda^2 (1 - 2\dot{\varphi}/\omega)^2} \tag{9.4, 8}$$

where p_a is the ambient pressure, L is the length of the bearing, r_1 is the shaft radius, $\varepsilon = |z|/\delta$ is the relative eccentricity of the shaft centre from the bearing centre, δ is the radial clearance in the bearing, and Λ is the compressibility number given by

$$\Lambda = 6\eta \omega r_1^2 p_a^{-1} \delta^2 \tag{9.4, 9}$$

where η is the coefficient of dynamic viscosity of the gas. The quantities ε, φ and z are related as follows:

$$z = \varepsilon\delta \exp (i\varphi) \, . \tag{9.4, 10}$$

For an incompressible medium the equations of motion of the system shown in Fig. 9.4, 1 can be rearranged to take the form

$$\left.\begin{array}{l} A z_1'' + b z_2'' - i\nu E(z_1' - z_2') + z_1 + K(2z_1' - i\nu z_1) = 0 \, , \\[4pt] b z_1'' + B z_2'' - i\nu E(z_2' - z_1') + z_2 - z_0 + K[2(z_2' - z_0') - i\nu(z_2 - z_0)] = 0 \, , \\[4pt] z_0'' - M\{z_2 - z_0 + K[2(z_2' - z_0') - i\nu(z_2 - z_0)]\} + q^2 \left(z_0 + k\,|z_0|\dfrac{z_0'}{|z_0'|}\right) = 0 \end{array}\right\} \tag{9.4, 11}$$

where

$$K = \frac{h}{m\omega_1} \, , \qquad M = \frac{m}{m_0} \, , \qquad q^2 = \frac{c_0/m_0}{c/m} \, .$$

K is the coefficient of the aerodynamic component, M the ratio between the masses, and q the tuning coefficient whose optimum value is to be established.

If the compressibility of the medium is taken into account, expedient introduction of the relative deflection

$$w = z_2 - z_0$$

and some rearrangement changes the equations of motion to

$$
\left.
\begin{aligned}
&Az_1'' + bz_2'' - i\nu E(z_1' - z_2') + z_1[1 + K(p_e - ip_\varphi)] = 0 , \\
&bz_1'' + Bz_2'' - i\nu E(z_2' - z_1') + w[1 + K(p_e - ip_\varphi)] = 0 , \\
&z_2'' - w'' + q^2[(z_2 - w) + k|z_2 - w|\frac{z_2' - w'}{|z_2' - w'|} - Mw[1 + K(p_e - ip_\varphi)] = 0
\end{aligned}
\right\}
$$

$$(9.4, 12)$$

where

$$
K = \frac{\pi p_a L r_1}{m\delta\omega_1^2} , \qquad p_e = \frac{\Lambda_0^2(\nu - 2\varphi')^2}{1 + \Lambda_0^2(\nu - 2\varphi')^2} ,
$$

$$
p_\varphi = \frac{\Lambda_0(\nu - 2\varphi')}{1 + \Lambda_0^2(\nu - 2\varphi')^2} , \qquad \Lambda_0 = 6\eta\omega_1 r_1 p_a^{-1}\delta^2 .
$$

As a first step, find the boundary values ν_s as functions of the coefficient of tuning q for the system described by (9.4, 12). Substituting in (9.4, 12) the solution

$$
z_1 = \exp(i\Omega\tau) , \qquad z_2 = (Z + iW)\exp(i\Omega\pi) , \qquad z_0 = (U + iV)\exp(i\Omega\tau)
$$

$$(9.4, 13)$$

leads to four algebraic equations determining Z, W, U and V as functions of Ω, that is,

$$
Z = \frac{(1 - F)}{H} , \qquad W = \frac{K_0}{H} , \qquad U = \frac{(X + K_0 Y)}{\Delta} , \qquad V = \frac{(Y - K_0 X)}{\Delta}
$$

$$(9.4, 14)$$

where

$$
K_0 = K(2\Omega - \nu) , \qquad F = \Omega^2 A - \nu\Omega E , \qquad G = \Omega^2 B - \nu\Omega E ,
$$

$$
H = \Omega^2 b - \nu\Omega E , \qquad \Delta = 1 + K_0^2 ,
$$

$$
X = -H - GZ + Z - K_0 W , \qquad Y = -GW + W + K_0 Z
$$

and two equations for the function $q(\Omega)$

$$
\left.
\begin{aligned}
q_{\mathrm{I}}^2 &= (U + kV)^{-1}\{\Omega^2 U + M[Z - U - K_0(W - V)]\} , \\
q_{\mathrm{II}}^2 &= (V + kU)^{-1}\{\Omega^2 V + M[W - V + K_0(Z - U)]\} .
\end{aligned}
\right\}
$$

$$(9.4, 15)$$

For different values of ν these equations yield two systems of curves, $q_{\mathrm{I}}(\Omega; \nu)$ and $q_{\mathrm{II}}(\Omega; \nu)$ the points of intersection of which describe the corresponding boundary values of ν_s as functions of q. The examples presented below were solved using a Hewlett-Packard 9830 A calculator with a graph plotter.

Several alternatives were examined to establish the effect of the centroid position ((a): $\alpha_1 = -0.25$, (b): $\alpha_1 = 0$, (c): $\alpha_1 = 0.5$, (d): $\alpha_1 = 1$, see Fig. 9.4, 3), the mass ratio M, of I/ml^2 and the gyroscopic action. Some of the results obtained are presented in diagrams showing the boundary value curves $\nu_s(q)$ (solid lines) and the curves $\nu_K(q)$ (dashed lines). The minimum value, $\nu_{s\min}$, is indicated by heavy solid line. The numbers in circles give the critical frequencies or the limit frequencies of initiation of self-excited vibration corresponding to the first, second, ... critical speed. In all the cases examined, the values of the parameters were:

$$
I/ml^2 = 1 , \qquad M = 3 , \qquad \gamma = 0.2 , \qquad k = 0.1 , \qquad K = 0.2
$$

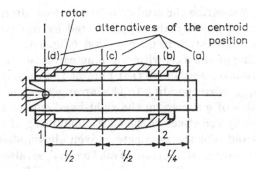

Fig. 9.4, 3

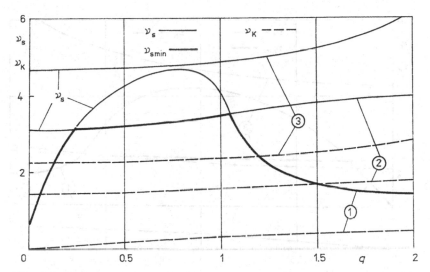

Fig. 9.4, 4

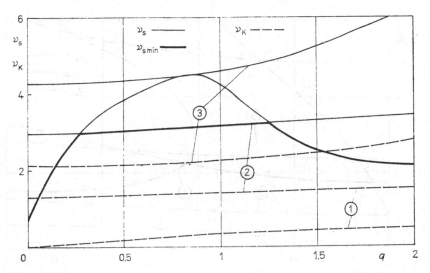

Fig. 9.4, 5

The diagrams in Figs. 9.4, 4 to 9.4., 7 describe the studied effect for four alternatives, (a) to (d); to provide further information, they are supplemented by records of the corresponding dependences of the self-excitation frequency $\Omega(q)$ (Fig. 9.4, 8). As the diagrams suggest, an elastic mounting of the axis with damping can raise the lowest limit of initiation of self-excited vibration well above that obtained in case of a rigid mounting ($q = \infty$). The value of $\nu_{s\,min}$ corresponding to this arrangement is slightly lower than $\nu_{s\,min}$ for the highest value of q shown on the right-hand side of the diagram. The increase of $\nu_{s\,min}$ produced by the effect of the elastic mounting of the axis depends on the position of the centroid. For the parameters given above, alternative (a) shows the greatest (several times the original) increase of all. The analysis also reveals that the maximum value of $\nu_{s\,min}$ grows larger with increasing mass ratio M, increasing coefficient of damping k and decreasing I/ml^2. The higher the values of M and I/ml^2,

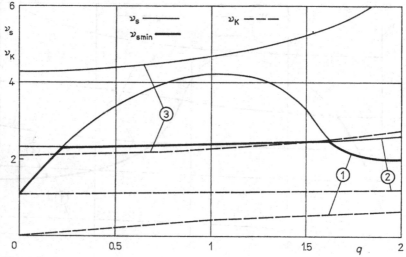

Fig. 9.4, 6

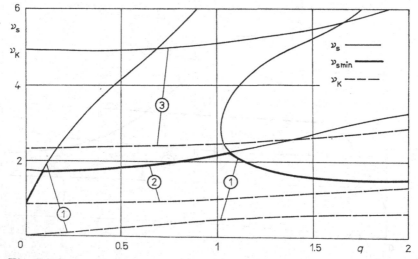

Fig. 9.4, 7

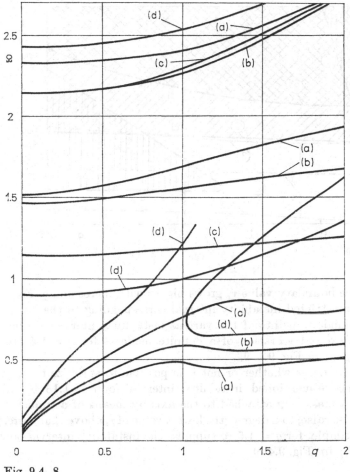

Fig. 9.4, 8

the more remote is the optimum position of the centroid from the centre of the span between the bearings.

The results of an analysis involving a compressible medium are shown in Fig. 9.4, 9. The diagram of the single example reported here (for additional examples refer to TONDL (1980a)) is drawn for γ, k, M, I/ml^2 used in the preceding analysis, for $K = \varLambda_0 = 1$ and alternative (b) ($\alpha_1 = 0$, $\alpha_2 = 1$).

If compressibility of the medium is considered, one can obtain both the limit of initiation and the limit of extinction of self-excited vibration of a definite mode (TONDL (1974a)). As can be seen in the diagram, this is what happened in the example being discussed. For greater clarity, different kinds of hatching are used for the various instability regions, and encircled numerals for the corresponding vibration modes. To facilitate a comparison, the dependence of the critical frequency ν_k on q (for the determination of ν_k, see TONDL (1980a)) and the correspondence with the different vibration modes are shown in dashed lines. The instability region belonging to the lowest natural frequency is seen to be divided in two parts. If the rotor speed of a system (having a given q) is raised, then — as long as $q < 1.18$ — the equilibrium condition becomes unstable and self-excited vibration starts to arise from a certain

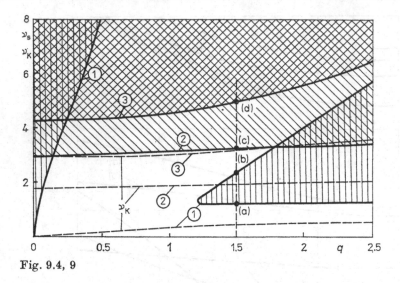

Fig. 9.4, 9

rotor speed up; the boundary value ν_s grows higher with increasing q. For $q > 1.18$, there also exists a restricted instability interval corresponding to the lowest natural frequency. The boundary points of the various instability intervals — intervals of occurrence of self-excited vibration of a definite mode — for $q = 1.5$ are marked by points (a), (b), . . . in Fig. 9.4, 9.

The question now arises whether or not it is possible to stabilize the equilibrium position in the above-mentioned instability interval for $q > 1.18$ by adding an absorber, such as a mass ring attached to the axis by means of a rubber ring (Fig. 9.4, 10) and thus also raise, to a degree, the lowest value of ν_s above that corresponding to $q < 1.18$; for example, for $q = 1.5$ to suppress the instability interval bounded by points (a) and (b) in Fig. 9.4, 9.

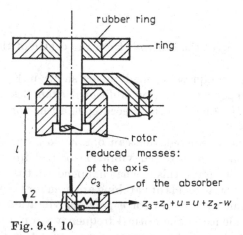

Fig. 9.4, 10

Denote by u the vector of the relative deflection of the absorber whose action is reduced to the plane of the second bearing (Fig. 9.4, 10) and write the corresponding system of differential equations of motion transformed to the dimensionless form (assuming, as before, an ideally balanced rotor) as follows:

$$
\left.
\begin{aligned}
& Az_1'' + bz_2'' - i\nu E(z_1' - z_2') + z_1[1 + K(p_e - ip_\varphi)] = 0 , \\
& bz_1'' + Bz_2'' - i\nu E(z_2' - z_1') + w[1 + K(p_e - ip_\varphi)] = 0 , \\
& z_2'' - w'' + q^2\left[z_2 - w + k\,|z_2 - w|\frac{z_2' - w'}{|z_2' - w'|}\right] \\
& \quad - Mw[1 + K(p_e - ip_\varphi)] + \mu(z_2'' - w'' + u'') = 0 , \\
& z_2'' - w'' + u'' + Q^2\left(u + k_0\,|u|\frac{u'}{|u'|}\right) = 0
\end{aligned}
\right\}
\tag{9.4, 16}
$$

where $\mu = m_3/m_0$ is the ratio between the reduced masses of absorber and axis, k_0 is the damping coefficient of the absorber, $Q = \left(\dfrac{c_3/m_3}{c/m}\right)^{1/2}$ is the tuning coefficient of the absorber, c_3 is the stiffness and m_3 the reduced mass of the absorber spring.

Substituting in (9.4, 16) the solution

$$
\left.
\begin{aligned}
& z_1 = \exp(i\Omega\tau) , \qquad z_2 = (Z + iY)\exp(i - \Omega\tau) , \\
& w = (W + iV)\exp(i\Omega\tau) , \qquad u = (U + iL)\exp(i - \Omega\tau)
\end{aligned}
\right\}
\tag{9.4, 17}
$$

yields the equations for subsequent determination of Z, Y, W, V as functions of Ω, viz.

$$
Z = (1 + KP_e - F)/H , \qquad Y = -KP_\varphi/H , \tag{9.4, 18}
$$

$$
\left.
\begin{aligned}
& W = \frac{1}{\Delta}[(1 + KP_e)(H + GZ) - KP_\varphi GY] , \\
& V = \frac{1}{\Delta}[(1 + KP_e)GY + KP_\varphi(H + GZ)]
\end{aligned}
\right\}
\tag{9.4, 19}
$$

where

$$
F = \Omega^2 A - \nu\Omega E , \qquad G = \Omega^2 B - \nu\Omega^2 E , \qquad H = \Omega^2 b + \nu\Omega E ,
$$

$$
\Delta = (1 + KP_e)^2 + (P_\varphi K)^2 ,
$$

$$
P_e = \frac{\Lambda_0^2(\nu - 2\Omega)^2}{1 + \Lambda_0^2(\nu - 2\Omega)^2} , \qquad P_\varphi = \frac{\Lambda_0(\nu - 2\Omega)}{1 + \Lambda_0^2(\nu - 2\Omega)^2}
$$

and the equations for obtaining U and L:

$$
\left.
\begin{aligned}
U = \{ & -(1 + \mu)\,\Omega^2(Z - W) + q^2[(Z - W) - k(Y - V)] \\
& - M[W(1 + KP_e) + VKP_\varphi]\}/\mu\Omega^2 , \\
L = \{ & -(1 + \mu)\,\Omega^2(Y - V) + q^2[(Y - V) + k(Z - W)] \\
& - M[V(1 + KP_e) - WKP_\varphi]\}/\mu\Omega^2 .
\end{aligned}
\right\}
\tag{9.4, 20}
$$

For ν changing step by step, the corresponding boundary values of Q (the points of intersection of curves $Q_{\mathrm{I}}(\Omega)$ and $Q_{\mathrm{II}}(\Omega)$) are obtained from the following equations:

$$
\left.
\begin{aligned}
& Q_{\mathrm{I}} = \Omega[(Z - W + U)/(U - k_0 L)]^{1/2} , \\
& Q_{\mathrm{II}} = \Omega[(Y - V + L)/(L + k_0 U)]^{1/2} .
\end{aligned}
\right\}
\tag{9.4, 21}
$$

Fig. 9.4, 11 shows the results obtained for a system corresponding to the basic system having the parameters given above, for $q = 1.5$, $\mu = 0.3$ and $k_0 = 0.2$. The boundary curves $\nu_s(Q)$ start (for $Q = 0$) from points (a), (b), (c), (d) (marked correspondingly as the points in Fig. 9.4, 9). It can be seen that in the range $0.53 < Q < 0.64$ the equilibrium position is wholly stabilized in the interval of ν in which a restricted instability interval (bounded by points (a), (b) in Fig. 9.4, 9) existed for the absorberless system. The boundary ν_s corresponding to the second natural frequency can be slightly increased only for $Q = 1.77$.

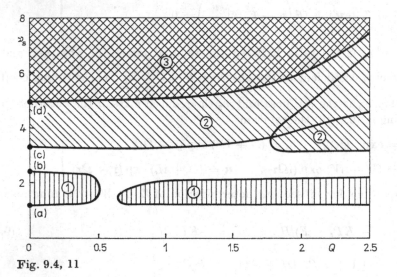

Fig. 9.4, 11

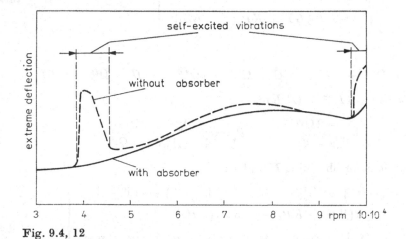

Fig. 9.4, 12

As the results of experiments made with models of cotton-yarn spinning machine spindles show, self-excited vibration of a definite mode is likely to be initiated within a restricted interval of the rotor speed. If this instability interval is not too broad, the vibration can be completely suppressed by application of a properly tuned absorber (Fig. 9.4, 12).

10. Vibration systems with narrow-band random excitation

10.1. Application of the quasi-static method

In what follows we investigate the vibrations of the system

$$y'' + \lambda y = 2\alpha\lambda y - by' - dy^2 y' - ey^3 - g(\tau) \cdot y \qquad (10.1, 1)$$

where now the parametric excitation $g(\tau)$ is assumed to be a stationary narrow-band random process with zero mean value,

$$g(\tau) = \gamma(\tau) \cos\left[2n\tau + \eta(\tau)\right] = \gamma(\tau) \cos \psi(\tau) \qquad (10.1, 2)$$

with the slowly varying excitation amplitude $\gamma = \gamma(\tau)$ and phase $\eta(\tau)$. Further, α is a frequency variation, b the coefficient of a linear and d of a non-linear damping and e that of a non-linear restoring force, while dashes denote derivatives with respect to the dimensionless time τ.

By analogy with the evaluation of amplitude formulae in case of deterministic excitation, our aim is now to determine the statistical *distribution* of the vibration amplitudes, the most important characteristic of a random vibration, as dependent on the system parameters.

The *quasi-static* method used by STRATONOVICH (1961) and LENNOX and KUAK (1976) for problems of forced vibrations and by Baxter (1971) for parametrically excited vibrations of bars involves the following procedure.

Corresponding to the averaging method (Section 2.5), we assume the main parametric resonance case

$$\lambda = n^2$$

and introduce the transformation

$$y = a \cos\varphi, \qquad y' = -na \sin\varphi, \qquad \varphi = n\tau + \vartheta \qquad (10.1, 3)$$

where $a = a(\tau)$ and $\vartheta = \vartheta(\tau)$ are slowly varying functions. Thus we get the equations in standard form

$$a' = -\frac{1}{n}\left[2n^2\alpha - g(\tau)\right] a \sin\varphi \cos\varphi - ba \sin^2\varphi$$

$$- da^3 \sin^2\varphi \cos^2\varphi + \frac{1}{n} ea^3 \sin\varphi \cos^3\varphi,$$

$$\vartheta' = -\frac{1}{n}\left[2n^2\alpha - g(\tau)\right] \cos^2\varphi - b \sin\varphi \cos\varphi$$

$$- da^2 \sin\varphi \cos^3\varphi + \frac{1}{n} ea^2 \cos^4\varphi$$

equivalent to (10.1, 1). Insertion of (10.1, 2) and trigonometric transformations give

$$a' = -n\alpha a \sin 2\varphi - \frac{1}{2} ba(1 - \cos 2\varphi) - \frac{1}{8} da^3(1 - \cos 4\varphi)$$

$$+ \frac{1}{8n} ea^3(2 \sin 2\varphi + \sin 4\varphi) + \frac{1}{4n} \gamma a[\sin (2\varphi - \psi) + \sin (2\varphi + \psi)],$$

$$\vartheta' = -n\alpha(1 + \cos 2\varphi) - \frac{1}{2} b \sin 2\varphi - \frac{1}{8} da^2(2 \sin 2\varphi + \sin 4\varphi)$$

$$+ \frac{1}{8n} ea^2(3 + 4 \cos 2\varphi + \cos 4\varphi)$$

$$+ \frac{1}{4n} \gamma[2 \cos \psi + \cos(2\varphi - \psi) + \cos(2\varphi + \psi)].$$

According to higher approximations of the averaging method (compare BOGOLJUBOV and MITROPOL'SKIJ (1963)), the dependence on φ, and hence on the time τ, of the quantities not containing γ on the right-hand sides has to be iteratively elimated by means of a suitable transformation

$$a = A + \varepsilon(A, \Phi), \qquad \vartheta = \Theta + \delta(A, \Phi), \qquad \Phi = n\tau + \Theta. \qquad (10.1, 4)$$

In contrast to the expansion with respect to a single small quantity applied by STRA-TONOVICH (1961) and BAXTER (1971), α, $\gamma(\tau)$, $\eta(\tau)$, and b are now assumed to be independent small quantities. The first iteration for ε, δ yields, when the small quantities of lowest order of magnitude are taken into account,

$$A' + n \frac{\partial \varepsilon}{\partial \Phi} = -n\alpha A \sin 2\Phi - \frac{1}{2} bA(1 - \cos 2\Phi) - \frac{1}{8} dA^3(1 - \cos 4\Phi)$$

$$+ \frac{1}{8n} eA^3(2 \sin 2\Phi + \sin 4\Phi) - \frac{1}{4n} \gamma A[\sin (2\Phi - \psi) + \sin (2\Phi + \psi)],$$

$$\Theta' + n \frac{\partial \delta}{\partial \Phi} = -n\alpha(1 + \cos 2\Phi) - \frac{1}{2} b \sin 2\Phi - \frac{1}{8} dA^2(2 \sin 2\Phi + \sin 4\Phi)$$

$$+ \frac{1}{8n} eA^2(3 + 4 \cos 2\Phi + \cos 4\Phi)$$

$$+ \frac{1}{4n} \gamma[2 \cos \psi + \cos (2\Phi - \psi) + \cos (2\Phi + \psi)].$$

The correction functions ε, δ are determined in such a way that they cancel on the right-hand sides the dependence on Φ in the terms not containing γ. Thus we are led to the differential equations

$$A' = - \frac{1}{2} bA - \frac{1}{8} dA^3 + \frac{1}{4n} \gamma A[\sin (2\Phi - \psi) + \sin (2\Phi + \psi)],$$

$$\Theta' = -n\alpha + \frac{3}{8n} eA^2 + \frac{1}{4n} \gamma[2 \cos \psi + \cos (2\Phi - \psi) + \cos (2\Phi + \psi)]$$

$$(10.1, 5)$$

and for the correction functions

$$n \frac{\partial \varepsilon}{\partial \Phi} = -n\alpha A \sin 2\Phi + \frac{1}{2} bA \cos 2\Phi + \frac{1}{8} dA^3 \cos 4\Phi$$

$$+ \frac{1}{8n} eA^3 (2 \sin 2\Phi + \sin 4\Phi) \,,$$

$$n \frac{\partial \delta}{\partial \Phi} = -n\alpha \cos 2\Phi - \frac{1}{2} b \sin 2\Phi - \frac{1}{8} dA^2 (2 \sin 2\Phi + \sin 4\Phi)$$

$$+ \frac{1}{8n} eA^2 (4 \cos 2\Phi + \cos 4\Phi)$$

from which we get, with arbitrary integration functions $\varepsilon_0(A)$, $\delta_0(A)$,

$$\varepsilon = \frac{1}{2} \alpha A \cos 2\Phi + \frac{1}{4n} bA \sin 2\Phi + \frac{1}{32n} dA^3 \sin 4\Phi$$

$$- \frac{1}{32n^2} eA^3 (4 \cos 2\Phi + \cos 4\Phi) + \varepsilon_0(A)$$

and

$$\delta = -\frac{1}{2} \alpha \sin 2\Phi + \frac{1}{4n} b \cos 2\Phi + \frac{1}{32n} dA^2 (4 \cos 2\Phi + \cos 4\Phi)$$

$$+ \frac{1}{32n^2} eA^2 (8 \sin 2\Phi + \sin 4\Phi) + \delta_0(A) \,.$$

We choose for simplicity $\varepsilon_0(A) = \delta_0(A) = 0$, as in STRATONOVICH (1961), p. 103 and BAXTER (1971), and in contrast to BOGOLJUBOV and MITROPOL'SKIJ (1963) where the correction functions are chosen such that the solution

$$y = a \cos \varphi = A \cos \Phi + (\varepsilon \cos \Phi - A\delta \sin \Phi) + \cdots$$

acquires, by the correction terms in parentheses, no additional terms to the basic harmonic,

$$\int_0^{2\pi} (\varepsilon \cos \Phi - A\delta \sin \Phi) \frac{\cos}{\sin} \Phi \, d\Phi = 0 \,.$$

For the original quantity y, a first approximation yields, according to (10.1, 3), (10.1, 4),

$$y = (A + \varepsilon) \cos (\Phi + \delta)$$

$$= A \cos \Phi + \frac{1}{2} \alpha A \cos \Phi + \frac{1}{4n} bA \sin \Phi + \frac{1}{32n} dA^3 (2 \sin \Phi - \sin 3\Phi)$$

$$- \frac{1}{32n^2} eA^3 (6 \cos \Phi - \cos 3\Phi) \,.$$

Following the quasi-static method, we assume the random functions change so slowly, that is, that the correlation time of the random excitation $g(\tau)$ is so large in comparison with the relaxation time of the amplitude — which is of the order of magnitude of $(nb)^{-1}$ — that so-called quasi-stable values of the amplitude and phase come to be established for every value of $\gamma(\tau)$ and $\eta(\tau)$. Therefore we assume that in (10.1, 5) the left-hand side derivatives A', Θ' vanish and that only the excitation terms with the arguments $2\Phi - \psi$, which vary slowly for small α, are taken into consideration.

The slowly varying amplitude γ as well as the phase η of the excitation being assumed constant, (10.1, 5) yields, by squaring and adding, besides $A = 0$,

$$\tfrac{1}{4}\gamma^2 = n^2(b + \tfrac{1}{4}dA^2)^2 + (2n^2\alpha - \tfrac{3}{4}eA^2)^2 \tag{10.1, 6}$$

from which follows the amplitude frequency formula

$$2n^2\alpha = \tfrac{3}{4}eA^2 \pm \sqrt{\tfrac{1}{4}\gamma^2 - n^2(b + \tfrac{1}{4}dA^2)^2} \,. \tag{10.1, 7}$$

Division of the two equations (10.1, 5) gives

$$\cot(2\Phi - \psi) = \frac{2n\alpha - \dfrac{3}{4n}eA^2}{b + \dfrac{1}{4}dA^2} \,.$$

10.2. Application of the integral equation method. Probability densities

The quasi-static method is based on the periodic solutions of (10.1, 1) for constant values of the slowly varying functions γ and η found in Section 10.1 by help of the second approximation of the averaging method. The integral equation method leads in a very straightforward way to these periodic solutions. Equation (10.1, 1) is an example of the general investigation in Chapter 5. For the following compare SCHMIDT and WENZEL (1984).

Assuming only a constant excitation amplitude and phase or, in other words, a harmonic excitation (10.1, 2) for an individual realization, the integral equation method leads to the amplitude equation (5.5, 1). This equation simplifies to (10.1, 7) if we neglect the second order excitation terms which allow an evaluation of the second resonance.

Real vibration amplitudes exist only if the amplitude-dependent threshold condition

$$\gamma \geqq 2n(b + \tfrac{1}{4}dA^2) \tag{10.2, 1}$$

holds. If the excitation amplitude is small enough to realize the sign of equality in (10.2, 1), then (10.1, 7) simplifies to the equation

$$\alpha = \frac{3}{8n^2}eA^2$$

for the backbone or skeleton curve, the geometric locus of all maximum values

$$A_{\max} = 2\sqrt{\frac{1}{d}\left(\frac{\gamma}{2n} - b\right)}$$

of the response curves for different excitation amplitudes γ (the dashed line in Fig. 10.2, 1).

The amplitude values A_v in the boundary points of the interval in which the resonance curves exist are determined by the condition

$$\frac{d\alpha}{dA} = 0$$

for vertical points of the resonance curves. If $e > 0$, the lower boundary point yields

$$\alpha = \alpha_v = -\frac{1}{2n^2}\sqrt{\frac{1}{4}\gamma^2 - n^2b^2}\,, \qquad A = A_v = 0$$

whereas the upper one leads to

$$\alpha_v = \frac{1}{2n^2}\left(\frac{\gamma}{2}\frac{\sqrt{9e^2 + n^2d^2}}{nd} - \frac{3eb}{d}\right), \qquad A_v^2 = \frac{4}{d}\left(-b + \frac{3e\gamma}{2n\sqrt{9e^2 + n^2d^2}}\right). \tag{10.2, 2}$$

Vice versa, if $e < 0$, the lower boundary point gives

$$\alpha_v = -\frac{1}{2n^2}\left(\frac{\gamma}{2}\frac{\sqrt{9e^2 + n^2d^2}}{nd} + \frac{3eb}{d}\right), \qquad A_v^2 = \frac{4}{d}\left(-b - \frac{3e\gamma}{2n\sqrt{9e^2 + n^2d^2}}\right), \tag{10.2, 3}$$

the upper one

$$\alpha_v = \frac{1}{2n^2}\sqrt{\frac{1}{4}\gamma^2 - n^2b^2}\,, \qquad A_v = 0\,.$$

Because $A_v \geq 0$ holds,

$$\frac{\gamma^2}{4} \gtreqless \frac{n^2b^2(9e^2 + n^2d^2)}{9e^2}$$

is the condition for the resonance curves to exist beyond the linear resonance interval

$$\left(-\frac{1}{2n^2}\sqrt{\frac{1}{4}\gamma^2 - n^2b^2}\,, \quad \frac{1}{2n^2}\sqrt{\frac{1}{4}\gamma^2 - n^2b^2}\right),$$

that is, for there to be an overhang of the resonance curves (Fig. 10.2, 1).

Now as before the threshold condition (10.2, 1) holds. For the relatively small excitation amplitudes

$$n^2b^2 < \frac{\gamma^2}{4} < \frac{n^2b^2(9e^2 + n^2d^2)}{9e^2}\,,$$

the resonance curves remain in the linear resonance interval, no overhang exists.

Insertion of α_v into formulae (10.2, 2), (10.2, 3) for A_v yields the equation

$$\alpha = \alpha_v = \left(\frac{3e}{8n^2} + \frac{d^2}{24e}\right)A_v^2 + \frac{bd}{6e} \tag{10.2, 4}$$

for a curve which reveals how the vertical boundary points of the resonance curve depend on α and which we denote as *vertical curve*. Fig. 10.2, 1 gives an example for

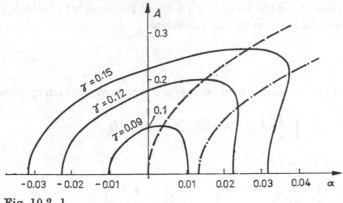

Fig. 10.2, 1

the group of resonance curves belonging to different values of γ, for the (dashed) backbone curve and the (dashed-dotted) vertical curve where $n = 1$, $e = 1$, $b = 0.04$ and $d = 2$, a very large value in order to make manifest the difference between vertical and backbone curve, which vanishes for $d \to 0$.

For every α, the corresponding value A_v is the smallest of all amplitude values belonging to the stable upper branches of the group of resonance curves.

The amplitude formula (10.1, 6) can, because of (10.2, 4), be written in the form

$$\frac{1}{4}\gamma^2 = n^2\left(b + \frac{1}{4}dA^2\right)^2 + \left[\frac{3}{4}e(A_v^2 - A^2) + \frac{n^2d^2}{12e}A_v^2 + \frac{n^2bd}{3e}\right]^2 \qquad (10.2, 5)$$

which reveals the dependence between the excitation amplitude γ, the resonance amplitude A and the minimum A_v of A. The derivative $\mathrm{d}\gamma^2/\mathrm{d}(A^2)$ shows that γ^2 monotonously depends on A^2.

Our aim is to determine the probability density $w(A)$ of the response amplitude A. To this end the probability density $v(\gamma)$ of the excitation amplitude must either be known (say, by experimental investigations) or has to be determined from the known probability density $u(g)$ of the excitation process $g(\tau)$. Examples and general methods of evaluation for the relationship between the probability densities $v(\gamma)$ and $u(g)$ have been given by KROPAČ (1972). We assume the rather general case of a Weibull distribution of the excitation *amplitude*, admitting additionally of a threshold value Γ, as the most comprehensive form to handle easily by analytical formulae:

$$v(\gamma) = \frac{k}{(\sqrt{2}\,\sigma)^k}(\gamma - \Gamma)^{k-1}\,\mathrm{e}^{-\left(\frac{\gamma - \Gamma}{\sqrt{2}\sigma}\right)^k}. \qquad (10.2, 6)$$

Here σ is the scale parameter and k the form parameter. With this often used distribution various forms of probability densities can be described as Fig. 10.2, 2 shows for $\sigma = 1/\sqrt{2}$.

In particular, for $k = 2$ and $\Gamma = 0$ the Weibull probability density simplifies to the Rayleigh probability density

$$v(\gamma) = \frac{\gamma}{\sigma^2}\mathrm{e}^{-\frac{\gamma^2}{2\sigma^2}} \qquad (10.2, 7)$$

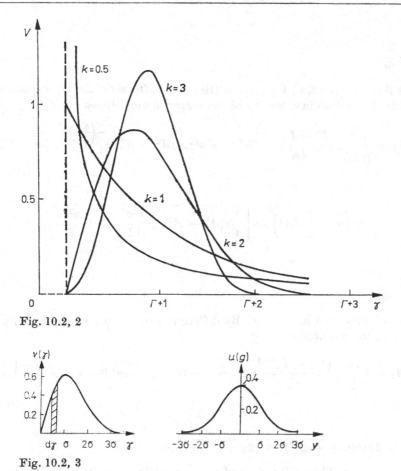

Fig. 10.2, 2

Fig. 10.2, 3

of the excitation *amplitudes*, which corresponds with the Gaussian probability density

$$u(g) = \frac{1}{\sqrt{2\pi}\sigma} e^{-\frac{g^2}{2\sigma^2}}$$

of the excitation *process* (Fig. 10.2, 3).

The probability density of the *response* amplitudes associated with strips $v(\gamma)\,\mathrm{d}\gamma$ of excitation amplitudes (Fig. 10.2, 3) can be obtained by the transformation

$$v(\gamma)\,\mathrm{d}\gamma = \mathrm{d}V(\gamma) = \mathrm{d}W(A) = w(A)\,\mathrm{d}A \tag{10.2, 8}$$

with the probability distributions V of the excitation amplitude and W of the response amplitude, if γ is a monotonic function of A (compare for instance STRATONOVICH (1961)). This condition holds for the problem at hand.

For the Weibull probability density of the excitation amplitudes (10.2, 6), the transformation (10.2, 8) leads for the response amplitudes to the probability distribution

$$W(A) = -e^{-\left(\frac{\gamma(A) - \Gamma}{\sqrt{2}\sigma}\right)^k}. \tag{10.2, 9}$$

We first assume $e > 0$.

In case

$$\alpha > \frac{bd}{6e},$$

we have to substitute (10.2, 5) for $\gamma(A)$ in (10.2, 9). Differentiation corresponding to (10.2, 8) yields the probability density of the response amplitudes

$$w(A) = \frac{k}{(\sqrt{2}\sigma)^k} \frac{(K - \Gamma)^{k-1}}{4K} (9e^2 + n^2d^2) A(A^2 - A_v^2) e^{-\left(\frac{K-\Gamma}{\sqrt{2}\sigma}\right)^k} \quad \text{for} \quad A \geq A_v$$

(10.2, 10)

where

$$K = 2\sqrt{n^2\left(b + \frac{1}{4} dA^2\right)^2 + \left[\frac{3}{4} e(A_v^2 - A^2) + \frac{n^2d^2}{12e} A_v^2 + \frac{n^2db}{3e}\right]^2}.$$

In case

$$\alpha \leq \frac{bd}{6e},$$

(10.1, 6) has to be inserted in (10.2, 9). By differentiating we get the probability density of the response amplitudes

$$w(A) = \frac{2k}{(\sqrt{2}\sigma)^k} \frac{(K_\alpha - \Gamma)^{k-1}}{K_\alpha} A\left[n^2(db - 6e\alpha) + \frac{1}{4} (n^2d^2 + 9e^2) A^2\right] e^{-\left(\frac{K_\alpha-\Gamma}{\sqrt{2}\sigma}\right)^k}$$

(10.2, 11)

where

$$K_\alpha = 2\sqrt{n^2(b + \frac{1}{4} dA^2)^2 + (2n^2\alpha - \frac{3}{4} eA^2)^2}.$$

Now assume $e < 0$. Then (10.1, 6) and the probability density formula (10.2, 11) are valid for $\alpha > \frac{bd}{6e}$. If $\alpha \leq \frac{bd}{6e}$, (10.2, 5) has to be used, and the probability density is (10.2, 10).

In the special case of a Rayleigh probability density (10.2, 7) of the excitation amplitudes, formula (10.2, 10) simplifies to

$$w(A) = \frac{9e^2 + n^2d^2}{2\sigma^2} A(A^2 - A_v^2) e^{-\frac{2L}{\sigma^2}} \quad \text{for} \quad A \geq A_v$$

(10.2, 12)

with

$$L = n^2\left(b + \frac{1}{4} dA^2\right)^2 + \left[\frac{3}{4} e(A_v^2 + A^2) + \frac{n^2d^2}{12e} A_v^2 + \frac{n^2bd}{3e}\right]^2$$

whereas formula (10.2, 11) becomes

$$w(A) = \frac{2}{\sigma^2} A\left[n^2(bd - 6e\alpha) + \frac{1}{4} (9e^2 + n^2d^2) A^2\right] e^{-\frac{2L_\alpha}{\sigma^2}}$$

(10.2, 13)

where

$$L_\alpha = n^2\left(b + \frac{1}{4} dA^2\right)^2 + \left(2n^2\alpha - \frac{3}{4} eA^2\right)^2.$$

The probability density (10.2, 12) disappears if A tends to infinity and if A tends from above to the minimum value A_v of the vertical curve or to zero. As amplitudes below the vertical curve do not appear, $w(A) = 0$ holds for $A < A_\mathrm{v}$.

The probability densities (10.2, 12) respectively (10.2, 13) are drawn in Fig. 10.2, 4 for the example $\sigma = 0.01$, $b = 0.002$, $e = 1$, $d = 0.2$ and different values of the fre-

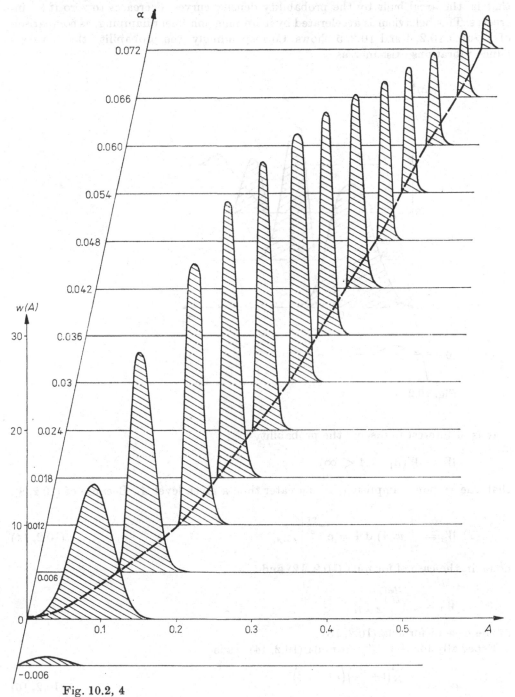

Fig. 10.2, 4

quency variation α, in Fig. 10.2, 5 for greater non-linear damping, $d = 1$. The vertical curves (10.2, 4) are marked by dashed lines, they represent the lower limit for real response amplitudes. For increasing positive frequency variation α, this lower limit increases, as too does the maximum of the probability density curves (which is called the most probable amplitude). The probability for positive response amplitudes, that is, the areal built by the probability density curves, decreases to zero if $|\alpha|$ increases. This behaviour is accelerated by increasing non-linear damping, as comparison of Figures 10.2, 4 and 10.2, 5 shows. Correspondingly, the probability that no resonance vibrations exist increases.

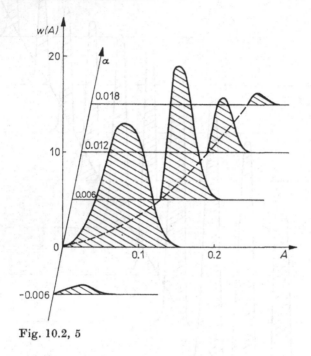

Fig. 10.2, 5

It is of interest to discuss the probability

$$W_1 = W(A_1 < A < \infty)$$

that the response amplitude A is greater than a given level A_1. Because of (10.2, 8),

$$W_1 = \int\limits_{A_1}^{\infty} w(A) \, \mathrm{d}A = \left. \mathrm{e}^{-\frac{2L}{\sigma^2}} \right|_{A = A_1} \tag{10.2, 14}$$

holds in the case of formula (10.2, 12) and

$$W_1 = \left. \mathrm{e}^{-\frac{2L\alpha}{\sigma^2}} \right|_{A = A_1}$$

in the case of formula (10.2, 13).

Especially for $A_1 = A_{\mathrm{v}}$, formula (10.2, 14) reads

$$W_{\mathrm{v}} = \mathrm{e}^{-\frac{2n^2}{\sigma^2}\left(1 + \frac{n^2 d^2}{9e^2}\right)\left(b + \frac{1}{4} dA_{\mathrm{v}}^2\right)^2} \tag{10.2, 15}$$

and gives the probability of there being any positive response amplitudes at all. Correspondingly, the probability for vanishing response amplitudes is

$$1 - W_v \,.$$

For the examples in Figures 10.2, 4 and 10.2, 5 and also for vanishing non-linear damping, $d = 0$, formula (10.2, 20) gives the probability of positive response amplitudes drawn in Fig. 10.2, 6. The corresponding curves for a greater linear damping, $b = 0.005$, are given in Fig. 10.2, 7. The figure shows that this probability decreases if linear or non-linear damping increases.

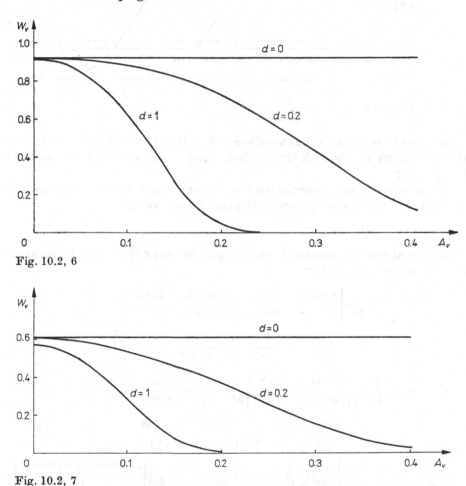

Fig. 10.2, 6

Fig. 10.2, 7

The probability (10.2, 14) that the amplitudes are greater than a given amplitude level A_1 is sketched in Fig. 10.2, 8 for the same values as in Fig. 10.2, 5, for $A_v = 0$ (that is, because of (10.2, 4), for $\alpha = 1/3000$), as well as for $A_v = 0.117$ (that is $\alpha = 0.006$) and $A_v = 0.167$ (that is $\alpha = 0.012$).

Vice versa, (10.2, 14) yields the amplitude A_1, amplitudes greater than which arise only with given probability p, as the (smallest positive) solution of the equation

$$n^2 \left(b + \frac{1}{4} d A_1^2 \right)^2 + \left[\frac{3}{4} e (A_v^2 - A_1^2) + \frac{n^2 d^2}{12 e} A_v^2 + \frac{n^2 b d}{3 e} \right]^2 = -\frac{\sigma^2}{2} \ln p \,.$$

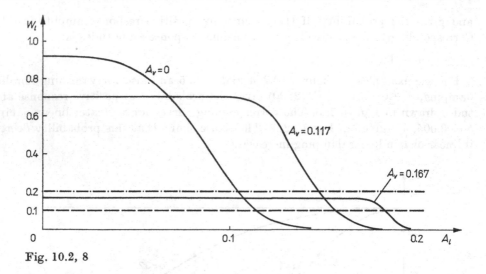

Fig. 10.2, 8

Examples of these amplitudes are given in Fig. 10.2, 8, at the points of intersection of the probability curves with the (dashed) line $p = 0.1$ and with the (dashed-dotted) line $p = 0.2$.

Up till now we have investigated the first or main parametric resonance. We now take into consideration a general k-th parametric resonance

$$\lambda = (kn)^2 .$$

The integral equation method leads (compare SCHMIDT (1975)) to the excitation parts of the iterative solutions

$$y_2 = -\frac{\gamma}{2n^2} \left[\frac{r \cos (k - 2) \, n\tau + s \sin (k - 2) \, n\tau}{\vartheta k^2 - (k - 2)^2} \right.$$
$$\left. + \frac{r \cos (k + 2) \, n\tau + s \sin (k + 2) \, n\tau}{k^2 - (k + 2)^2} \right],$$

$$y_3 = \frac{\gamma^2}{4n^4} \left\{ \frac{r \cos (k - 4) \, n\tau + s \sin (k - 4) \, n\tau}{[\vartheta k^2 - (k - 2)^2] \, [\vartheta k^2 - (k - 4)^2]} \right.$$
$$+ \frac{r \cos (k + 4) \, n\tau + s \sin (k + 4) \, n\tau}{[k^2 - (k + 2)^2] \, [k^2 - (k + 4)^2]}$$
$$\left. - \left[\frac{1}{\vartheta k^2 - (k - 2)^2} + \frac{1}{k^2 - (k + 2)^2} \right] \frac{r \cos kn\tau + s \sin kn\tau}{k^2} \right\},$$

$$\ldots$$

Insertion into the periodicity equations yields the *amplitude frequency formula*

$$2n^2\alpha = \frac{3e}{4} A^2 + G\gamma^2 + O(\gamma^4) \pm \sqrt{\sum_{\nu=1}^{\infty} \delta_\nu^k G_\nu^2 \gamma^{2\nu} - n^2 \left(b + \frac{1}{4} dA^2 \right)^2} \qquad (10.2, 16)$$

where

$$G = \frac{1}{4n^2} \left[\frac{1}{\vartheta k^2 - (k - 2)^2} + \frac{1}{k^2 - (k + 2)^2} \right]$$

and

$$G_\nu = \frac{1}{2^\nu n^{2\nu-2}[\nu^2 - (\nu-2)^2]\,[\nu^2 - (\nu-4)^2]\,\ldots[\nu^2 - (4-\nu)^2]\,[\nu^2 - (2-\nu)^2]}$$

with $\nu - 1$ factors in brackets in the denominator, especially

$$G_1 = \frac{1}{2}, \qquad G_2 = \frac{1}{16n^2}, \qquad G_3 = \frac{1}{512n^4}.$$

The term $O(\gamma^4)$ is not specified because it is not needed in what follows. Formula (10.2, 16) includes (10.1, 7) for $k = 1$. The threshold condition, generalizing (10.2, 1), is now

$$G_k \gamma^k \gtreqqless n\left(b + \frac{1}{4}\,dA^2\right),$$

and instead of (10.2, 5) the formula

$$G_k^2 \gamma^{2k} = n^2\left(b + \frac{1}{4}\,dA^2\right)^2 + \left[\frac{3}{4}\,e(A_\mathrm{v}^2 - A^2) + \frac{n^2 d^2}{12e}\,A_\mathrm{v}^2 + \frac{n^2 b d}{3e}\right]^2$$

holds.

Confining attention, for simplicity, to a Rayleigh probability density (10.2, 7) of the excitation amplitudes, we get

$$W(A) = -e^{-\frac{L^{1/k}}{2\sigma^2 G_k^{2/k}}}$$

and instead of (10.2, 12)

$$w(A) = \frac{(9e^2 + n^2 d^2)\,A(A^2 - A_\mathrm{v}^2)}{8k\sigma^2 G_k^{2/k} L^{1-1/k}}\,e^{-\frac{L^{1/k}}{2\sigma^2 G_k^{2/k}}}.$$

In the same way formulae corresponding to (10.2, 11), (10.2, 13) can be found, which can be discussed quite analogously. It can be shown that the probability densities for $k > 1$ already tend to zero for much smaller amplitudes than for $k = 1$.

11. Vibration systems with broad-band random excitation

11.1. The amplitude probability density

In Chapter 10 we have assumed a narrow-band random excitation (10.1, 2) for which the bandwidth of the spectrum is much smaller than $(nb)^{-1}$. In this chapter we consider the opposite extreme, that the bandwidth of the excitation spectrum is much greater than $(nb)^{-1}$. Here the apparatus of Marcov process theory is applicable (compare for instance STRATONOVICH (1961), M. F. DIMENTBERG (1980)). Stratonovich assumes approximately white noise excitation processes and thus gets results depending on the excitation spectral density for special values. Following MITROPOL'SKIJ and KOLOMIEC (1976), WEDIG (1978), M. F. DIMENTBERG (1980) and other recent investigations, we consider ideal white noise excitation processes, generalized derivations of Wiener processes, and are thus able to use the elegant mathematical tool of Ito equation calculus, developed for instance in GICHMAN and SKOROCHOD (1968, 1975), CHAS'MIN-SKIJ (1969) or ARNOLD (1973).

As a rather general equation of the type (5.1, 1), we investigate the vibration equation

$$\left.\begin{aligned}
x'' + \omega^2 x &= F, \\
F = F(x, x', x'', t) &= f\dot{\xi}(t) - g\dot{\xi}(t)\, x - bx' - dx^2 x' - ex^3 - hx(x'^2 + xx'')
\end{aligned}\right\}$$

$$(11.1, 1)$$

where $\dot{\xi}(t)$ is a white noise random process, f, g are the coefficients of forced and parametric excitation respectively, $b \geqq 0$ the coefficient of a linear and $d > 0$ that of a non-linear damping, e the coefficient of a non-linear restoring force and h that of a non-linear inertia force (compare, for the occurance of such forces, BOLOTIN (1956) and Chapter 12). The coefficients f, g and b are assumed small in the sense of the iterative procedure. For vanishing forced excitation, equations of this type have been investigated by STRATONOVICH (1961), NIKOLAENKO (1967), BAXTER (1971), SCHMIDT (1978), and DIMENTBERG, ISIKOV and MODEL (1981) and others; for vanishing parametric excitation see KÜHNLENZ (1979). Simultaneous forced and parametric excitation has already been considered by DIMENTBERG and GORBUNOV (1975) and MODEL (1978 a, b). For the following results compare SCHMIDT (1981 a).

Corresponding to the investigation of narrow-band excitations in Section 10.1, we may introduce by

$$x = a \cos \varphi, \qquad x' = -\omega a \sin \varphi$$

the amplitude $a > 0$ and the phase

$$\varphi = \omega t + \vartheta$$

as random functions of time, which leads to the standard form

$$-\omega a' = F \sin \varphi, \qquad -\omega a \vartheta' = F \cos \varphi.$$

We understand these equations as physical or Stratonovich equations, compare for instance ARNOLD (1973). If they are written in the form

$$dy_i = m_i \, dt + n_i \, d\xi, \qquad i = 1, 2$$

with $y_1 = a$, $y_2 = \vartheta$ and $d\xi = \dot{\xi}(t) \, dt$, the corresponding Ito equations are

$$dy_i = (m_i + \mu_i) \, dt + n_i \, d\xi, \qquad \mu_i = \frac{1}{2} \sum_{j=1}^{2} \frac{dn_i}{dy_j} n_j.$$

The additional Ito terms can be found to be

$$\mu_1 = \frac{f^2}{4\omega^2 a} (1 + \cos 2\varphi) - \frac{fg}{4\omega^2} (3 \cos \varphi + \cos 3\varphi)$$

$$+ \frac{g^2 a}{16\omega^2} (3 + 4 \cos 2\varphi + \cos 4\varphi),$$

$$\mu_2 = -\frac{f^2}{2\omega^2 a^2} \sin 2\varphi + \frac{fg}{2\omega^2 a} (\sin \varphi + \sin 3\varphi) - \frac{g^2}{8\omega^2} (2 \sin 2\varphi + \sin 4\varphi).$$

Thus the Ito equations can be written, after trigonometric transformations, in the form

$$da = p \, dt + q \, d\xi, \qquad d\vartheta = r \, dt + s \, d\xi \qquad\qquad (11.1, 2)$$

where

$$p = -\frac{ba}{2} + \frac{ba}{2} \cos 2\varphi - \frac{da^3}{8} + \frac{da^3}{8} \cos 4\varphi + \frac{ea^3}{4\omega} \sin 2\varphi + \frac{ea^3}{8\omega} \sin 4\varphi$$

$$- \frac{\omega h a^3}{4} \sin 4\varphi + \frac{f^2}{4\omega^2 a} + \frac{f^2}{4\omega^2 a} \cos 2\varphi - \frac{3fg}{4\omega^2} \cos \varphi - \frac{fg}{4\omega^2} \cos 3\varphi$$

$$+ \frac{3g^2 a}{16\omega^2} + \frac{g^2 a}{4\omega^2} \cos 2\varphi,$$

$$q = -\frac{f}{\omega} \sin \varphi + \frac{ga}{2\omega} \sin 2\varphi,$$

$$r = -\frac{b}{2} \sin 2\varphi - \frac{da^2}{4} \sin 2\varphi - \frac{da^2}{8} \sin 4\varphi + \frac{3ea^2}{8\omega} + \frac{ea^2}{2\omega} \cos 2\varphi + \frac{ea^2}{8\omega} \cos 4\varphi$$

$$- \frac{\omega h a^2}{4} - \frac{\omega h a^2}{2} \cos 2\varphi - \frac{\omega h a^2}{4} \cos 4\varphi - \frac{f^2}{2\omega^2 a^2} \sin 2\varphi + \frac{fg}{2\omega^2 a} \sin \varphi$$

$$+ \frac{fg}{2\omega^2 a} \sin 3\varphi - \frac{g^2}{4\omega^2} \sin 2\varphi - \frac{g^2}{8\omega^2} \sin 4\varphi,$$

$$s = -\frac{f}{\omega a} \cos \varphi + \frac{g}{2\omega} + \frac{g}{2\omega} \cos 2\varphi.$$

The expressions in p and r depending on f and g correspond to the additional Ito terms. In F, in the sense of the iteration method x'' is replaced by $-\omega^2 a \cos \varphi$.

As the basis for iteratively solving this equation, introduce, corresponding with Section 10.1, a transformation of amplitude and phase

$$a = A + \varepsilon(A, \Phi), \qquad \varphi = \Phi + \delta(A, \Phi) \tag{11.1, 3}$$

with small correction functions ε and δ which have to be suitably chosen. For the transformed phases, let

$$\Phi = \omega t + \Phi, \quad \text{that is,} \quad \vartheta = \Theta + \delta(A, \Phi). \tag{11.1, 4}$$

Our goal is successively to determine the stochastic differentials for the random processes A and Θ, written in the form

$$dA = P \, dt + Q \, d\xi, \qquad d\Theta = R \, dt + S \, d\xi. \tag{11.1, 5}$$

The stochastic differential of the process $\varepsilon = \varepsilon(A, \Phi)$ is, following of Ito's formula (ARNOLD (1973)),

$$d\varepsilon = \frac{\partial \varepsilon}{\partial A} dA + \frac{\partial \varepsilon}{\partial \Phi} d\Phi + \left(\frac{1}{2} \frac{\partial^2 \varepsilon}{\partial A^2} Q^2 + \frac{\partial^2 \varepsilon}{\partial A \, \partial \Phi} QS + \frac{1}{2} \frac{\partial^2 \varepsilon}{\partial \Phi^2} S^2 \right) dt.$$

Therefore (11.1, 3) yields with (11.1, 4), (11.1, 5), (11.1, 2),

$$dA = \left[-\omega \frac{\partial \varepsilon}{\partial \Phi} - \frac{\partial \varepsilon}{\partial A} P - \frac{\partial \varepsilon}{\partial \Phi} R - \frac{1}{2} \frac{\partial^2 \varepsilon}{\partial A^2} Q^2 \right.$$
$$\left. - \frac{\partial^2 \varepsilon}{\partial A \, \partial \Phi} QS - \frac{1}{2} \frac{\partial^2 \varepsilon}{\partial \Phi^2} S^2 + p(A + \varepsilon, \Phi + \delta) \right] dt$$
$$+ \left[-\frac{\partial \varepsilon}{\partial A} Q - \frac{\partial \varepsilon}{\partial \Phi} S + q(A + \varepsilon, \Phi + \delta) \right] d\xi. \tag{11.1, 6}$$

Analogously we get

$$d\Theta = \left[-\omega \frac{\partial \delta}{\partial \Phi} - \frac{\partial \delta}{\partial A} P - \frac{\partial \delta}{\partial \Phi} R - \frac{1}{2} \frac{\partial^2 \delta}{\partial A^2} Q^2 \right.$$
$$\left. - \frac{\partial^2 \delta}{\partial A \, \partial \Phi} QS - \frac{1}{2} \frac{\partial^2 \delta}{\partial \Phi^2} S^2 + r(A + \varepsilon, \Phi + \delta) \right] dt$$
$$+ \left[-\frac{\partial \delta}{\partial A} Q - \frac{\partial \delta}{\partial \Phi} S + s(A + \varepsilon, \Phi + \delta) \right] d\xi.$$

With the notation (11.1, 5) the Fokker Planck Kolmogorov equation for the two-dimensional probability density $w(A, \Theta; t)$ of amplitude and phase reads

$$\frac{\partial w}{\partial t} + \frac{\partial (Pw)}{\partial A} + \frac{\partial (Rw)}{\partial \Theta} = \frac{1}{2} \frac{\partial^2 (Q^2 w)}{\partial A^2} + \frac{\partial^2 (QSw)}{\partial A \, \partial \Theta} + \frac{1}{2} \frac{\partial^2 (S^2 w)}{\partial \Theta^2} \tag{11.1, 7}$$

as is shown, for instance, in MITROPOL'SKIJ and KOLOMIEC (1976). We confine ourselves to the important stationary case $\partial w/\partial t = 0$. But even then no closed solutions of this equation are known. Therefore we use the averaging method for iteratively evaluating the solution of (11.1, 7). For discussion on, and generalization of, the averaging method compare EBELING and ENGEL-HERBERT (1982) and EBELING, HERZEL, RICHERT, and SCHIMANSKY-GEIER (1986).

In the *first approximation*, which has already been evaluated by DIMENTBERG and GORBUNOV (1975) and MODELL (1978) for $f = 0$, the averaging method without correction functions is used. In other words, for the solution of (11.1, 7) in P, R, Q^2, QS,

S^2 consider only the terms already found in (11.1, 2) and from these only the ones which do not oscillate, that is, which do not depend on Φ and which we can therefore get by averaging with respect to Φ. Correspondingly (STRATONOVICH (1961)), we can assume also that the probability density is a function of the amplitude only. Thus (11.1, 7) simplifies to

$$\frac{d\Psi}{dA} = 0$$

where

$$\Psi = Pw - \frac{1}{2}\frac{d}{dA}(Q^2 w)$$

and

$$P = P_1 = \frac{f^2}{4\omega^2 A} + \left(\frac{3g^2}{16\omega^2} - \frac{b}{2}\right)A - \frac{d}{8}A^3,$$

$$Q^2 = Q_1^2 = \frac{f^2}{2\omega^2} + \frac{g^2 A^2}{8\omega^2}.$$

Assuming $\Psi \to 0$, $w \to 0$ for $A \to \infty$, and therefore that $\Psi \equiv 0$, we arrive at, if we use the abbreviation

$$J = \frac{2P - \dfrac{dQ^2}{dA}}{Q^2}, \tag{11.1, 8}$$

the probability density

$$w = C\,e^{\int J\,dA}.$$

For vanishing forced excitation, $f - 0$, comes

$$w = CA^{1 - \frac{8\omega^2 b}{g^2}}\,e^{-\frac{\omega^2 d}{g^2}A^2} \tag{11.1, 9}$$

and for vanishing parametric excitation, $g = 0$,

$$w = CA\,e^{-\frac{\omega^2 b}{f^2}A^2 - \frac{\omega^2 d}{8f^2}A^4}. \tag{11.1, 10}$$

In case $g \neq 0$ a partial fraction decomposition yields

$$J = \frac{4f^2 + (g^2 - 8\omega^2 b)A^2 - 2\omega^2 dA^4}{A(4f^2 + g^2 A^2)}, \tag{11.1, 11}$$

$$= -\frac{2\omega^2 d}{g^2}A + \frac{4f^2 + \left(g^2 - 8\omega^2 b + \dfrac{8\omega^2 df^2}{g^2}\right)A^2}{A(4f^2 + g^2 A^2)} \tag{11.1, 12}$$

$$= -\frac{2\omega^2 d}{g^2}A + \frac{1}{A} - 8\omega^2\left(b - \frac{df^2}{g^2}\right)\frac{A}{4f^2 + g^2 A^2}$$

therefore

$$\int J\,dA = -\frac{\omega^2 d}{g^2}A^2 + \ln A - \frac{4\omega^2}{g^2}\left(b - \frac{df^2}{g^2}\right)\ln\left(1 + \frac{g^2 A^2}{4f^2}\right) \tag{11.1, 13}$$

and

$$w = \overline{C} A \left(A^2 + \frac{4f^2}{g^2} \right)^{-\frac{4\omega^2 b}{g^2} + \frac{4\omega^2 df^2}{g^4}} e^{-\frac{\omega^2 d}{g^2} A^2} \tag{11.1, 14}$$

with

$$\overline{C} = \left(\frac{4f^2}{g^2} \right)^{\frac{4\omega^2 b}{g^2} - \frac{4\omega^2 df^2}{g^4}} C .$$

The constants C respectively $\overline{C}$ have to be determined from the normalization condition

$$\int_0^\infty w \, \mathrm{d}A = 1 . \tag{11.1, 15}$$

Formula (11.1, 10) is not included in (11.1, 14) for $g \to 0$, in connection with the subtraction and addition of terms in (11.1, 12) — the first one respectively the last one on the right-hand side — which tend to infinity for $g \to 0$. But if we heuristically develop in (11.1, 13)

$$\ln \left(1 + \frac{g^2 A^2}{4f^2} \right)$$

into a power series we get for

$$\frac{g^2 A^2}{4f^2} < 1$$

the formula

$$w = CA \, e^{-\frac{\omega^2 b}{f^2} A^2 - \omega^2 \left(d - \frac{bg^2}{f^2} \right) \left[\frac{1}{2} \frac{A^4}{4f^2} - \frac{1}{3} \frac{g^2 A^6}{(4f^2)^2} + \frac{1}{4} \frac{g^4 A^8}{(4f^2)^3} - + \ldots \right]} \tag{11.1, 16}$$

which now does contain (11.1, 10) for $g \to 0$. An exact verification of (11.1, 16) has to start from (11.1, 11), to develop

$$\frac{1}{4f^2 + g^2 A^2} = \frac{1}{4f^2} \frac{1}{1 + \dfrac{g^2 A^2}{4f^2}}$$

into a power series and to integrate term by term.

The formulae thus found reveal how the excitation coefficients f, g as well as the damping coefficients b, d influence the probability density. Non-linear damping, as well as linear damping for vanishing parametric excitation, cause the exponential decrease of the probability density and hence the existence of a positive normalization constant. But the influence of restoring and inertia force non-linearities with the coefficients e and h respectively remains undetermined.

In order to determine this influence, we apply in what follows the *second approximation* of the averaging method. For the evaluation of the second approximation $P = P_1 + P_2$, first determine in the expression for P, that is, in the first expression of (11.1, 6) in brackets, the term of highest order of magnitude $-\omega(\partial\varepsilon/\partial\Phi)$ such as to compensate the terms depending not on ε, δ, but on Φ (which are omitted in first approximation by averaging). Integration yields, up to an arbitrary function of the amplitude chosen for simplicity to be equal to zero (compare p. 325)

$$
\varepsilon = \frac{bA}{4\omega}\sin 2\Phi + \frac{dA^3}{32\omega}\sin 4\Phi - \frac{eA^3}{8\omega^2}\cos 2\Phi - \frac{eA^3}{32\omega^2}\cos 4\Phi
$$

$$
+ \frac{hA^3}{16}\cos 4\Phi + \frac{f^2}{8\omega^3 A}\sin 2\Phi - \frac{3fg}{4\omega^3}\sin \Phi - \frac{fg}{12\omega^3}\sin 3\Phi
$$

$$
+ \frac{g^2 A}{8\omega^3}\sin 2\Phi + \frac{g^2 A}{64\omega^3}\sin 4\Phi \, . \tag{11.1, 17}
$$

Analogously, the counterbalance of the terms in R depending not on ε, δ but on Φ by $-\omega(\partial\delta/\partial\Phi)$ leads to

$$
\delta = \frac{b}{4\omega}\cos 2\Phi + \frac{dA^2}{8\omega}\cos 2\Phi + \frac{dA^2}{32\omega}\cos 4\Phi + \frac{eA^2}{4\omega^2}\sin 2\Phi
$$

$$
+ \frac{eA^2}{32\omega^2}\sin 4\Phi - \frac{hA^2}{4}\sin 2\Phi - \frac{hA^2}{16}\sin 4\Phi + \frac{f^2}{4\omega^3 A^2}\cos 2\Phi
$$

$$
- \frac{fg}{2\omega^3 A}\cos \Phi - \frac{fg}{6\omega^3 A}\cos 3\Phi + \frac{g^2}{8\omega^3}\cos 2\Phi + \frac{g^2}{32\omega^3}\cos 4\Phi \, .
$$

$$\tag{11.1, 18}$$

The correction functions ε and δ produce numerous further terms in P not dependent on Φ. If we insert in the first expression of (11.1, 6) in brackets the first approximation of P, R, Q, S and transform the last expression as

$$
(A + \varepsilon)^3 \cos 4(\Phi + \delta) = A^3 \cos 4\Phi + 3A^2\varepsilon \cos 4\Phi - 4A^3\delta \sin 4\Phi \, ,
$$

we get for the additional terms of second approximation:

$$
P_2 = -\frac{\partial\varepsilon}{\partial A}P_1 - \frac{\partial\varepsilon}{\partial\Phi}R_1 - \frac{1}{2}\frac{\partial^2\varepsilon}{\partial A^2}Q_1^2 - \frac{\partial^2\varepsilon}{\partial A\,\partial\Phi}Q_1 S_1 - \frac{1}{2}\frac{\partial^2\varepsilon}{\partial\Phi^2}S_1^2
$$

$$
- \frac{b}{2}\varepsilon - \frac{b}{2}\varepsilon \cos 2\Phi - bA\delta \sin 2\Phi - \frac{3dA^2}{8}\varepsilon + \frac{3dA^2}{8}\varepsilon \cos 4\Phi
$$

$$
- \frac{dA^3}{2}\delta \sin 4\Phi + \frac{3eA^2}{4\omega}\varepsilon \sin 2\Phi + \frac{eA^3}{2\omega}\delta \cos 2\Phi + \frac{3eA^2}{8\omega}\varepsilon \sin 4\Phi
$$

$$
+ \frac{eA^3}{2\omega}\delta \cos 4\Phi - \frac{3\omega hA^2}{4}\varepsilon \sin 4\Phi - \omega hA^3\delta \cos 4\Phi
$$

$$
- \frac{f^2}{4\omega^2 A^2}\varepsilon - \frac{f^2}{4\omega^2 A^2}\varepsilon \cos 2\Phi - \frac{f^2}{2\omega^2 A}\delta \sin 2\Phi
$$

$$
+ \frac{3fg}{4\omega^2}\delta \sin \Phi + \frac{3fg}{4\omega^2}\delta \sin 3\Phi + \frac{3g^2}{16\omega^2}\varepsilon + \frac{g^2}{4\omega^2}\varepsilon \cos 2\Phi
$$

$$
- \frac{g^2 A}{2\omega^2}\delta \sin 2\Phi + \frac{g^2}{16\omega^2}\varepsilon \cos 4\Phi - \frac{g^2 A}{4\omega^2}\delta \sin 4\Phi \, ,
$$

where again only the averaged terms (that is, the terms which do not depend on Φ) are taken into consideration. The terms now neglected by averaging are of a higher order of magnitude with respect to the small parameters than the terms neglected in first approximation. Using (11.1, 17), (11.1, 18), we can show that

$$P_2 = \frac{hbA^3}{8} + \frac{deA^5}{32\omega^2} - \frac{9ef^2A}{32\omega^4} + \frac{hf^2A}{16\omega^2} - \frac{25eg^2A^3}{128\omega^4} + \frac{9hg^2A^3}{64\omega^2}.$$

In the same way, (11.1, 6) leads to the additional expression of second order

$$Q_2 = - \frac{\partial\varepsilon}{\partial A}Q_1 - \frac{\partial\varepsilon}{\partial\Phi}S_1 - \frac{f}{\omega}\delta\cos\Phi + \frac{g\varepsilon}{2\omega}\sin 2\Phi + \frac{gA}{\omega}\delta\cos 2\Phi$$

from which we get, by using (11.1, 17), (11.1, 18) and averaging,

$$Q^2 = Q_1^2 + 2Q_1Q_2 + Q_2^2$$

where

$$2Q_1Q_2 + Q_2^2 = - \frac{3ef^2A^2}{16\omega^4} - \frac{hf^2A^2}{8\omega^2} - \frac{5eg^2A^4}{64\omega^4} + \frac{hg^2A^4}{32\omega^2}.$$

When we insert $P = P_1 + P_2$ and Q^2, the integrand (11.1, 8) reads

$$J = \frac{-lA^6 - MA^4 - NA^2 + G}{A(rA^4 + SA^2 + G)}$$

with the abbreviations

$$G = \frac{f^2}{2\omega^2}, \qquad M = \frac{d}{4} + m, \qquad N = b - \frac{g^2}{8\omega^2} + n, \qquad S = \frac{g^2}{8\omega^2} + s$$

where small letters indicate terms of second order:

$$l = - \frac{de}{16\omega^2}, \qquad m = - \frac{hb}{4} + \frac{5eg^2}{64\omega^4} - \frac{5hg^2}{32\omega^2}, \qquad n = \frac{3ef^2}{16\omega^4} - \frac{3ef^2}{8\omega^2},$$

$$r = - \frac{5eg^2}{64\omega^4} + \frac{hg^2}{32\omega^2}, \qquad s = - \frac{3ef^2}{16\omega^4} - \frac{hf^2}{8\omega^2}$$

(r, s not to be confused with the coefficients of the Ito equations (11.1, 2)).

If we consider a non-vanishing parametric excitation, $g \neq 0$, decompose into partial fractions, set (neglecting, as above, terms of higher order)

$$\frac{1}{rA^4 + SA^2 + G} = \frac{1}{SA^2 + G} - \frac{rA^4}{(SA^2 + G)^2}$$

and assume, because of the first approximation, $S > 0$, we come to

$$J = -2UA - 4VA^3 + \frac{1}{A} + \frac{2TA}{A^2 + \dfrac{G}{S}} + \frac{2ZA}{\left(A^2 + \dfrac{G}{S}\right)^2},$$

using the abbreviations

$$U = \frac{M}{2S} - \frac{Nr + Gl}{2S^2} + \frac{G(M + r)r}{S^3}, \qquad V = \frac{lS - (M + r)r}{4S^2},$$

$$T = - \frac{1}{2} - \frac{N}{2S} + \frac{G(M - r)}{2S^2} - \frac{G(Gl + 2Nr)}{2S^3} + \frac{3G^2(M + r)r}{2S^4},$$

$$Z = G^2r\frac{NS + S^2 - G(M + r)}{2S^5}.$$

The probability density yields

$$w = \bar{C} A \left(A^2 + \frac{G}{S} \right)^T e^{-UA^2 - VA^4 - \frac{z}{A + G/S}}.$$ (11.1, 19)

This reduces to (11.1, 14) if the additional terms of second approximation written with small letters are put equal to zero.

In case of vanishing forced excitation, $f = 0$, (11.1, 19) simplifies to the formula

$$w = \bar{C} A^{2T+1} e^{-UA^2 - VA^4}, \qquad 2T + 1 = -\frac{N}{S} = 1 - \frac{8\omega^2 b}{g^2}.$$ (11.1, 20)

A comparison with (11.1, 9) shows that the second approximation leads to additional terms only in the exponential expression.

All probability density formulae derived contain only the second power of the forced and parametric excitation coefficients f, g, from which it follows that the signs of these coefficients do not influence the probability density.

11.2. Statistical properties of the vibrations

The normalization constant can be evaluated from (11.1, 15) by numerical integration. Analytical formulae can be found in the following two special cases.

First special case. Vanishing forced excitation, $f = 0$. By use of the integrals

$$\int_0^\infty x^{\nu-1} e^{-\varrho x - \sigma x^2} \, dx = (2\sigma)^{-\nu/2} \Gamma(\nu) e^{\varrho^2/8\sigma} D_{-\nu}\left(\frac{\varrho}{\sqrt{2\sigma}} \right) \qquad (\sigma > 0, \nu > 0)$$ (11.2, 1)

with the parabolic cylinder function D and the Gamma function Γ and

$$\int_0^\infty x^{\nu-1} e^{-\varrho x} \, dx = \varrho^{-\nu} \Gamma(\nu) \qquad (\varrho > 0, \nu > 0)$$ (11.2, 2)

(compare for instance GRADŠTEJN and RYŽIK (1971), p. 351 and 331) we can find

$$\frac{1}{C} = \int_0^\infty A^{-N/S} e^{-UA^2 - VA^4} \, dA$$

$$= \begin{cases} \dfrac{1}{2} (2V)^{(N/4S) - 1/4} \Gamma\left(\dfrac{1}{2} - \dfrac{N}{2S} \right) e^{U^2/8V} D_{(N/2S) - 1/2}\left(\dfrac{U}{\sqrt{2V}} \right) & \text{for} \quad V > 0, \\[3mm] \dfrac{1}{2} U^{(N/2S) - 1/2} \Gamma\left(\dfrac{1}{2} - \dfrac{N}{2S} \right) & \text{for} \quad V = 0, \quad U > 0 \end{cases}$$ (11.2, 3)

under the condition that the argument of the Gamma function is positive, in other words, that the threshold condition

$$g^2 > 4\omega^2 b$$ (11.2, 4)

for parametric excitation to exceed a certain damping threshold value holds.

Second special case. Non-vanishing forced excitation, but $V = Z = 0$, especially the first approximation. If we substitute

$$x = A^2 + \frac{G}{S}$$

and use the integral

$$\int_\tau^\infty x^{\nu-1} e^{-\varrho x} \, dx = \varrho^{-\nu} \Gamma(\nu, \varrho\tau) \qquad (\tau > 0, \varrho > 0), \tag{11.2, 5}$$

$\Gamma(\nu, \mu)$ being the incomplete Gamma function (compare for instance GRADŠTEJN and RYŽIK (1971), p. 331) we get

$$\frac{1}{\overline{C}} = \frac{1}{2} e^{(GU)/S} \int_{G/S}^\infty x^T e^{-Ux} \, dx = \frac{1}{2} e^{(GU)/S} U^{-T-1} \Gamma\left(T + 1, \frac{GU}{S}\right), \tag{11.2, 6}$$

where the conditions $G/S > 0$ and $U > 0$ can be assumed valid because of the first approximation. By means of the formula

$$\Gamma(T + 1, z) = \Gamma(T + 1) - \sum_{n=0}^\infty \frac{(-1)^n z^{T+n+1}}{n!(T + n + 1)} \tag{11.2, 7}$$

in GRADŠTEJN and RYŽIK (1971), p. 955, we can substitute for the incomplete Gamma function the (simple) Gamma function if $T \neq -1, -2, \dots$ For $f \to 0$, formula (11.2, 6) simplifies because of (11.2, 7) to the second formula (11.2, 3).

The derivative of the probability density (11.1, 19) is

$$\frac{dw}{dA} = \overline{C}\left(A^2 + \frac{G}{S}\right)^{T-2}\left[1 - 2UA^2 - 4VA^4\right)\left(A^2 + \frac{G}{S}\right)^2$$

$$+ 2TA^2\left(A^2 + \frac{G}{S}\right) + 2ZA^2\right] e^{-UA^2 - VA^4 - \frac{Z}{A^2 + G/S}}. \tag{11.2, 8}$$

The condition for extreme values, the vanishing of the expression in brackets, is an equation of fourth degree in A^2.

In the *first special case* of vanishing forced excitation, the threshold condition (11.2, 4) has to be fulfilled, otherwise the probability of positive amplitudes is zero and no vibration exists. We have

$$w(0) = \infty \quad \text{for} \quad g^2 < 8\omega^2 b$$

and

$$w(0) = \overline{C} \quad \text{for} \quad g^2 = 8\omega^2 b,$$

in both cases no extreme value of the probability density exists and it decreases monotonicly for A increasing (left curves in Figures 11.2, 3 and 11.2, 4). Corresponding to STRATONOVICH (1961), we speak of an "undevelopped vibration".

In contrast,

$$w(0) = 0 \quad \text{for} \quad g^2 > 8\omega^2 b.$$

In this case a maximum of the probability density exists, that is, there is a most probable value A_m which is determined, because of (11.2, 8), by

$$
\left.
\begin{aligned}
4VA_m^2 &= -U + \sqrt{U^2 + 4\left(1 - \frac{8\omega^2 b}{g^2}\right)V} \quad \text{for} \quad V > 0\,, \\[2mm]
A_m^2 &= \frac{g^2 - 8\omega^2 b}{2g^2 U} = \frac{1}{2\omega^2}\,\frac{g^2 - 8\omega^2 b}{d + 4m + 4\left(1 - \dfrac{8\omega^2 b}{g^2}\right)r} \\[2mm]
&\qquad\qquad\qquad \text{for} \quad V = 0\,, \qquad U > 0\,.
\end{aligned}
\right\}
\qquad (11.2,\,9)
$$

Here the condition $U > 0$ is valid because of the predominant expressions of first approximation. Since $T < 0$, the derivative of the probability density tends to infinity for $A \to 0$ so that the probability density already assumes great values for small amplitudes (half developed vibration).

For *non-vanishing forced excitation*, (11.1, 19) always yields $w(0) = 0$ so that one or two maxima of the probability density exist. The derivative of the probability density is finite, equal to

$$
\overline{C}\left(\frac{G}{S}\right)^T e^{-(ZS)/G} \quad \text{for} \quad A = 0
$$

so that the probability density assumes great values only for greater amplitudes (fully developped vibration).

In the *second special case* $V = Z = 0$, exactly one maximum follows from (11.2, 8), that is, a most probable amplitude A_m is given by

$$
2UA_m^2 = Y + \sqrt{Y^2 + \frac{2GU}{S}}
$$

if

$$
Y = T + \frac{1}{2} - \frac{GU}{S}
$$

is written as an abbreviation. For $Y = 0$, that is, in first approximation if

$$
g^2(g^2 - 8\omega^2 b) = 8\omega^2 df^2
$$

then in particular

$$
A_m^2 = \sqrt{\frac{G}{2SU}} \to \frac{\sqrt{2}\,|f|}{\omega\sqrt{d}} = \frac{|g|\sqrt{g^2 - 8\omega^2 b}}{2\omega^2 d}
$$

where the arrow indicates reduction to first approximation. For

$$
Y^2 \gg \left|\frac{2GU}{S}\right|
$$

we get by series expansion

$$
A_m^2 = \frac{G}{2S|Y|} +
\begin{cases}
\dfrac{Y}{U}\,, & \text{if} \quad Y > 0\,, \\[3mm]
0\,, & \text{if} \quad Y < 0
\end{cases}
$$

$$
\to \frac{4f^2}{|g^2 - 8\omega^2 b|} +
\begin{cases}
\dfrac{g^2 - 8\omega^2 b}{2\omega^2 d}\,, & \text{if} \quad g^2 > 8\omega^2 b\,, \\[3mm]
0\,, & \text{if} \quad g^2 < 8\omega^2 b\,;
\end{cases}
$$

that is, for $f \to 0$ the most probable amplitude tends to zero if $g^2 < 8\omega^2 b$ and to the positive value given by (11.2, 9) if $g^2 > 8\omega^2 b$.

Examples for the probability density (11.1, 19) are given in Figures 11.2, 1 to 11.2, 4 where $\omega = 1$, $e = 4$, $h = 0$ and $b = 1/2000$ have been chosen and the normalization constant has been evaluated by numerical integration. For $f = 0.01$ and $g = 0.06$ (fully developed vibration), Fig. 11.2, 1 shows how a greater non-linear damping d diminishes the probability of greater amplitudes; Fig. 11.2, 2 gives for $d = 0.2$ the difference between first (dashed line) and second approximation (full line); the second approximation also somewhat diminishes the probability of greater amplitudes. Fig. 11.2, 3 shows (for $d = 1$ and $g = 0.06$) how the probability of greater amplitudes increases with forced excitation (left curve $f = 0$, undeveloped vibration,

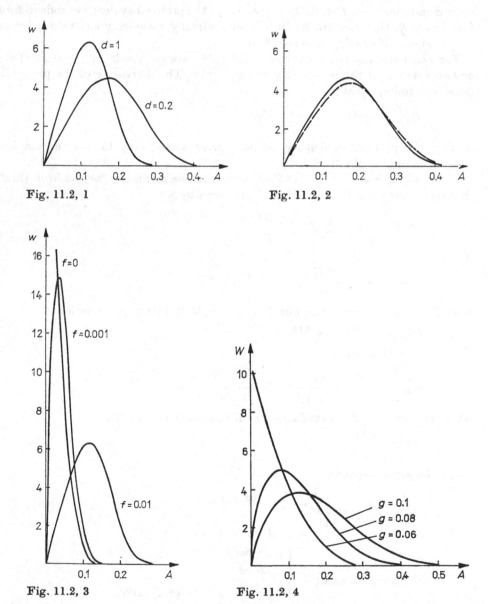

Fig. 11.2, 1

Fig. 11.2, 2

Fig. 11.2, 3

Fig. 11.2, 4

fully developed vibration for the other curves). An analogous influence of parametric excitation reveals Fig. 11.2, 4 for $d = 0.2$, $f = 0$ and $g = 0.06$ (undevelopped vibration), $g = 0.08$ respectively $g = 0.1$ (half developped vibrations).

The *moments*

$$m_k = \int\limits_0^\infty A_k w(A)\, \mathrm{d}A\,, \qquad k = 1, 2, 3, \dots$$

of the amplitude can be evaluated by help of the probability density.

For the *first special case* of vanishing forced excitation we get the formula

$$m_k = \begin{cases} \dfrac{\Gamma\left(\dfrac{k+1}{2} - \dfrac{N}{2S}\right) D_{\frac{N}{2S} - \frac{k+1}{2}}\left(\dfrac{U}{\sqrt{2V}}\right)}{(2V)^{k/4}\, \Gamma\left(\dfrac{1}{2} - \dfrac{N}{2S}\right) D_{\frac{N}{2S} - \frac{1}{2}}\left(\dfrac{U}{\sqrt{2V}}\right)}\,, & \text{if } V > 0\,, \\[2em] \dfrac{\Gamma\left(\dfrac{k+1}{2} - \dfrac{N}{2S}\right)}{U^{k/2}\, \Gamma\left(\dfrac{1}{2} - \dfrac{N}{2S}\right)}\,, & \text{if } V = 0\,, \quad U > 0\,. \end{cases}$$

The second moment or the mean square amplitude is because of $\Gamma(x+1) = x\Gamma(x)$

$$m_2 = \begin{cases} \dfrac{\left(\dfrac{1}{2} - \dfrac{N}{2S}\right) D_{\frac{N}{2S} - \frac{3}{2}}\left(\dfrac{U}{\sqrt{2V}}\right)}{\sqrt{2V}\, D_{\frac{N}{2S} - \frac{1}{2}}\left(\dfrac{U}{\sqrt{2V}}\right)}\,, & \text{if } V > 0\,, \\[2em] \dfrac{S - N}{2SU} = \dfrac{1}{\omega^2}\, \dfrac{g^2 - 4\omega^2 b}{d + 4m + 4\left(1 - \dfrac{8\omega^2 b}{g^2}\right) r}\,, & \text{if } V = 0\,, \quad U > 0 \end{cases}$$

$$(11.2, 10)$$

from where follows the rms value $\sqrt{m_2}$.

As for $r\gamma = 0$

$$\frac{N}{2S} - \frac{k+1}{2} = -\frac{k}{2} - 1 + \frac{4\omega^2 b}{g^2} = \begin{cases} -2 & \text{if } b = 0\,, \quad k = 2\,, \\ -1 & \text{if } b = 0\,, \quad k = 0 \end{cases}$$

and because of GRADŠTEJN and RYŽIK (1971), p. 1081 and 1108,

$$D_{-1}(z) = e^{z^2/4} \sqrt{\frac{\pi}{2}} \left[1 - \Phi\left(\frac{z}{\sqrt{2}}\right)\right], \qquad (11.2, 11)$$

$$D_{-2}(z) = -e^{z^2/4} \sqrt{\frac{\pi}{2}} \left\{z\left[1 - \Phi\left(\frac{z}{\sqrt{2}}\right)\right] - \sqrt{\frac{2}{\pi}}\, e^{-z^2/2}\right\} \qquad (11.2, 12)$$

hold, in the case of vanishing linear damping and $V > 0$ the mean square amplitude can be expressed only by means of the Γ function and the error function

$$\Phi(x) = \frac{2}{\sqrt{\pi}} \int\limits_0^x e^{-\xi^2}\, \mathrm{d}\xi\,. \qquad (11.2, 13)$$

Because the argument $U/\sqrt{2V}$ of the parabolic cylinder function is of the order of magnitude $1/g$, the asymptotic formula (compare MAGNUS and OBERHETTINGER (1943) p. 92)

$$D_\nu(z) \sim \mathrm{e}^{-z^2/4}\, z^\nu \left[1 - \frac{\nu(\nu - 1)}{2z^2} + \frac{\nu(\nu - 1)\,(\nu - 2)\,(\nu - 3)}{2 \cdot 4z^2} - + \ldots \right]$$

$$\text{for} \quad z \gg 1\,, \quad z \gg |\nu|$$

can also be used for the evaluation of the moments.

For $V = 0$, $U > 0$, formula (11.2, 10) immediately shows the influence of the system parameters on the mean square amplitude. For example, the influence of predominant order of magnitude of the parametric excitation g and the linear damping b appears in the numerator which is

equal to g^2 for $b = 0$,

equal to $\dfrac{g^2}{2}$ for $b = \dfrac{g^2}{8\omega^2}$ and

equal to zero for $b = \dfrac{g^2}{4\omega^2}$.

The dependence on non-linear restoring force e and non-linear damping d, say, represents the formula, written for $h = 0$, $g^2 = 5\omega^2 b$:

$$m_2 = \frac{b}{d + \dfrac{85eb}{64\omega^2}}\,. \tag{11.2, 14}$$

Examples of the diminishing of the mean square amplitude by the non-linear restoring force and the non-linear damping are given in Fig. 11.2, 5 for $b = 0.01$, $\omega = 1$. In case $e = 10$, $b = 0.01d$, the influence of e and by this of the second approximation causes a diminishing of the mean square amplitude by about 12 per cent.

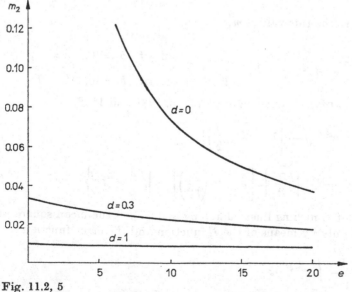

Fig. 11.2, 5

The first moment or the amplitude mean value is for $V = 0$, $U > 0$

$$m_1 = \frac{|g|}{\omega} \frac{\Gamma\left(\dfrac{3}{2} - \dfrac{4\omega^2 b}{g^2}\right)}{\Gamma\left(1 - \dfrac{4\omega^2 b}{g^2}\right) \sqrt{d + 4m + 4\left(1 - \dfrac{8\omega^2 b}{g^2}\right) r}}$$

from which there follows with (11.2, 10) the dispersion

$$D = m_2 - m_1^2$$

of the amplitude:

$$D = \frac{1}{\omega^2} \frac{1}{d + 4m + 4\left(1 - \dfrac{8\omega^2 b}{g^2}\right) r} \left[g^2 - 4\omega^2 b - \frac{g^2 \Gamma^2\left(\dfrac{3}{2} - \dfrac{4\omega^2 b}{g^2}\right)}{\Gamma^2\left(1 - \dfrac{4\omega^2 b}{g^2}\right)} \right].$$

In the *second special case* $V = Z = 0$ and non-vanishing forced excitation we derive, by using the integral

$$\int_0^\infty x^{\nu-1}(x + \tau)^{-\sigma} e^{-\varrho x}\, dx = \tau^{\frac{\nu-\sigma-1}{2}}\, \varrho^{\frac{\sigma-\nu-1}{2}}\, e^{\frac{\tau\varrho}{2}}\, \Gamma(\nu)\, W_{\frac{1-\nu-\sigma}{2}, \frac{\nu-\sigma}{2}}(\tau\varrho)$$

$$(\varrho > 0, \nu > 0)$$

(compare for instance GRADŠTEJN and RYŽIK (1971), p. 333) and $G/S > 0$, $U > 0$, the formula

$$m_k = \left(\frac{G}{SU}\right)^{k/4} \Gamma\left(\frac{k}{2} + 1\right) \frac{W_{T/2-k/4,\, T/2+1/2+k/4}\left(\dfrac{GU}{S}\right)}{W_{T/2,\, T/2+1/2}\left(\dfrac{GU}{S}\right)}$$

where $W_{\mu\nu}(z)$ are the Whittaker functions. The argument of these functions is in the first approximation

$$\frac{GU}{S} = \frac{4\omega^2 d f^2}{g^4}.$$

This can often be assumed as large in comparison with 1 so that the asymptotic development (MAGNUS and OBERHETTINGER (1943), p. 89)

$$W_{\mu\nu}(z) \sim e^{-z^{1/2}} z^\mu \left(1 + \sum_{n=1}^{\infty} \frac{[\nu^2 - (\mu - \frac{1}{2})^2]\,[\nu^2 - (\mu - \frac{3}{2})^2] \cdots [\nu^2 - (\mu - n + \frac{1}{2})^2]}{n!\, z^n}\right)$$

$$(11.2, 15)$$

can be used. The first two terms of this development lead to the approximative formula

$$m_k = \frac{\Gamma\left(\dfrac{k}{2} + 1\right)}{U^{k/2}} \frac{GU + \dfrac{k+2}{2} ST}{GU + ST}$$

where the second summands in the second fraction and by this the second fraction as a whole stem from the second term of the asymptotic development (11.2, 15). In particular, the mean square amplitude, using the expression for the first approximation, yields

$$m_2 = \frac{1}{U} \frac{GU + 2ST}{GU + ST} \rightarrow \frac{g^2}{\omega^2 d} \frac{3df^2 - 2g^2 b}{2df^2 - g^2 b} .$$

A comparison with the opposite limiting case $f = 0$ of formula (11.2, 10) for, say, vanishing linear damping, $b = 0$, now (for predominant forced excitation f) results in one and a half times the value following from (11.2, 10) for $f = 0$.

The *probability*

$$W_l = W(A_l < A < \infty)$$

of exceeding a given amplitude level A_l is

$$W_l = \int\limits_{A_l}^{\infty} w(A) \, \mathrm{d}A .$$

In the second special case $V = Z = 0$ (vanishing forced excitation, $f = 0$, included), especially for the first approximation we can find, by analogy to (11.2, 6), that

$$W_l = \frac{\overline{C}}{2} \, \mathrm{e}^{(GU)/S} \, U^{-T-1} \Gamma\left(T + 1, A_l^2 U + \frac{GU}{S}\right).$$

Insertion of (11.2, 6) yields

$$W_l = \frac{\Gamma\left(T + 1, A_l^2 U + \dfrac{GU}{S}\right)}{\Gamma\left(T + 1, \dfrac{GU}{S}\right)} . \tag{11.2, 16}$$

In evaluating W_l, formula (11.2, 7) can be used.

In two examples, (11.2, 16) simplifies significantly. For $T = 0$, that is in first approximation for

$$df^2 = bg^2 \tag{11.2, 17}$$

because $\Gamma(1, z) = \mathrm{e}^{-z}$ the formula

$$W_l = W_{l1} = \mathrm{e}^{-A_l^2 U} \tag{11.2, 18}$$

holds. On the other hand, for $T = -1/2$, that is in first approximation for

$$df^2 = \left(b - \frac{g^2}{8\omega^2}\right) f^2 ,$$

because

$$\Gamma\left(\tfrac{1}{2}, z\right) = \sqrt{\pi} - \sqrt{\pi} \, \Phi(\sqrt{z})$$

the formula

$$W_l = W_{l2} = \frac{1 - \Phi\left(\sqrt{A_l^2 U + \dfrac{GU}{S}}\right)}{1 - \Phi\left(\sqrt{\dfrac{GU}{S}}\right)}$$

holds.

A comparison of these two examples clarifies the influence of different system parameters. For vanishing forced excitation, $f = 0$, the last formula simplifies to

$$W_{l2} = 1 - \Phi(A_l \sqrt{\overline{U}}) , \tag{11.2, 19}$$

while the condition $T = -1/2$ simplifies in the first approximation to

$$b = \frac{g^2}{8\omega^2} , \qquad f = 0 . \tag{11.2, 20}$$

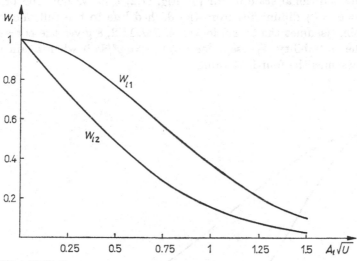

Fig. 11.2, 6

The curves given by (11.2, 18) for $T = 0$ and by (11.2, 19) for $T = -1/2$, $f = 0$ are drawn in Fig. 11.2, 6 in general form, without confinement to the first approximation. In the first approximation we get

$$U = \frac{\omega^2 d}{g^2} , \tag{11.2, 21}$$

and the conditions for W_{l2} are (11.2, 20), whereas the condition (11.2, 17) for W_{l1} holds (say) for $b = f = 0$ or for

$$b = \frac{g^2}{8\omega^2} , \qquad f^2 = \frac{g^4}{8\omega^2 d} . \tag{11.2, 22}$$

In other words, W_{l2} shows, in comparison with W_{l1}, the *decrease* of the probability of exceeding A_l if $T = 0$ changes to $T = -1/2$, in the first approximation if the linear damping increases from zero to the value given by (11.2, 22) and if the forced excitation vanishes. On the other hand, W_{l1} gives, in comparison with W_{l2}, the *increase* of the probability for exceeding A_l if particularly in first approximation the forced excitation increases from zero to the value given by (11.2, 22) and

$$b = \frac{g^2}{8\omega^2}$$

holds.

The influence of non-linear damping d manifests itself — in the first approximation and for vanishing forced excitation, $f = 0$ — only in (11.2, 21), that is, in a modification of the abscissa scale. For instance, if the non-linear damping increases to the p-fold value, U increases correspondingly, so that the probability of exceeding the diminished level

$$\frac{A_l}{\sqrt{p}}$$

remains the same; or, in other words, for a fixed level A_l, the probability of exceeding diminishes by the additional scale factor $\sqrt{p}$. Fig. 11.2, 7 shows how the probability W_{l1} (say, for $f = b = 0$) diminishes from the dashed line to the full one when the non-linear damping assumes the four-fold value. Fig. 11.2, 8 gives the corresponding diminution of the probability W_{l2} (say, for $f = 0$, $b = g^2/8\omega^2$) when again the non-linear damping assumes the four-fold value.

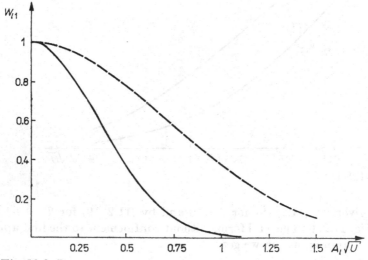

Fig. 11.2, 7

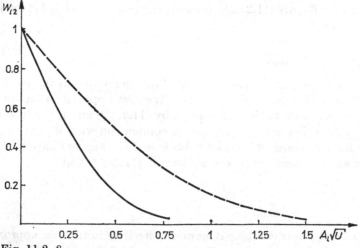

Fig. 11.2, 8

In order to discuss the influence of the other non-linearities, say of non-linear inertia with the coefficient h, by means of the formulae of the second approximation, we have first to take into consideration the conditions $V = Z = 0$ of (11.2, 10). The second condition leads, approximatively, to the condition (11.2, 11) for W_{h1}. Choose for instance $b = f = 0$, so that instead of (11.2, 15) the formula

$$U = \frac{\omega^2 d}{g^2} - \frac{h}{2}$$

holds. The condition $V = 0$ yields

$$3e = 2\omega^2 h \ .$$

Consequently, non-linear inertia, as well as non-linear damping, leads to an additional scale factor. Fig. 11.2, 9 presents an example for a scale factor 0.9 which follows for instance for non-linear inertia $h = 9.5$ and $\omega = 1$, $d = 1/4$, $g = 0.1$ and which causes an increase of the probability W_{h1} from the values of the dashed line to those of the full line.

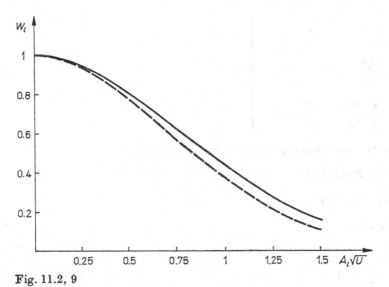

Fig. 11.2, 9

11.3. Non-stationary probability density, transition probability density and two-dimensional probability density

On the basis of the stationary probability densities thus found, we consider in this section the dependence of the probability density on time, in other words, the first expression in the Fokker Planck Kolmogorov equation (11.1, 7). We use a method introduced by STRATONOVICH (1961), compare also Nikolaenko (1967).

Corresponding with Section 11.1, write the Fokker Planck Kolmogorov equation in the form

$$\frac{\partial w}{\partial t} = -\frac{\partial (Pw)}{\partial A} + \frac{1}{2}\frac{\partial^2(Q^2 w)}{\partial A^2} = -\frac{\partial \Psi}{\partial A} \ .$$

First, assume the probability density as a product,

$$w(A, t) = \overline{w}(A) \, \varphi(t) \, . \tag{11.3, 1}$$

Separation of variables yields

$$\frac{\dfrac{\mathrm{d}\varphi}{\mathrm{d}t}}{\varphi} = \frac{1}{\overline{w}} \left[-\frac{\mathrm{d}(P\overline{w})}{\mathrm{d}A} + \frac{1}{2} \frac{\mathrm{d}^2(Q^2\overline{w})}{\mathrm{d}A^2} \right] = \text{const} \, . \tag{11.3, 2}$$

If we write $-\lambda$ for the constant, we get

$$\varphi = \mathrm{e}^{-\lambda(t - t_0)}$$

as well as the differential equation

$$H \frac{\mathrm{d}^2\overline{w}}{\mathrm{d}A^2} + I \frac{\mathrm{d}\overline{w}}{\mathrm{d}A} + K\overline{w} = 0 \tag{11.3, 3}$$

for $\overline{w}$, using the abbreviations

$$\left. \begin{aligned} H &= \frac{1}{2} Q^2 \, , \\[2mm] I &= \frac{\mathrm{d}Q^2}{\mathrm{d}A} - P \, , \\[2mm] K &= \frac{1}{2} \frac{\mathrm{d}^2Q^2}{\mathrm{d}A^2} - \frac{\mathrm{d}P}{\mathrm{d}A} + \lambda \, . \end{aligned} \right\} \tag{11.3, 4}$$

This differential equation can be simplified by writing

$$\overline{w}(A) = E(A) \, v(A)$$

and introducing the variable

$$\alpha = \varkappa A^2 \, , \tag{11.3, 5}$$

that is,

$$\frac{\mathrm{d}}{\mathrm{d}A} = 2\varkappa A \frac{\mathrm{d}}{\mathrm{d}\alpha} \, ,$$

$$\frac{\mathrm{d}^2}{\mathrm{d}A^2} = 2\varkappa \frac{\mathrm{d}}{\mathrm{d}\alpha} + 4\varkappa^2 A^2 \frac{\mathrm{d}^2}{\mathrm{d}\alpha^2}$$

with a constant $\varkappa$ which we will fix later on. If we use the notation

$$v(A) = \overline{v}(\alpha) \, ,$$

we come to the differential equation

$$4\varkappa^2 A^2 HE \frac{\mathrm{d}^2\overline{v}}{\mathrm{d}\alpha^2} + 2\varkappa \left(2AH \frac{\mathrm{d}E}{\mathrm{d}A} + AIE + HE \right) \frac{\mathrm{d}\overline{v}}{\mathrm{d}\alpha}$$

$$+ \left(H \frac{\mathrm{d}^2E}{\mathrm{d}A^2} + I \frac{\mathrm{d}E}{\mathrm{d}A} + KE \right) \overline{v} = 0 \, . \tag{11.3, 6}$$

Choose E in such a way that the coefficient of $d\bar{v}/d\alpha$ vanishes,

$$2AH\frac{dE}{dA} = -(AI + H)E,\qquad\qquad(11.3, 7)$$

from where follows, assuming $AH \neq 0$, the expression

$$E = e^{-\int\left(\frac{T}{2H}+\frac{1}{2A}\right)dA}\qquad\qquad(11.3, 8)$$

for E. Substituting in (11.3, 6) the derivatives of E by (11.3, 7) and the formula derived by differentiation of (11.3, 7), and considering the abbreviations (11.3, 4) and (11.3, 5), we get the differential equation

$$4\alpha^2\frac{d^2\bar{v}}{d\alpha^2} + \Pi\bar{v} = 0\qquad\qquad(11.3, 9)$$

using the abbreviation

$$\Pi = \frac{3}{4} + \frac{A^2P\dfrac{dQ^2}{dA} - A^2P^2}{Q^4} - \frac{A^2\dfrac{dP}{dA} - 2A^2\lambda}{Q^2}.$$

Insertion of the formulae derived in Section 11.1 reveals the dependence of the expression Π on the amplitude,

$$\Pi = \frac{3}{4} + \frac{1}{2}\frac{G + (4\lambda + N')A^2 + 3M'A^4 + 5lA^6}{G + SA^2 + rA^4} + \frac{1}{4(G + SA^2 + rA^4)^2}$$
$$\times [-G^2 + 2G(2S + N')A^2 + (2GM' - N'^2 - 4N'S + 8Gr)A^4$$
$$- 2(M'N' + 2M'S - Gl + 4N'r)A^6 - (M'^2 + 2N'l + 4Sl + 8M'r)A^8$$
$$- 2(M' + 4r)lA^{10} - l^2A^{12}]$$

using the former abbreviations and

$$M' = \frac{d}{4} - \frac{hb}{4} + \frac{25eg^2}{64\omega^4} - \frac{9hg^2}{32\omega^2},$$

$$N' = b - \frac{3g^2}{8\omega^2} + \frac{9ef^2}{16\omega^4} - \frac{hf^2}{8\omega^2}$$

differing from the corresponding expressions M, N in second respectively first approximation.

If forced excitation does not vanish, we can, in connection with the predominance of small amplitudes A, in general assume that

$$G \gg SA^2 + rA^4.$$

Developing the denominators of Π into series leads to the expression

$$\Pi = 1 + \frac{N' + S + 2\lambda}{G}A^2 - \frac{N'^2 + 9S^2 + 10N'S - 8GM' - 8Gr + 8S\lambda}{4G^2}A^4$$
$$+ \frac{6G^2l - GM'N' - 7GM'S - 7GN'r - 15GSr + N'^2S + 7N'S^2 + 2S^3 - 4Gr\lambda)}{2G^3}$$
$$\times A^6 + O(A^8).\qquad\qquad(11.3, 10)$$

If we neglect the amplitude terms of sixth and higher degree in (11.3, 10) and choose the free constant $\varkappa$ such that

$$\varkappa = \frac{1}{2G}\sqrt{N'^2 + 9S^2 + 10N'S - 8GM' - 8Gr + 8S\lambda}\,,$$

we get the differential equation (11.3, 9) for $\bar{v}$ in form of the Whittaker equation

$$4\alpha^2 \frac{d^2\bar{v}}{d\alpha^2} = (\alpha^2 - 4p\alpha + 4q^2 - 1)\,\bar{v} \qquad\qquad (11.3, 11)$$

with

$$p = \frac{N' + S + 2\lambda}{4G\varkappa}\,, \qquad q = 0\,.$$

Following the eigenfunction method of STRATONOVICH (1961), we use the eigenvalues

$$\lambda = \lambda_\nu \qquad (\nu = 0, 1, 2, \ldots)$$

and corresponding eigenfunctions

$$\bar{w} = w_\nu(A) \qquad (\nu = 0, 1, 2, \ldots)$$

for given boundary conditions, in our case vanishing values for $A = 0$ and $A \to \infty$. The first eigenfunction for $\lambda_0 = 0$ is the stationary probability density

$$w_0(A) = w_{\text{stat}}(A) \qquad\qquad (11.3, 12)$$

as comparison of (11.3, 2) with the stationary case of Section 11.1 shows.

The unstationary probability density can be written in the form of the eigenfunction development

$$w(A, t) = C_0 w_0(A) + \sum_{\nu=1}^{\infty} C_\nu w_\nu(A)\,\mathrm{e}^{-\lambda\nu(t-t_0)} \qquad\qquad (11.3, 13)$$

generalizing the assumption (11.3, 1) of one product. Integration of (11.3, 3) leads, for $\lambda_\nu \neq 0$, to

$$\int_0^\infty w_\nu(A)\,\mathrm{d}A = 0\,.$$

Under the assumption that no two eigenfunctions have the same eigenvalue,

$$\lambda_\mu \neq \lambda_\nu \quad \text{for} \quad \mu \neq \nu\,,$$

we shall now establish an orthogonality relation. The differential equation (11.3, 3) reads

$$\frac{\partial}{\partial A}\,\Psi[w_\mu] = \lambda_\mu w_\mu\,,$$

$$\frac{\partial}{\partial A}\,\Psi[w_\nu] = \lambda_\nu w_\nu$$

for two different eigenfunctions. Multiplication by w_ν/w_0 respectively w_μ/w_0 and integration yields

$$\left.\begin{array}{l}\displaystyle\int \frac{w_\nu}{w_0}\frac{\partial}{\partial A}\,\Psi[w_\mu]\,\mathrm{d}A = \lambda_\mu \int \frac{w_\nu w_\mu}{w_0}\,\mathrm{d}A\,,\\[3mm]\displaystyle\int \frac{w_\mu}{w_0}\frac{\partial}{\partial A}\,\Psi[w_\nu]\,\mathrm{d}A = \lambda_\nu \int \frac{w_\mu w_\nu}{w_0}\,\mathrm{d}A\,.\end{array}\right\} \tag{11.3, 14}$$

The expression

$$J = \int \tau\frac{\partial}{\partial A}\,\Psi[w_0\sigma]\,\mathrm{d}A$$

$$= \int \tau\frac{\partial}{\partial A}\,(Pw_0\sigma)\,\mathrm{d}A - \frac{1}{2}\int \tau\frac{\partial^2}{\partial A^2}\,(Q^2 w_0\sigma)\,\mathrm{d}A\,, \tag{11.3, 15}$$

σ and τ being arbitrary functions, can be written in the form

$$J = \int \tau\sigma\frac{\partial}{\partial A}\,(Pw_0)\,\mathrm{d}A + \int \tau\,Pw_0\frac{\partial\sigma}{\partial A}\,\mathrm{d}A - \frac{1}{2}\int \tau\sigma\frac{\partial^2}{\partial A^2}\,(Q^2 w_0)\,\mathrm{d}A$$

$$- \int \tau\frac{\partial\sigma}{\partial A}\frac{\partial}{\partial A}\,(Q^2 w_0)\,\mathrm{d}A - \frac{1}{2}\int \tau Q^2 w_0\frac{\partial^2\sigma}{\partial A^2}\,\mathrm{d}A\,.$$

The first and the third expression cancel out because w_0 is the stationary solution. In the same way, the fourth expression can be written as double the second one with negative sign. Thus we get

$$J = -\int \tau Pw_0\frac{\partial\sigma}{\partial A}\,\mathrm{d}A - \frac{1}{2}\int \tau Q^2 w_0\frac{\partial^2\sigma}{\partial A^2}\,\mathrm{d}A\,.$$

Integration by parts yields

$$J = \int \frac{\partial}{\partial A}\,(\tau Pw_0)\,\sigma\,\mathrm{d}A - \frac{1}{2}\int \frac{\partial^2}{\partial A^2}\,(\tau Q^2 w_0)\,\sigma\,\mathrm{d}A$$

$$= \int \frac{\partial\Psi[w_0\tau]}{\partial A}\,\sigma\,\mathrm{d}A \tag{11.3, 16}$$

if we assume

$$\tau\frac{\partial\sigma}{\partial A} = \sigma\frac{\partial\tau}{\partial A}$$

or

$$\tau\Psi[w_0\sigma] = \sigma\Psi[w_0\tau]$$

on the boundary. When we choose

$$\sigma = \frac{w_\mu}{w_0}\,,\qquad \tau = \frac{w_\nu}{w_0}\,,$$

(11.3, 15) and (11.3, 16) show that the left-hand sides of the two equations (11.3, 14) are equal. The right-hand sides give the equation

$$(\lambda_\mu - \lambda_\nu)\int \frac{w_\mu w_\nu}{w_0}\,\mathrm{d}A = 0\,,$$

that is, the orthogonality of the eigenfunctions belonging to different eigenvalues with weight $1/w_0$. Normalizing the eigenfunctions by

$$\int \frac{w_\nu^2}{w_0} \, dA = 1 ,$$

we can write the orthogonality relation in the form

$$\int \frac{w_\mu w_\nu}{w_0} \, dA = \delta_\mu^\nu .$$

Multiplication of the eigenfunction development (11.3, 13) by w_μ/w_0, putting $t = t_0$, integrating and using the orthogonality relation gives

$$C_\mu = \int \frac{w(A, t_0) \, w_\mu(A)}{w_0(A)} \, dA .$$

When the initial distribution is the Dirac delta function,

$$w(A, t_0) = \delta(A - A_0) ,$$

then the eigenfunction development (11.3, 13) is the transition probability density

$$p_{tt_0}(A, A_0) = \sum_{\nu=0}^{\infty} \frac{w_\nu(A) \, w_\nu(A_0)}{w_0(A_0)} \, e^{-\lambda\nu(t-t_0)}$$

where $\lambda_0 = 0$. By means of the transition probability density and the initial distribution, all distributions of finite order can be evaluated. In the case of a stationary initial distribution, the two-dimensional probability density

$$w_\tau(A, A_0) = p_{t, t-\tau}(A, A_0) \, w_0(A_0)$$

satisfies the simple formula

$$w_\tau(A, A_0) = \sum_{\nu=0}^{\infty} w_\nu(A) \, w_\nu(A_0) \, e^{-\lambda\bar{\nu}|\tau|} , \qquad \tau = t - t_0 . \tag{11.3, 17}$$

The Whittaker equation (11.3, 11) which we have found here has for
$2q$ not equal to an integer
the system of linearly independent solutions

$$M_{p, q}(\alpha) = \alpha^{q+1/2} \, e^{-\alpha/2} \, {}_1F_1(q - p + \tfrac{1}{2}, 2q + 1, \alpha) ,$$

$$M_{p, -q}(\alpha) = \alpha^{-q+1/2} \, e^{-\alpha/2} \, {}_1F_1(-q - p + \tfrac{1}{2}, -2q + 1, \alpha)$$

where ${}_1F_1$ is given by

$$_1F_1(r, s, \alpha) = 1 + \sum_{\nu=1}^{\infty} \frac{r(r + 1) \dots (r + \nu - 1) \alpha^\nu}{s(s + 1) \dots (s + \nu - 1) \nu!} ,$$

the so-called Pochhammer function or confluent hypergeometric function, which can be expressed in the form

$$_1F_1(r, s, \alpha) = \frac{\Gamma(s) \, \Gamma(1 - r)}{\Gamma(s - r)} \, L_{-r}^{(s-1)}(\alpha)$$

by generalized Laguerre functions. A system of solutions also linearly independent for $2q$ equal to an integer is given by the Whittaker functions

$$W_{p,q}(\alpha) = \frac{\Gamma(-2q)}{\Gamma(\frac{1}{2} - q - p)} M_{p,q}(\alpha) + \frac{\Gamma(2q)}{\Gamma(\frac{1}{2} + q - p)} M_{p,-q}(\alpha).$$

We can transform the Whittaker equation (11.3, 11) by

$$\bar{v}(\alpha) = \alpha^{q+1/2}\, e^{-\alpha/2}\, u(\alpha)$$

to the differential equation

$$\alpha \frac{d^2 u}{d\alpha^2} + (2q + 1 - \alpha)\frac{du}{d\alpha} + \left(p - q - \frac{1}{2}\right) u = 0$$

which has (KAMKE (1959)), for natural numbers $2q + 1$ and the boundary conditions that u is limited for $\alpha \to 0$ and not greater than a power of α for $\alpha \to \infty$, the eigenvalues $p - q - 1/2 = 0, 1, 2, \ldots$ and the eigenfunctions $L_n^{(2q)}(\alpha)$.

In our case

$$p = \frac{N' + S + 2\lambda}{4G\varkappa}, \qquad q = 0$$

the eigenvalue equation yields

$$\lambda_n = (2n + 1)\, G\varkappa - \frac{N'}{2} - \frac{S}{2}, \qquad n = 0, 1, 2, \ldots \tag{11.3, 18}$$

or, after inserting the formulae found above,

$$\lambda_n = \left(n + \frac{1}{2}\right)\sqrt{b^2 - \frac{df^2}{\omega^2} + \frac{bg^2}{2\omega^2} - \frac{3g^4}{16\omega^4} + \frac{g^2}{\omega^2}\lambda_n}$$

$$+ \frac{g^2}{8\omega^2} - \frac{b}{2} + O(f^2, \sqrt{bf}, \sqrt{bfg}, \sqrt{fg^3}). \tag{11.3, 19}$$

The solutions are, in the approximation of the terms written explicitely in (11.3, 19), given by

$$\lambda_n = \frac{2n^2 + 2n + 1}{4\omega^2}\, g^2 - \frac{b}{2}$$

$$\pm \sqrt{\left(n + \frac{1}{2}\right)^2 b^2 - \left(n + \frac{1}{2}\right)^2 \frac{df^2}{\omega^2} + \frac{(4n^3 + 8n^2 + 5n + 1)\, n}{16\omega^4}\, g^4}.$$

We obtain real values λ_n (and $\varkappa$ because of (11.3, 18)) if forced excitation and non-linear damping are small enough for the radicand to be non-negative,

$$df^2 \leqq \omega^2 b^2 + \frac{(4n^3 + 8n^2 + 5n + 1)\, n}{4(2n + 1)^2\, \omega^2}\, g^4. \tag{11.3, 20}$$

This inequality holds for every n if it holds for $n = 0$ in which case it reads

$$df^2 \leqq \omega^2 b^2.$$

At least one of the two values λ_n for any integer n is positive if the inequality

$$df^2 < \frac{4n(n+1)}{(2n+1)^2}\,\omega^2 b^2 + \frac{2n^2+2n+1}{(2n+1)^2}\,bg^2 - \frac{3n^2+3n+1}{4(2n+1)^2\,\omega^2}\,g^4$$

or the inequality (with this inequality also (11.3, 20) is met)

$$g^2 > \frac{2\omega^2 b}{2n^2+2n+1}$$

is valid. The last one also holds for every n is it holds for $n = 0$:

$$g^2 > 2\omega^2 b\,.$$

In the case of vanishing parametric excitation, $g = 0$, (11.3, 19) yields

$$\lambda_n = \left(n+\frac{1}{2}\right)\sqrt{b^2 - \frac{df^2}{\omega^2}} - \frac{b}{2}\,. \tag{11.3, 21}$$

For linear damping, $d = 0$,

$$\lambda_n = nb\,. \tag{11.3, 22}$$

Equation (11.3, 22) shows that λ_n is proportional to n, (11.3, 21) shows approximately the same, from which it follows that higher terms in the development (11.3, 13) quickly decrease as t increases. In general, $\lambda_n > 0$ holds so that the non-stationary solutions tend to the stationary ones after some time.

If forced excitation vanishes, $\gamma = 0$, the coefficient Π of the transformed differential equation (11.3, 9) reads, after developing the denominators into series and neglecting amplitude terms of sixth and higher degrees, in the form

$$\Pi = \frac{3S^2 - 2N'S - N'^2 + 8S\lambda}{4S^2} + \frac{M'S^2 - M'N'S + N'^2 r + 7N'Sr - 4Sr\lambda}{2S^3}\,A^2$$

$$+ \frac{-M'^2S^2 - 6M'S^2 r - 2N'S^2 l + 4M'N'Sr + 6S^3 l + 6N'Sr^2 - 3N'^2 r^2 + 8Sr^2\lambda}{4S^4}\,A^4\,.$$

The choice of

$$\varkappa = \frac{1}{2S^2}\sqrt{M'^2S^2 + 6M'S^2 r + 2N'S^2 l - 4M'N'Sr - 6S^3 l - 6N'Sr^2 + 3N'^2 r^2 - 8Sr^2\lambda}$$

(where the radicand is positive because the first term is of highest order of magnitude) leads again to the Whittaker equation (11.3, 11), where now

$$q^2 = \left(\frac{1}{2} - \frac{2\omega^2 b}{g^2}\right)^2 - \frac{4\omega^2}{g^2}\lambda$$

and, by developing into series and neglecting terms of higher degree,

$$p = 1 - \frac{2\omega^2 b}{g^2} + p'$$

with

$$p' = \frac{8eb^2}{dg^2} - \frac{6\omega^2 hb^2}{dg^2} - \frac{5eb}{\omega^2 d} + \frac{3hb}{d} + \frac{eg^2}{2\omega^4 d}$$

which yields the eigenvalue equation

$$4\omega^2\lambda_n = -n(n-1)\,g^2 - 4n\omega^2 b + [(2n-1)\,g^2 + 4\omega^2 b]\,p'\,.$$

The condition $\lambda_n > 0$ holds, for instance, for $g^2 \gg b$ (linear damping small in comparison with parametric excitation) and

$$eg^2 > \frac{2n(n-1)}{2n-1}\,\omega^4 d$$

or for $g^2 \gg b$, vanishing non-linear restoring forces ($e = 0$) and

$$\frac{b}{d} > \frac{n(n-1)}{3(2n-1)\,h}\,,$$

that is, always for $n = 0$ and $n = 1$, for

$$\frac{b}{d} > \frac{2}{9h}\quad\text{if}\quad n = 2\,,$$

and so on.

Following (11.3, 8), (11.3, 4) and (11.1, 8), the function E can be given in a more explicit form by

$$E = \mathrm{e}^{-\int\left(\frac{\frac{\mathrm{d}Q^2}{\mathrm{d}A} - P}{Q^2} + \frac{1}{2A}\right)\mathrm{d}A}$$

$$= \mathrm{e}^{\frac{1}{2}\int\left(\frac{2P - \frac{\mathrm{d}Q^2}{\mathrm{d}A}}{Q^2} - \frac{\frac{\mathrm{d}Q^2}{\mathrm{d}A}}{Q^2} - \frac{1}{A}\right)\mathrm{d}A}$$

$$= \sqrt{\frac{w_{\mathrm{stat}}}{CAQ^2}}\,.$$

So we can derive explicit formulae for the transition probability density and by this for different probability densities of finite order, for instance in the case of prevailing forced excitation the formula

$$w_\tau(A, A_0) = \frac{\varkappa}{C}\sqrt{\frac{AA_0 w_{\mathrm{stat}}(A)\,w_{\mathrm{stat}}(A_0)}{Q^2(A)\,Q^2(A_0)}}\,\mathrm{e}^{-\frac{\varkappa}{2}(A^2 + A_0^2)}$$

$$\times \sum_{\nu=0}^{\infty} L_\nu^{(0)}(\varkappa A^2)\,L_\nu^{(0)}(\varkappa A_0^2)\,\mathrm{e}^{-\lambda_\nu|\tau|}$$

for the two-dimensional probability density (11.3, 17) in form of a Laguerre function development.

This way the Fokker Planck Kolmogorov equation can lead to explicit results on multi-dimensional amplitude distributions in rather complex non-linear vibration problems.

12. Systems with autoparametric coupling

12.1. Basic properties

As an example of coupled vibrations in a system with two degrees of freedom system, we shall investigate systems with autoparametric coupling. In such systems a forced excitation of one mode acts, because of the non-linear connection between different modes, as a parametric excitation of another mode. We shall consider in particular the autoparametric vibration absorber system investigated by HAXTON and BARR (1972) and IBRAHIM and ROBERTS (1976, 1977); compare SCHMIDT and SCHULZ (1982). It consists of a main mass and a weightless beam with a concentrated end mass mounted on the main mass (Fig. 12.1, 1).

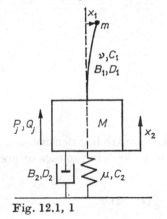

Fig. 12.1, 1

Forced vertical vibrations of the main mass impose forced axial vibrations and, in case of a parametric resonance, additional lateral vibrations on the beam which react on the main mass through inertial non-linearities to give a vibration absorber effect.

The equations of motion of various systems with autoparametric coupling are of the form

$$\ddot{x}_1 \times \Omega_1^2 x_1 = -B_1 \dot{x}_1 - D_1 x_1^2 \dot{x}_1 - C_1 x_1^3 - E x_1 (\dot{x}_1^2 + x_1 \ddot{x}_1)$$
$$+ J x_1 x_2 + K_1 x_1 \ddot{x}_2 + L_1 (\dot{x}_2^2 + x_2 \ddot{x}_2)$$
$$- \sum_{j=1}^{\infty} (G_j \cos j\omega t + H_j \sin j\omega t) x_1 ,$$

$$\ddot{x}_2 + \Omega_2^2 x_2 = \sum_{j=1}^{\infty} (P_j \cos j\omega t + Q_j \sin j\omega t) - B_2 \dot{x}_2 - D_2 x_2^2 \dot{x}_2$$
$$- C_2 x_2^3 + F x_2^2 + K_2 \ddot{x}_1 x_2 + L_2 (\dot{x}_1^2 + x_1 \ddot{x}_1) .$$

For the case of the autoparametric vibration absorber, these equations follow easily from the Lagrangian formalism, where x_1 is the lateral displacement of the absorber mass m, x_2 the vertical displacement of the main mass M,

$$\Omega_1^2 = \frac{\nu}{m}, \qquad \Omega_2^2 = \frac{\mu}{M + m}$$

with the lateral spring stiffness ν and the vertical spring stiffness μ, with B_i, D_i, C_i the linear and non-linear damping and the non-linear restoring forces acting on the absorber mass and the main mass respectively, E the coefficient of non-linear inertia forces, K_1, L_2 the coefficients of coupling terms, P_j, Q_j the Fourier coefficients of vertical forces, and setting

$$F = \dot{G}_j = H_j = J = K_2 = L_1 = 0 . \tag{12.1, 1}$$

The dimensionless equations corresponding to (6.1, 8) are, with (12.1, 1),

$$y_1'' + \lambda_1 y_1 = \lambda_1 \alpha_1 y_1 - \frac{B_1}{\omega} y_1' - \frac{D_1}{\omega} y_1^2 y_1' - \frac{C_1}{\omega^2} y_1^3 - E y_1 (y_1'^2 + y_1 y_1'') + K_1 y_1 y_1'' ,$$

$$y_2'' + \lambda_2 y_2 = \frac{1}{\omega^2} \sum_{j=1}^{\infty} (P_j \cos j\tau + Q_j \sin j\tau) + \lambda_2 \alpha_2 y_2$$

$$- \frac{B_2}{\omega} y_2' - \frac{D_2}{\omega} y_2^2 y_2' - \frac{C_2}{\omega_2} y_2^3 + L_2 (y_1'^2 + y_1 y_1'') .$$

In the non-resonance case neither λ_1 nor λ_2 is the square of an integer, and the first approximative solution turns out to be

$$y_{110} = 0 , \qquad y_{210} = \frac{1}{\omega^2} \sum_{j=1}^{\infty} \frac{P_j \cos j\tau + Q_j \sin j\tau}{\vartheta_\lambda^{j^2} \lambda_2 - j^2} .$$

Now we assume a two-fold resonance

$$\lambda_1 = n^2 , \qquad \lambda_2 = n_2^2$$

with integers n, n_2. Instead of (6.1, 7) now

$$\alpha_1 = \frac{\omega - \omega_1}{\omega_1} , \qquad \alpha_2 = \frac{\omega - \omega_2}{\omega_2} \approx \frac{\omega - \omega_2}{\omega_1} \tag{12.1, 2}$$

holds with two fixed frequencies ω_1, ω_2 in the neighbourhood of ω. As we shall show, the case of a single resonance is included in the following analysis. The first approximation is

$$y_{11} = r \cos n\tau + s \sin n\tau ,$$

$$y_{21} = r_2 \cos n_2\tau + s_2 \sin n_2\tau + y_{210} .$$

This leads to the second approximation for y_2

$$y_{22} = y_{21} - \alpha_2(r_2 \cos n_2\tau + s_2 \sin n_2\tau)$$

$$+ \frac{B_2}{n_2\omega}(s_2 \cos n_2\tau - r_2 \sin n_2\tau) + \frac{D_2}{4n_2\omega}(r_2^2 + s_2^2)(s_2 \cos n_2\tau - r_2 \sin n_2\tau)$$

$$+ \frac{D_2}{32n_2\omega}[s_2(3r_2^2 - s_2^2) \cos 3n_2\tau + r_2(3s_2^2 - r_2^2) \sin 3n_2\tau]$$

$$+ \frac{3C_2}{4n_2^2\omega^2}(r_2^2 + s_2^2)(r_2 \cos n_2\tau + s_2 \sin n_2\tau)$$

$$+ \frac{C_2}{32n_2^2\omega^2}[r_2(r_2^2 - 3s_2^2) \cos 3n_2\tau + s_2(3r_2^2 - s_2^2) \sin 3n_2\tau]$$

$$+ \frac{n^2 L_2}{4n^2 - \vartheta_{n_2}^{2n} n_2^2}[(r^2 - s^2) \cos 2n\tau + 2rs \sin 2n\tau].$$

Using this, the second approximation of y_1 turns out to be

$$y_{12} = y_{11} - \alpha_1(r \cos n\tau + s \sin n\tau) + \frac{B_1}{n\omega}(s \cos n\tau - r \sin n\tau)$$

$$+ \frac{D_1}{4n\omega}(r^2 + s^2)(s \cos n\tau - r \sin n\tau)$$

$$+ \frac{D_1}{32n\omega}[s(3r^2 - s^2) \cos 3n\tau + r(3s^2 - r^2) \sin 3n\tau]$$

$$+ \frac{3C_1}{4n^2\omega^2}(r^2 + s^2)(r \cos n\tau + s \sin n\tau)$$

$$+ \frac{C_1}{32n^2\omega^2}[r(r^2 - 3s^2) \cos 3n\tau + s(3r^2 - s^2) \sin 3n\tau]$$

$$- \frac{E}{2}(r^2 + s^2)(r \cos n\tau + s \sin n\tau)$$

$$- \frac{E}{16}[r(r^2 - 3s^2) \cos 3n\tau + s(3r^2 - s^2) \sin 3n\tau]$$

$$- \frac{K_1}{2\omega^2}\sum_{j=1}^{\infty}\frac{j^2}{\vartheta_{n_2}^{j} n_2^2 - j^2}\left[\frac{(rP_j - sQ_j)\cos(n+j)\tau + (sP_j + rQ_j)\sin(n+j)\tau}{n^2 - (n+j)^2}\right.$$

$$\left. + \frac{(rP_j + sQ_j)\cos(n-j)\tau + (sP_j - rQ_j)\sin(n-j)\tau}{\vartheta_j^{2n} n^2 - (n-j)^2}\right]$$

$$+ \frac{n_2^2(1 - \alpha_2) K_1}{2[(n+n_2)^2 - n^2]}[(rr_2 - ss_2)\cos(n+n_2)\tau$$

$$+ (sr_2 + rs_2)\sin(n+n_2)\tau]$$

$$+ \frac{n_2^2(1 - \alpha_2) K_1}{2[(n-n_2)^2 - \vartheta_{n_2}^{2n} n^2]}[(rr_2 + ss_2)\cos(n-n_2)\tau$$

$$+ (sr_2 - rs_2)\sin(n-n_2)\tau]$$

$$+ \frac{n_2 B_2 K_1}{2[(n+n_2)^2 - n^2]\,\omega} \big[(rs_2 + sr_2)\cos(n+n_2)\tau$$

$$+ (ss_2 - rr_2)\sin(n+n_2)\tau\big]$$

$$+ \frac{n_2 B_2 K_1}{2[(n-n_2)^2 - \vartheta_{n_2}^{2n} n^2]\,\omega} \big[(rs_2 - sr_2)\cos(n-n_2)\tau$$

$$+ (ss_2 + rr_2)\sin(n-n_2)\tau\big]$$

$$+ \frac{n_2 D_2 K_1(r_2^2 + s_2^2)}{8[(n+n_2)^2 - n^2]\,\omega} \big[(rs_2 + sr_2)\cos(n+n_2)\tau$$

$$+ (ss_2 - rr_2)\sin(n+n_2)\big]\tau$$

$$+ \frac{n_2 D_2 K_1(r_2^2 + s_2^2)}{8[(n-n_2)^2 - \vartheta_{n_2}^{2n} n^2]\,\omega} \big[(rs_2 - sr_2)\cos(n-n_2)\tau$$

$$+ (ss_2 + rr_2)\sin(n-n_2)\tau\big]$$

$$+ \frac{3C_2 K_1(r_2^2 + s_2^2)}{8[(n+n_2)^2 - n^2]\,\omega^2} \big[(rr_2 - ss_2)\cos(n+n_2)\tau$$

$$+ (sr_2 + rs_2)\sin(n+n_2)\tau\big]$$

$$+ \frac{3C_2 K_1(r_2^2 + s_2^2)}{8[(n-n_2)^2 - \vartheta_{n_2}^{2n} n^2]\,\omega^2} \big[(rr_2 + ss_2)\cos(n-n_2)\tau$$

$$+ (sr_2 - rs_2)\sin(n-n_2)\tau\big]$$

$$+ \frac{9n_2 D_2 K_1}{64[(n+3n_2)^2 - n^2]\,\omega} \big\{[(rs_2(3r_2^2 - s_2^2) - sr_2(3s_2^2 - r_2^2)]\cos(n+3n_2)\tau$$

$$+ [ss_2(3r_2^2 - s_2^2) + rr_2(3s_2^2 - r_2^2)]\sin(n+3n_2)\tau\big\}$$

$$+ \frac{9n_2 D_2 K_1}{64[(n-3n_2)^2 - \vartheta_{3n_2}^{2n} n^2]\,\omega} \big\{[rs_2(3r_2^2 - s_2^2) + sr_2(3s_2^2 - r_2^2)]$$

$$\times \cos(n-3n_2)\tau + [ss_2(3r_2^2 - s_2^2)$$

$$- rr_2(3s_2^2 - r_2^2)]\sin(n-3n_2)\tau\big\}$$

$$+ \frac{9C_2 K_1}{64[(n+3n_2)^2 - n^2]\,\omega^2} \big\{[rr_2(r_2^2 - 3s_2^2) - ss_2(3r_2^2 - s_2^2)]\cos(n+3n_2)\tau$$

$$+ [sr_2(r_2^2 - 3s_2^2) + rs_2(3r_2^2 - s_2^2)]$$

$$\times \sin(n+3n_2)\tau\big\}$$

$$+ \frac{9C_2 K_1}{64[(n-3n_2)^2 - \vartheta_{3n_2}^{2n} n^2]\,\omega^2} \big\{[rr_2(r_2^2 - 3s_2^2) + ss_2(3r_2^2 - s_2^2)]$$

$$\times \cos(n-3n_2)\tau + [sr_2(r_2^2 - 3s_2^2)$$

$$- rs_2(3r_2^2 - s_2^2)]\sin(n-3n_2)\tau\big\}$$

$$+ \frac{2n^2 K_1 L_2(r^2 + s^2)}{4n^2 - \vartheta_{n_2}^{2n} n_2^2}(r\cos n\tau + s\sin n\tau)$$

$$+ \frac{n^2 K_1 L_2}{4(4n^2 - \vartheta_{n_2}^{2n} n_2^2)}\big[r(r^2 - 3s^2)\cos 3n\tau + s(3r^2 - s^2)\sin 3n\tau\big].$$

It is not easy to interpret these approximations, which include much information about the time dependence of the solutions. The periodicity equations (2.3, 3) stemming from them are, in vector form,

$$
a_1 \begin{pmatrix} r \\ s \end{pmatrix} - b_1 \begin{pmatrix} s \\ -r \end{pmatrix} + p_1 \begin{pmatrix} r \\ -s \end{pmatrix} + q_1 \begin{pmatrix} s \\ r \end{pmatrix}
$$

$$
- k \left[r_2 \begin{pmatrix} r \\ -s \end{pmatrix} + s_2 \begin{pmatrix} s \\ r \end{pmatrix} \right] + b \left[r_2 \begin{pmatrix} s \\ r \end{pmatrix} + s_2 \begin{pmatrix} -r \\ s \end{pmatrix} \right]
$$

$$
+ d \left[s_2 (s_2^2 - 3r_2^2) \begin{pmatrix} r \\ -s \end{pmatrix} + r_2 (r_2^2 - 3s_2^2) \begin{pmatrix} s \\ r \end{pmatrix} \right]
$$

$$
+ c \left[r_2 (3s_2^2 - r_2^2) \begin{pmatrix} r \\ -s \end{pmatrix} + s_2 (s_2^2 - 3r_2^2) \begin{pmatrix} s \\ r \end{pmatrix} \right] = 0 \tag{12.1, 3}
$$

and

$$
a_2 \begin{pmatrix} r_2 \\ s_2 \end{pmatrix} - b_2 \begin{pmatrix} s_2 \\ -r_2 \end{pmatrix} - l \begin{pmatrix} \dfrac{r^2 - s^2}{2rs} \end{pmatrix} + \begin{pmatrix} p_2 \\ q_2 \end{pmatrix} = 0 \tag{12.1, 4}
$$

with the abbreviations

$$
a_1 = n^2 \alpha_1 - \left(\frac{3C_1}{4\omega^2} - \frac{n^2 E}{2} + \frac{2n^4 K_1 L_2}{4n^2 - \vartheta_{n_1}^{2n} n_2^2} \right) A^2 ,
$$

$$
a_2 = n_2^2 \alpha_2 - \frac{3C^2}{4\omega^2} A_2^2 ,
$$

$$
b_1 = \frac{n}{\omega} \left(B_1 + \frac{D_1}{4} A^2 \right), \qquad b_2 = \frac{n_2}{\omega} \left(B_2 + \frac{D_2}{4} A_2^2 \right), \tag{12.1, 5}
$$

$$
p_1 = \frac{2n^2 K_1}{(4n^2 - \vartheta_{n_1}^{2n} n_2^2) \omega^2} P_{2n} , \qquad p_2 = \frac{1}{\omega^2} P_{n_2} ,
$$

$$
q_1 = \frac{2n^2 K_1}{(4n^2 - \vartheta_{n_1}^{2n} n_2^2) \omega^2} Q_{2n} , \qquad q_2 = \frac{1}{\omega^2} Q_{n_2} ,
$$

$$
k = \delta_{n_2}^{2n} \left(2n^2 - \frac{a_2}{2} \right) K_1 , \qquad b = \delta_{n_2}^{2n} \frac{b_2 K_1}{2} ,
$$

$$
l = \delta_{n_2}^{2n} n^2 L_2 , \qquad d = \delta_{3n_2}^{2n} \frac{3n D_2 K_1}{32\omega} ,
$$

$$
c = \delta_{3n_2}^{2n} \frac{9C_2 K_1}{64\omega^2} ,
$$

where

$$
A^2 = r^2 + s^2 , \qquad A_2^2 = r_2^2 + s_2^2
$$

are the squares of the partial amplitudes (more exactly: the squares of the resonance parts of the first approximation of the partial amplitudes) in direction y_1 and y_2 respectively.

The equations (12.1, 4) are linear with respect to r_2, s_2 and their solution is

$$\left.\begin{array}{l} Br_2 = a_2 l(r^2 - s^2) + 2b_2 lrs - a_2 p_2 - b_2 q_2\,, \\ Bs_2 = 2a_2 lrs - b_2 l(r^2 - s^2) - a_2 q_2 + b_2 p_2\,, \end{array}\right\} \tag{12.1, 6}$$

using the abbreviation

$$B = a_2^2 + b_2^2\,.$$

Squaring the equations (12.1, 6), adding and dividing by B leads to

$$BA_2^2 = l^2 A^4 + p_2^2 + q_2^2 - 2lp_2(r^2 - s^2) - 4lq_2 rs\,. \tag{12.1, 7}$$

The equations (12.1, 3) and (12.1, 4), complicated as they are, already show the coupling between the vertical vibrations of the main mass, represented by r_2, s_2, A_2, and the lateral absorber vibrations, represented by r, s, A. If $l = 0$, that is, if the resonance condition $n_2 = 2n$ does not hold, the vertical vibration is (in the approximation at hand) not influenced by the lateral vibration, but it always influences the lateral vibration. The latter influence is given already by L_2, n_2 and the forced excitation terms P_{2n}, Q_{2n} in a_1, p_1, q_1, but it is complicated by the linear r_2, s_2 terms in (12.1, 3), if k, b do not disappear, that is, if $n_2 = 2n$ and $K_1 \neq 0$, and by the non-linear r_2, s_2 terms, if c, d not disappear, that is, if $3n_2 = 2n$ and K_1, C_2 respectively D_2 are not zero.

The process of evaluating the first approximations also enables us to recognize the behaviour of higher approximations. The coupling term $K_1 y_{12} y_{21}''$ leads to terms containing $\genfrac{}{}{0pt}{}{\cos}{\sin} (3n - n_2)\tau$ which add to the periodicity equations (12.1, 3) not only for $n_2 = 2n$, but also for $n_2 = 4n$. In detail, this term leads to

$$\begin{aligned} y_{1\,\mathrm{add}} = {} & \frac{n_2^2 D_1 K_1}{64n[(3n - n_2)^2 - \vartheta_{(3n-n_2)}^{n_2} n^2]\,\omega} \{[s(3r^2 - s^2)\,r_2 + r(3s^2 - r^2)\,s_2] \\ & \times \cos(3n - n_2)\,\tau \\ & + [r(3s^2 - r^2)\,r_2 + s(s^2 - 3r^2)\,s_2] \\ & \times \sin(3n - n_2)\,\tau\} \\ & + \frac{n_2^2(C_1 - 2n^2\omega^2 E)\,K_1}{64n^2[(3n - n_2)^2 - \vartheta_{(3n-n_2)}^{n_2} n^2]\,\omega^2} \{[r(r^2 - 3s^2)\,r_2 + s(3r^2 - s^2)\,s_2] \\ & \times \cos(3n - n_2)\,\tau \\ & + [s(3r^2 - s^2)\,r_2 + r(3s^2 - r^2)\,s_2] \\ & \times \sin(3n - n_2)\,\tau\} \end{aligned}$$

from which there follow the additional terms

$$\begin{aligned} & -\delta_{n_2}^{2n}\frac{nD_1 K_1}{16\omega}\left[r_2\binom{s(3r^2 - s^2)}{r(3s^2 - r^2)} + s_2\binom{r(3s^2 - r^2)}{s(s^2 - 3r^2)}\right] \\ & +\delta_{n_2}^{2n}\frac{(2n^2\omega^2 E - C_1)\,K_1}{16\omega^2}\left[r_2\binom{r(r^2 - 3s^2)}{s(3r^2 - s^2)} + s_2\binom{s(3r^2 - s^2)}{r(3s^2 - r^2)}\right] \\ & -\delta_{n_2}^{4n}\frac{nD_1 K_1}{4\omega}\left[r_2\binom{s(3r^2 - s^2)}{r(r^2 - 3s^2)} + s_2\binom{r(3s^2 - r^2)}{s(3r^2 - s^2)}\right] \\ & +\delta_{n_2}^{4n}\frac{(2n^2\,\omega^2 E - C_1)\,K_1}{4\omega^2}\left[r_2\binom{r(r^2 - 3s^2)}{s(s^2 - 3r^2)} + s_2\binom{s(3r^2 - s^2)}{r(r^2 - 3s^2)}\right] \end{aligned} \tag{12.1, 8}$$

on the left-hand side of (12.1, 3).

Insertion of (12.1, 6) into (12.1, 3), (12.1, 8) gives two coupled non-linear equations for r and s which can be solved numerically. A formula for the amplitude A only of the absorber vibrations can be found if we

a) exclude the resonances $3n_2 = 2n$ and $n_2 = 4n$, or

b) restrict ourselves to the approximation (12.1, 3), (12.1, 4) and assume the non-linear restoring and damping forces in vertical direction to vanish,

$$C_2 = D_2 = 0 \,, \quad \text{that ist} \quad c = d = 0 \,. \tag{12.1, 9}$$

In what follows, we assume one of these two restrictions as given. Then the insertion of (12.1, 6) into (12.1, 3) mentioned above yields

$$(Ba_1 - a_2klA^2 + b_2blA^2)\binom{r}{s}$$

$$+ (Bp_1 + a_2kp_2 + a_2bq_2 + b_2kq_2 - b_2bp_2)\binom{r}{-s}$$

$$+ (Bq_1 + a_2kq_2 - a_2bp_2 - b_2kp_2 - b_2bq_2)\binom{s}{r}$$

$$- (Bb_1 + a_2blA^2) + b_2klA^2)\binom{s}{-r} = 0 \,, \tag{12.1, 10}$$

two equations which contain r, s, in contrast to (12.1, 4), besides the linear factors in form of A_2^2 only.

The determinant condition for a non-vanishing solution r, s reads

$$(Ba_1 - a_2klA^2 + b_2blA^2)^2 + (Bb_1 + a_2blA^2 + b_2klA^2)^2$$
$$= (Bp_1 + a_2kp_2 + a_2bq_2 + b_2kq_2 - b_2bp_2)^2$$
$$+ (Bq_1 + a_2kq_2 - a_2bp_2 - b_2kp_2 - b_2bq_2)^2 \,. \tag{12.1, 11}$$

This is a biquadratic equation in A the coefficients of which are of fourth order in A^2 in the case of restriction a) and independent of A_2 in the case (12.1, 9) of the second restriction b), which we assume in what follows.

Abbreviating

$$a_1 = a - eA^2 \,, \quad b_1 = f + gA^2$$

instead of (12.1, 5), we get the solution of (12.1, 11)

$$[(Be + a_2kl - b_2bl)^2 + (Bg + a_2bl + b_2kl)^2] \, A^2$$
$$= Ba(Be + a_2kl - b_2bl) - Bf(Bg + a_2bl + b_2kl)$$
$$\pm \{ -B^2[f(Be + a_2kl - b_2bl) + a(Bg + a_2bl + b_2kl)]^2$$
$$+ [(Be + a_2kl - b_2bl)^2 + (Bg + a_2bl + b_2kl)^2]$$
$$\times [(Bp_1 + a_2kp_2 + a_2bq_2 + b_2kq_2 - b_2bp_2)^2$$
$$+ (Bq_1 + a_2kq_2 - a_2bp_2 - b_2kp_2 - b_2bq_2)^2] \}^{1/2} \,. \tag{12.1, 12}$$

We shall first discus the case $n_2 \neq 2n$ (no internal resonance). Because $b = k = l = 0$, (12.1, 12) simplifies to the amplitude formula

$$(e^2 + g^2) \, A^2 = ae - fg \pm \sqrt{(e^2 + g^2)(p_1^2 + q_1^2) - (ef + ag)^2} \tag{12.1, 13}$$

which again shows that the motion of the main mass influences the absorber amplitudes only by means of the values L_2, P_{2n}, Q_{2n}, n_2 in e and by p_1, q_1.

The equation $A = 0$ gives the resonance interval

$$a_\pm = \pm \sqrt{p_1^2 + q_1^2 - f^2}$$

which is real if linear damping is smaller than an excitation threshold value,

$$f < \sqrt{p_1^2 + q_1^2} \, .$$

The resonance curves given by (12.1, 13) coalesce, that is, determine a maximum amplitude only for positive non-linear damping g.

Fig. 12.1, 2 gives resonance curves corresponding to (12.1, 13) for $e = -0.65$, $a = \alpha_1$, $f = 0.001$, $p_1 = 0.002$, $q_1 = 0$ and different values of non-linear damping g, Fig. 12.1, 3 analogously for $g = 0.1$ and different values of forced excitation p_1. It shows that increasing non-linear damping diminishes, and forced excitation increases, the amplitudes.

Formula (12.1, 7) for the main mass amplitude now reads

$$BA_2^2 = p_2^2 + q_2^2$$

or, in the original coefficients,

$$n_2^2 \omega^2 (n_2^2 \omega^2 \alpha_2^2 + B_2^2) \, A_2^2 = P_{n_2}^2 + Q_{n_2}^2 \, , \qquad (12.1, 14)$$

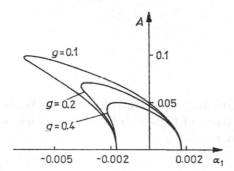

Fig. 12.1, 2

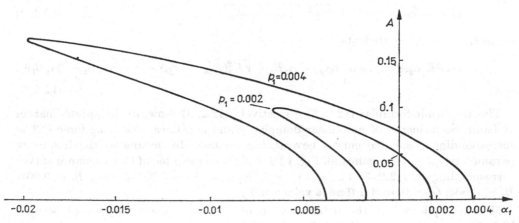

Fig. 12.1, 3

and is independent of the absorber motion. Fig. 12.1, 4 shows an example for

$$B = 9(9\alpha_2^2 + 16 \cdot 10^{-6}), \qquad q_2 = 0$$

and different values of p_1. The resonance curves given by (12.1, 14) have the maximum value

$$A_{2\,max} = \frac{1}{n_2 \omega B_2} \sqrt{P_{n_2}^2 + Q_{n_2}^2} \tag{12.1, 15}$$

for $\alpha_2 = 0$.

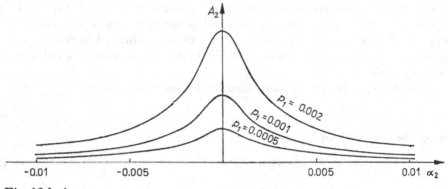

Fig. 12.1, 4

12.2. Internal resonance

For internal resonance, $n_2 = 2n$, the absorber equation (12.1, 12) can be simplified if we take into consideration only the terms of highest order of magnitude, in particular if we omit the non-linear damping terms,

$$Bg \ll b_2 kl, \tag{12.2, 1}$$

thus giving

$$klA^2 = aa_2 - fb_2 \pm \sqrt{k^2(p_2^2 + q_2^2) - (ab_2 + fa_2)^2} \tag{12.2, 2}$$

or, in the original coefficients,

$$n^2\omega^2 K_1 L_2 A^2 = 2n^2\omega^2\alpha_1\alpha_2 - B_1 B_2 \pm \sqrt{K_1^2(P_{2n}^2 + Q_{2n}^2) - n^2\omega^2(\alpha_1 B_2 + 2\alpha_2 B_1)^2}. \tag{12.2, 3}$$

The amplitude formulae (12.2, 2) respectively (12.2, 3) show, in the approximation at hand, no influence of non-linear damping g and non-linear restoring force e. The corresponding resonance curves nevertheless coalesce by means of the frequency parameters α_1, α_2 in the radicand. Fig. 12.2, 1 gives an example of the resonance curves corresponding to (12.2, 3) for $n = \omega = 1$, $K_1 = 3$, $L_2 = 1/3$, $\alpha_1 = \alpha_2$, $B_1 = 0.001$, $B_2 = 0.004$, $Q_2 = 0$, and different values of P_2.

More complicated is the determination of the main mass amplitudes because 12.1, 7) contains not only A but also $r^2 - s^2$ and rs.

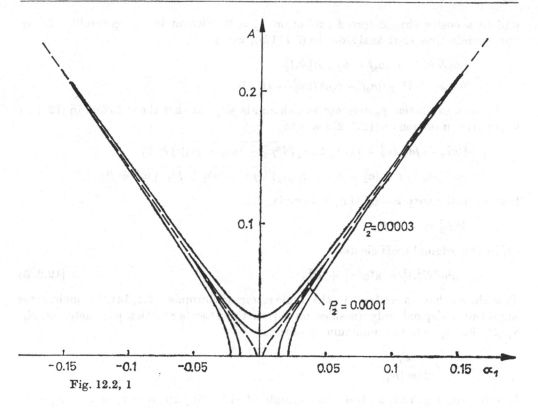

Fig. 12.2, 1

If we abbreviate (12.1, 10) by

$$u \begin{pmatrix} r \\ s \end{pmatrix} + v \begin{pmatrix} r \\ -s \end{pmatrix} + V \begin{pmatrix} s \\ r \end{pmatrix} - U \begin{pmatrix} s \\ -r \end{pmatrix} = 0 \,,$$

that is

$$(u + v)\, r = (U - V)\, s \,, \qquad (V + U)\, r = (v - u)\, s \,, \qquad (12.2,\,4)$$

we get by multiplying these equations with each other the equation

$$(u + v)\,(V + U)\, r^2 - (v - u)\,(U - V)\, s^2 = 0$$

or, writing the left-hand side in the form $xA^2 + y(r^2 - s^2)$,

$$(uV + vU)\,(r^2 - s^2) = -(uU + vV)\, A^2 \,.$$

Multiplying (12.2, 4) by $(u - v)\, r$ and $(u + v)\, s$ respectively and adding yields

$$2(uV + vU)\, rs = (v^2 - u^2)\, A^2 \,.$$

This gives (12.1, 7) as an equation for A_2 and A only:

$$(uV + vU)\, BA_2^2 = (uV + vU)\,(l^2 A^4 + p_2^2 + q_2^2)$$
$$+ 2(uU + vV)\, l p_2 A^2 + 2(u^2 - v^2)\, l q_2 A^2 \,.$$

Confining attention to the terms of highest order of magnitude

$$u = Ba - a_2 kl A^2 \,, \qquad U = Bf + b_2 kl A^2 \,,$$
$$v = a_2 k p_2 + b_2 k q_2 \,, \qquad V = a_2 k q_2 - b_2 k p_2$$

and to a cosine-shaped forced excitation, $q_2 = 0$ (without loss of generality, by an appropriate time shift analogous to (5.1, 12)), we get

$$[2a_2b_2klA^2 + (a_2f - b_2a) B] kA_2^2$$
$$= 2aflBA^2 + (a_2f - b_2a) (kp_2^2 - kl^2A^4) .$$

If forced excitation p_2 does not vanish and is so great that the radicand in (12.2, 2) is positive, insertion of (12.2, 2) leads to

$$[(ab_2 + fa_2) (a_2^2 - b_2^2) \pm 2a_2b_2 \sqrt{k^2p_2^2 - (ab_2 + fa_2)^2}] k^2A_2^2$$
$$= [(ab_2 + fa_2) (a_2^2 - b_2^2) \pm 2a_2b_2 \sqrt{k^2p_2^2 - (ab_2 + fa_2)^2}] (a^2 + f^2) ,$$

that is, to the surprisingly simple formula

$$k^2A_2^2 = a^2 + f^2 ,$$

or, in the original coefficients,

$$4n^2\omega^2K_1^2A_2^2 = n^2\omega^2\alpha_1^2 + B_1^2 . \tag{12.2, 5}$$

This shows that, in contrast to the single resonance formula (12.1, 14), the main mass amplitudes depend only (besides on ω) on the absorber motion parameters n, K_1, α_1, B_1. For $\alpha_1 = 0$, the minimum amplitude

$$A_2 = \frac{B_1}{2n\omega |K_1|}$$

is assumed. Fig. 12.2, 2 shows an example of (12.2, 5) with $n = 1$, $\omega = 1$, $K_1 = 3$, and $B_1 = 0.001$.

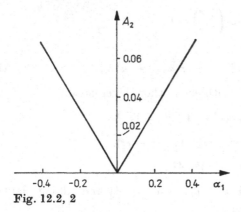

Fig. 12.2, 2

To discuss the amplitude formulae (12.2, 3) and (12.2, 5), we shall introduce the *detuning*

$$\varepsilon = \alpha_1 - \alpha_2$$

which, as the approximative equations (12.1, 2) show, does *not* depend on the variable frequency ω, but only on the fixed frequencies ω_1, ω_2:

$$\varepsilon = \frac{\omega_2 - \omega_1}{\omega_1} .$$

Formula (12.2, 3) can be written in the form

$$n^2\omega^2 K_1 L_2 A^2 = 2n^2\omega^2\left(\alpha_1 - \frac{\varepsilon}{2}\right)^2 - \frac{1}{2}n^2\omega^2\varepsilon^2 - B_1 B_2 \pm \sqrt{\;}, \qquad (12.2, 6)$$

with

$$\sqrt{\;} = \sqrt{K_1^2(P_{2n}^2 + Q_{2n}^2) - n^2\omega^2[(2B_1 + B_2)\,\alpha_1 - 2B_1\varepsilon]^2}\;.$$

The resonance curves given by (12.2, 6) are approximately equidistant — in the vertical direction — to the vertical curves (compare Section 10.2.)

$$n^2\omega^2 K_1 L_1 A_v^2 = 2n^2\omega^2\left(\alpha_1 - \frac{\varepsilon}{2}\right)^2 - \frac{1}{2}n^2\omega^2\varepsilon^2 - B_1 B_2 \qquad (12.2, 7)$$

for which $\sqrt{\;} = 0$ and which are, because of $K_1 L_1 > 0$, hyperbolae meeting the α_1 axis at

$$\alpha_{1\pm} = \frac{\varepsilon}{2} \pm \sqrt{\frac{\varepsilon^2}{4} + \frac{B_1 B_2}{2n^2\omega^2}}$$

(dashed curves in Fig. 12.2, 1). Fig. 12.2, 3 gives an example of (12.2, 6) corresponding to Fig. 12.2, 1 and $P_2 = 0.0003$.

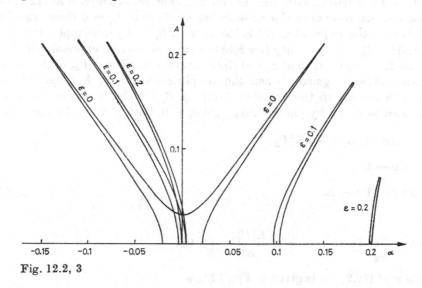

Fig. 12.2, 3

Differentiating (12.2, 6) with respect to α_1 and putting $dA/d\alpha_1 = 0$ gives an equation of fourth degree in α_1 as well as in ε for the amplitude extreme values:

$$[4n^2\omega^2(2\alpha_1 - \varepsilon)^2 + (2B_1 + B_2)^2]\,[(2B_1 + B_2)\alpha_1 - 2B_1\varepsilon]^2$$
$$= 4K_1^2(P_{2n}^2 + Q_{2n}^2)\,(2\alpha_1 - \varepsilon)^2\;.$$

Vertical tangents for the resonance curves occur for $d\alpha_1/dA = 0$, that is, for $\sqrt{\;} = 0$ from which it follows that

$$(2B_1 + B_2)\alpha_1 = 2B_1\varepsilon \pm \frac{K_1}{n\omega}\sqrt{P_{2n}^2 + Q_{2n}^2}\;.$$

In particular for vanishing detuning, $\varepsilon = 0$, amplitude extreme values appear for

$$\alpha_1 = \pm \frac{1}{n\omega(2B_1 + B_2)} \sqrt{K_1^2(P_{2n}^2 + Q_{2n}^2) - \left(B_1 + \frac{B_2}{2}\right)^4},$$

vertical tangents (coalescence of the resonance curves) for

$$\alpha_1 = \pm \frac{K_1}{n\omega(2B_1 + B_2)} \sqrt{P_{2n}^2 + Q_{2n}^2}. \tag{12.2, 8}$$

The maximum main mass amplitudes can be evaluated from (12.2, 5) (for $Q_{2n} = 0$) with (12.2, 8):

$$A_{2\,max}^2 = \left[\frac{P_{2n}}{2n\omega(2B_1 + B_2)} \pm \frac{B_1\varepsilon}{K_1(2B_1 + B_2)}\right]^2 + \frac{B_1^2}{4n^2\omega^2K_1^2}.$$

For non-vanishing detuning, one of these values is greater than the value

$$A_{2\,max}^2 = \frac{P_{2n}^2}{4n^2\omega^2(2B_1 + B_2)^2} + \frac{B_1^2}{4n^2\omega^2K_1^2} \tag{12.2, 9}$$

for $\varepsilon = 0$. A comparison with the maximum main mass amplitude (12.1, 15) for single, not internal resonance (for no absorber at all) and $Q_{n_2} = 0$ shows equality for $B_1 = 0$ whereas the expression (12.1, 15) is for $2B_1 = B_2$ quadruple, for $B_1 = B_2$ ninefold and for $B_1 = 2B_2$ twenty-five-fold the value of the first expression in (12.2, 9).

Given the damping coefficient B_2 and the excitation coefficient P_{2n} of the main mass the absorber effect augments when the coupling coefficient K_1 augments and is maximal with respect to the absorber damping B_1 for the minimum value $A_{2\,max}$. The latter can be found by putting $dA_{2\,max}/dB_1 = 0$, that is, from the equation

$$B_1(2B_1 + B_2)^3 = 2K_1^2P_{2n}^2$$

which simplifies to

$$\delta(2\delta + 1)^3 = 2\varkappa \tag{12.2, 10}$$

with

$$\delta = \frac{B_1}{B_2} \ (> 0), \qquad \varkappa = \frac{K_1^2 P_{2n}^2}{B_2^4} \ (> 0).$$

The solution of (12.2, 10) is given in Fig. 12.2, 4.

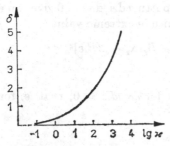

Fig. 12.2, 4

If we do not exclude the *influence of non-linear damping* by (12.2, 1), we can find instead of (12.2, 2)

$$(k^2l^2 + 2b_2klg + Bg^2) A^2 = (aa_2 - fb_2) kl - Bfg$$
$$\pm \sqrt{(k^2l^2 + 2b_2klg + Bg^2) k^2(p_2^2 + q_2^2) - [(ab_2 + fa_2) kl + aBg]^2} .$$

In the original coefficients, the condition $\sqrt{} = 0$ for vertical tangents reads

$$\left[2n^2K_1L_2(\alpha_1 B_2 + 2\alpha_2 B_1) + \left(2n^2\alpha_2^2 + \frac{B_2^2}{2\omega^2}\right) \alpha_1 D_1 \right]^2$$
$$\leqq \left[4n^2K_1^2L_2^2 + \frac{2K_1L_2B_2D_1}{\omega^2} + \left(\alpha_2^2 + \frac{B_2^2}{4n^2\omega^2}\right) \frac{D_1^2}{\omega^2} (P_{2n}^2 + Q_{2n}^2) \right],$$

an equation of sixth order in α_1, α_2. For vanishing detuning ε, it simplifies to an equation of third order in α_1^2:

$$[2n^2K_1L_2(2B_1 + B_2) + \left(2n^2\alpha_1^2 + \frac{B_2^2}{2\omega^2}\right) D_1]^2 \alpha_1^2$$
$$\leqq \left[4n^2K_1^2L_2^2 + \frac{2K_1L_2B_2D_1}{\omega^2} + \left(\alpha_1^2 + \frac{B_2^2}{4n^2\omega^2}\right) \frac{D_1^2}{\omega^2} \right] \frac{K_1^2}{\omega^2} (P_{2n}^2 + Q_{2n}^2) .$$

An estimation for small values of the non-linear damping coefficient D_1 shows that in comparison with linear damping, $D_1 = 0$, the interval $(-\alpha_1, \alpha_1)$ of real amplitudes diminishes if the excitation coefficients P_{2n}, Q_{2n} are not too small,

$$2K_1^2(P_{2n}^2 + Q_{2n}^2) > (2B_1 + B_2)^2 B_1 B_2 .$$

In the case of non-linear damping, the formula for the main mass amplitudes becomes very complicated and reveals, as formula (12.2, 5) for linear damping would already do if we had considered a higher approximation, a dependence on the forced excitation.

It should be noted here that the at first sight surprising result, that the main mass amplitudes do *not* depend on the forced excitation of the main mass, is not really so surprising because the α_1 interval for the two-mode vibration with the main mass amplitude (12.2, 5) and by this the *maximum* amplitudes A_2 do depend on the forced excitation.

12.3. Narrow-band random excitation

In what follows we shall assume, as in Chapter 10, that not a periodic excitation, but a stationary narrow-band random excitation with mean value zero

$$P(t) = \pi(t) \cos [2n\omega t + \delta(t)]$$

acts on the main mass. The excitation amplitude $\pi(t)$ and the excitation phase $\delta(t)$ are slowly varying functions, the process $P(t)$ is assumed Gaussian distributed, so that the excitation amplitude is Rayleigh distributed, with probability density

$$v(\pi) = \frac{\pi}{\sigma^2} e^{-\frac{\pi^2}{2\sigma^2}} .$$

A generalisation to Weibull distributed excitation amplitudes is possible as in Chapter 10.

The probability density of the response amplitudes associated with strips $w(\pi)\,\mathrm{d}\pi$ of excitation amplitudes can be found by the transformation

$$v(\pi)\,\mathrm{d}\pi = \mathrm{d}V(\pi) = \mathrm{d}W(A) = w(A)\,\mathrm{d}A \tag{12.3, 1}$$

using the fact that π is a monotonic function of A. The amplitude formula (12.2, 3) can now be written in the form

$$K_1^2\pi^2 = n^2\omega^2(\alpha_1 B_2 + 2\alpha_2 B_1)^2 + (n^2\omega^2 K_1 L_2 A^2 - 2n^2\omega^2\alpha_1\alpha_2 + B_1 B_2)^2$$

so that (12.3, 1) yields

$$\mathrm{d}W(A) = -\mathrm{e}^{-\varphi}$$

with

$$\varphi = \frac{1}{2K_1^2\sigma^2}\left[n^2\omega^2(\alpha_1 B_2 + 2\alpha_2 B_1)^2 + (n^2\omega^2 K_1 L_2 A^2 - 2n^2\omega^2\alpha_1\alpha_2 + B_1 B_2)^2\right].$$

By differentiation we get the probability density of the vibration amplitude

$$w(A) = \frac{2n^2\omega^2 L_2 A}{K_1\sigma^2}\left(n^2\omega^2 K_1 L_2 A^2 - 2n^2\omega^2\alpha_1\alpha_2 + B_1 B_2\right)\mathrm{e}^{-\varphi} \tag{12.3, 2}$$

or, because of (12.2, 7),

$$w(A) = \frac{2n^4\omega^4 L_2^2}{\sigma^2} A(A^2 - A_v^2)\,\mathrm{e}^{-\varphi}. \tag{12.3, 3}$$

The factor in parentheses vanishes on the vertical curve $A = A_v$. Because amplitudes belonging to parts of the resonance curve below the vertical curve do not appear, $A \gtrless A_v$, we have to assume $v(A) = 0$ for negative values of the above factor in parentheses as in Section 10.2, so that always $w(A) \gtrless 0$ holds. Correspondingly, the same expression in parentheses in $\mathrm{d}W(A)$ has to be omitted if it is negative.

The whole factor preceding $\mathrm{e}^{-\varphi}$ in (12.3, 3) ensures that the probability density disappears if the amplitudes approach $A_v > 0$ or zero whereas $\mathrm{e}^{-\varphi}$ causes the probability density to disappear for great amplitudes.

Formula (12.3, 2) for the probability density of the vibration amplitude enables us to evaluate the *moments*

$$m_k = \int_0^\infty A^k w(A)\,\mathrm{d}A, \qquad k = 1, 2, 3, \ldots$$

of the amplitude. Abbreviating in what follows (12.3, 2) by

$$w(A) = c_1 A(c_2 A^2 + c_3)\,\mathrm{e}^{-c_4[(c_2 A^2 + c_3)^2 + c_5]},$$

where $c_4 > 0$ and the signs of c_1 and c_2 are chosen so that $c_3 \gtrless 0$, we get, writing

$$A^2 = x \quad \text{and} \quad \frac{c_1}{2}\mathrm{e}^{-c_4(c_3^2 + c_5)} = c_6,$$

the equation

$$m_k = c_6 \int_0^\infty x^{k/2}(c_2 x + c_3)\,\mathrm{e}^{-c_2^2 c_4 x^2 - 2c_2 c_3 c_4 x}\,\mathrm{d}x$$

and, using formula (11.2, 1),

$$m_k = c_2 c_6 (2c_2^2 c_4)^{-(k/4)-1} \, \Gamma\!\left(\frac{k}{2} + 2\right) e^{(1/2)c_3^2 c_4} \, D_{-(k/2)-2}\!\left(c_3 \sqrt{2c_4}\right)$$

$$+ \, c_3 c_6 (2c_2^2 c_4)^{-(k/4)-1/2} \, \Gamma\!\left(\frac{k}{2} + 1\right) e^{(1/2)c_3^2 c_4} \, D_{-(k/2)-1}\!\left(c_3 \sqrt{2c_4}\right)$$

with the parabolic cylinder function D and the Gamma function Γ. The seond moment of the amplitude yields

$$m_2 = \frac{c_6}{c_2^2 \sqrt{2c_4^3}} \, e^{(1/2)c_3^2 c_4} \, D_{-3}\!\left(c_3 \sqrt{2c_4}\right) + \frac{c_3 c_6}{2c_2^2 c_4} \, e^{(1/2)c_3^3 c_4} \, D_{-2}\!\left(c_3 \sqrt{2c_4}\right)$$

where (11.2, 12) and, because of the recurrence relation for parabolic cylinder functions (compare GRADŠTEJN and RYŽIK (1971), p. 1080)

$$D_{p+1}(z) - z D_p(z) + p D_{p-1}(z) = 0$$

and (11.2, 11),

$$D_{-3}(z) = \sqrt{\frac{\pi}{8}} \, (1 + z^2) \, e^{(1/4)z^2} \left[1 - \Phi\!\left(\frac{z}{\sqrt{2}}\right)\right] - \frac{z}{2} \, e^{-(1/4)z^2}$$

holds with the error function

$$\Phi(y) = \frac{2}{\sqrt{\pi}} \int_0^y e^{-t^2} \, \mathrm{d}t \, .$$

The probability that the resonance amplitudes A are greater than a given level A_l is

$$W_l = W(A_l < A < \infty) \, .$$

From (12.3, 1) it follows that

$$W_l = \int_{A_l}^{\infty} w(A) \, \mathrm{d}A \leqq 1 \, .$$

Because $\alpha_2 = \alpha_1 - \varepsilon$, we get

$$W_l = \exp\!\left[-\frac{1}{2K_1^2 \sigma^2} \{n^2 \omega^2 [(2B_1 + B_2)\alpha_1 - 2B_1 \varepsilon]^2 \right.$$

$$\left. + [n^2 \omega^2 K_1 L_2 A_l^2 - 2n^2 \omega^2 \alpha_1(\alpha_1 - \varepsilon) + B_1 B_2]^2\}\right] \quad (12.3, 4)$$

for

$$n^2 \omega^2 K_1 L_2 A_l^2 \geqq 2n^2 \omega^2 \alpha_1(\alpha_1 - \varepsilon) - B_1 B_2$$

and

$$W_l = e^{-\frac{n^2 \omega^2}{2K_1^2 \sigma^2} [(2B_1 + B_2)\alpha_1 - 2B_1 \varepsilon]^2} \quad (12.3, 5)$$

for

$$n^2 \omega^2 K_1 L_2 A_l^2 < 2n^2 \omega^2 \alpha_1(\alpha_1 - \varepsilon) - B_1 B_2 \, .$$

that is, for

$$2n^2\omega^2\alpha_1(\alpha_1 - \varepsilon) > B_1 B_2$$

and sufficiently small A_l, independently of A_l.

For $A_l = 0$, there follows the *probability of there being non-vanishing amplitudes at all*

$$W_0 = \exp\left[-\frac{1}{2K_1^2\sigma^2}\{n^2\omega^2[(2B_1 + B_2)\alpha_1 - 2B_1\varepsilon]^2 \right.$$
$$\left. + [2n^2\omega^2\alpha_1(\alpha_1 - \varepsilon) - B_1 B_2]^2\}\right] \tag{12.3, 6}$$

for

$$2n^2\omega^2\alpha_1(\alpha_1 - \varepsilon) < B_1 B_2$$

and

$$W_0 = \varepsilon^{-\frac{n^2\omega^2}{2K_1^2\sigma^2}[(2B_1 + B_2)\alpha_1 - 2B_1\varepsilon]^2} \tag{12.3, 7}$$

for

$$2n^2\omega^2\alpha_1(\alpha_1 - \varepsilon) \geqq B_1 B_2$$

We shall discuss the probability W_l. Formula (12.3, 5) states that the maximum value $W_l = 1$ would appear for

$$(2B_1 + B_2)\alpha_1 = 2B_1\varepsilon$$

(but then (12.3, 5) does not hold because $A_l^2 < 0$ would follow) and that W_l monotonously diminishes if α_1 removes from this value. In every case, $W_l < 1$ holds so that the probability of vanishing amplitude A is not zero.

The condition

$$\frac{\mathrm{d}W_l}{\mathrm{d}\alpha} = 0$$

for extreme values of W_l leads, if we use formula (12.3, 4), to a cubic equation in α_1. In order to permit an analytical discussion, we confine ourselves to vanishing detuning, $\varepsilon = 0$. In that case, the three extreme values are for $\alpha_1 = \alpha_{10} = 0$

$$W_{l0} = \mathrm{e}^{-\frac{1}{2K_1^2\sigma^2}(n^2\omega^2 K_1 L_2 A_l^2 + B_1 B_2)^2} \tag{12.3, 8}$$

and — for

$$2\alpha_1^2 = 2\alpha_{1m}^2 = K_1 L_2 A_l^2 - \frac{1}{n^2\omega^2}\left(B_1^2 + \frac{1}{4}B_2^2\right)$$

(only real and different from $\alpha_{10} = 0$, if

$$n^2\omega^2 K_1 L_2 A_l^2 > B_1^2 + \tfrac{1}{4}B_2^2 \tag{12.3, 9}$$

holds) —

$$W_{lm} = \mathrm{e}^{-\frac{(B_1 + B_2/2)^2}{2K_1^2\sigma^2}[2n^2\omega^2 K_1 L_2 A_l^2 - (B_1 - B_2/2)^2]} \tag{12.3, 10}$$

It can be shown that $W_{l0} < W_{lm}$ because of (12.3, 8), so that W_l has a minimum W_{l0} for $\alpha_1 = 0$ and maxima W_{lm} for $\alpha_1 = \pm\alpha_{1m}$. These maxima appear because the condition for the validity of (12.3, 4) holds for $\alpha_1 = \pm\alpha_{1m}$.

If the condition (12.3, 8) for maxima in $\pm\alpha_{1m} \neq 0$ does not hold, W_{l0} is the maximum value of W_l, as the factor

$$B_1^2 + \tfrac{1}{4}B_2^2 - n^2\omega^2 K_1 L_2 A_l^2$$

of α_1^2 in (12.3, 4) shows.

Vice versa, the amplitude level A_l, amplitudes above which have only the (perhaps small) probability W_l, follows from (12.3, 10)

$$n^2\omega^2 K_1 L_2 A_l^2 = -\frac{K_1^2\sigma^2 \ln W_l}{\left(B_1 + \dfrac{B_2}{2}\right)^2} + \frac{1}{2}\left(B_1 - \frac{B_2}{2}\right)^2, \tag{12.3, 11}$$

if

$$n^2\omega^2 K_1 L_2 A_l^2 > B_1^2 + \tfrac{1}{4}B_2^2$$

and from (12.3, 8)

$$n^2\omega^2 K_1 L_2 A_l^2 = \sqrt{-2K_1^2\sigma^2 \ln W_l} - B_1 B_2, \tag{12.3, 12}$$

if

$$n^2\omega^2 K_1 L_2 A_l^2 \leq B_1^2 + \tfrac{1}{4}B_2^2.$$

If the equality sign in the last condition holds, the formulae coincide. They show the dependence of A_l on the different parameters. Whereas (12.3, 12) causes a diminution of A_l for any increase of the damping coefficients B_1 or B_2, (12.3, 11) leads to a minimum A_l for a certain damping ratio which follows from

$$\left(\frac{B_2}{2B_1} - 1\right)\left(\frac{B_2}{2B_1} + 1\right)^3 = -\frac{2K_1^2\sigma^2 \ln W_l}{B_1^4} \tag{12.3, 13}$$

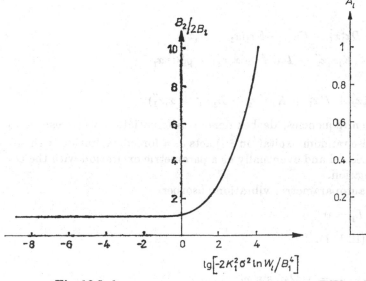

Fig. 12.3, 1

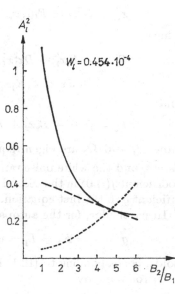

Fig. 12.3, 2

for (say) B_1 given. Fig. 12.3, 1 gives the solution of (12.3, 13). Examples of the dependence of A_l on B_2/B_1 represented by (12.3, 11) (full lines) and (12.3, 12) (dashed lines) are shown in Figures 12.3, 2 to 12.3, 4 for $n^2\omega^2 K_1 L_2 = 1$, $K_1^2\sigma^2 = 10^{-6}$, $B_1 = 0.02$ and $W_l = 0.0000454$, $W_l = 0.00674$ and $W_l = 0.0821$ respectively. The dotted line marks the condition for either formula (12.3, 11) or formula (12.3, 12): on the left-hand side of the dotted line the equation (12.3, 11) holds, given by the full line, whereas on the right-hand side (12.3, 12) is valid, given by the dashed line. For $W_l = 0.0821$, no real amplitude A_l results for $B_2/B_1 > 5.59$ because then the probability of there being non-vanishing amplitudes at all is smaller than W_l.

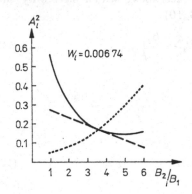

Fig. 12.3, 3

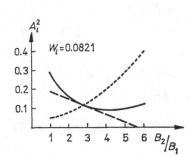

Fig. 12.3, 4

12.4. Broad-band random excitation

By analogy with Section 12.1, we shall investigate the system of two stochastic differential equations

$$x_i'' + \Omega_i^2 x_i = F_i(t, x_\nu, x_\nu', x_\nu'') , \qquad i = 1, 2 ; \qquad \nu = \{1, 2\} \qquad (12.4, 1)$$

where

$$F_1 = -B_1 x_1' - D x_1^2 x_1' - C x_1^3 - E x_1(x_1'^2 + x_1 x_1'')$$
$$+ J x_1 x_2 + K_1 x_1 x_2'' + L_1(x_2'^2 + x_2 x_2'') - g\dot{\xi}(t) x_1$$

and

$$F_2 = f\dot{\xi}(t) - B_2 x_2' + H x_1^2 + K_2 x_1'' x_2 + L_2(x_1'^2 + x_1 x_1'') .$$

Now Ω_1 and Ω_2 are eigenfrequencies, dashes denote differentiation with respect to time t, and the white noise random excitation $\dot{\xi}(t)$ acts as a forced excitation with the coefficient $f(\neq 0)$ in the second and eventually as a parametric excitation with the coefficient g in the first equation.

In particular, for the autoparametric vibration absorber,

$$g = J = K_2 = L_1 = 0$$

holds, corresponding to (12.1, 1).

Introducing by

$$x_i = a_i \cos \varphi_i , \qquad x_i' = -\Omega_i a_i \sin \varphi_i$$

partial amplitudes $a_i > 0$, and phases

$$\varphi_i = \Omega_i t + \vartheta_i$$

as random functions of time, we get the differential equations (12.4, 1) in standard form

$$-\Omega_i a_i' = F_i \sin \varphi_i , \qquad -\Omega_i a_i \vartheta_i' = F_i \cos \varphi_i . \qquad (12.4, 2)$$

In the expressions (12.4, 2) for F_i, we can replace — up to higher approximations, which are generally neglected — because of (12.5, 1)

$$x_i'' = -\Omega_i^2 x_i + \delta_i^2 f \dot{\xi}(t) . \qquad (12.4, 3)$$

The second part on the right-hand side is taken into consideration only in the term $K_1 x_1 x_2''$ of highest order of magnitude, thus (12.4, 2) reads

$$
\begin{aligned}
-\Omega_1 a_1' = {} & \Omega_1 a_1 (B_1 + D a_1^2 \cos^2 \varphi_1) \sin^2 \varphi_1 - C a_1^3 \sin \varphi_1 \cos^3 \varphi_1 \\
& - \Omega_1^2 E a_1^3 (\sin^2 \varphi_1 - \cos^2 \varphi_1) \sin \varphi_1 \cos \varphi_1 \\
& + J a_1 a_2 \sin \varphi_1 \cos \varphi_1 \cos \varphi_2 - \Omega_2^2 K_1 a_1 a_2 \sin \varphi_1 \cos \varphi_1 \cos \varphi_2 \\
& + \Omega_2^2 L_1 a_2^2 (\sin^2 \varphi_2 - \cos^2 \varphi_2) \sin \varphi_1 \\
& - (g - K_1 f) a_1 \dot{\xi}(t) \sin \varphi_1 \cos \varphi_1 ,
\end{aligned}
$$

$$
\begin{aligned}
-\Omega_2 a_2' = {} & \Omega_2 B_2 a_2 \sin^2 \varphi_2 + H a_1^2 \cos^2 \varphi_1 \sin \varphi_2 \\
& - \Omega_1^2 K_2 a_1 a_2 \cos \varphi_1 \sin \varphi_2 \cos \varphi_2 \\
& + \Omega_1^2 L_2 a_1^2 (\sin^2 \varphi_1 - \cos^2 \varphi_1) \sin \varphi_2 + f \dot{\xi}(t) \sin \varphi_2 ,
\end{aligned}
$$

$$
\begin{aligned}
-\Omega_1 a_1 \vartheta_1' = {} & \Omega_1 a_1 (B_1 + D a_1^2 \cos^2 \varphi_1) \sin \varphi_1 \cos \varphi_1 - C a_1^3 \cos^4 \varphi_1 \\
& - \Omega_1^2 E a_1^3 (\sin^2 \varphi_1 - \cos^2 \varphi_1) \cos^2 \varphi_1 \\
& + J a_1 a_2 \cos^2 \varphi_1 \cos \varphi_2 - \Omega_2^2 K_1 a_1 a_2 \cos^2 \varphi_1 \cos \varphi_2 \\
& + \Omega_2^2 L_1 a_2^2 \cos \varphi_1 (\sin^2 \varphi_2 - \cos^2 \varphi_2) - (g - K_1 f) a_1 \dot{\xi}(t) \cos^2 \varphi_1 ,
\end{aligned}
$$

$$
\begin{aligned}
-\Omega_2 a_2 \vartheta_2' = {} & \Omega_2 B_2 a_2 \sin \varphi_2 \cos \varphi_2 + H a_1^2 \cos^2 \varphi_1 \cos \varphi_2 \\
& - \Omega_2^2 K_2 a_1 a_2 \cos \varphi_1 \cos^2 \varphi_2 \\
& + \Omega_1^2 L_2 a_1^2 (\sin^2 \varphi_1 - \cos^2 \varphi_1) \cos \varphi_2 + f \dot{\xi}(t) \cos \varphi_2 .
\end{aligned}
$$

We understand these equations as physical or Stratonovich ones. If we abbreviate them in the form

$$dy_k = m_k \, dt + n_k \, d\xi , \qquad k = 1, 2, 3, 4$$

where

$$y_i = a_i , \qquad y_{i+2} = \vartheta_i \qquad (i = 1, 2)$$

and

$$d\xi = \dot{\xi}(t) \, dt ,$$

the corresponding Ito equations can be written

$$dy_k = (m_k + \mu_k) \, dt + n_k \, d\xi , \qquad \mu_k = \frac{1}{2} \sum_{l=1}^{4} \frac{\partial n_k}{\partial y_l} n_l \qquad (12.4, 4)$$

where the additional Ito terms are now

$$\mu_1 = \frac{(g - K_1 f)^2 a_1}{2\Omega_1^2} \cos^4 \varphi_1 , \qquad \mu_2 = \frac{f^2}{2\Omega_2^2 a_2} \cos^2 \varphi_2 ,$$

$$\mu_3 = \frac{(g - K_1 f)^2}{4\Omega_1^2} (1 + \cos 2\varphi_1) \sin 2\varphi_1 , \qquad \mu_4 = -\frac{f^2}{2\Omega_2^2 a_2^2} \sin 2\varphi_2 .$$

After trigonometric transformations, the Ito equations (12.4, 4) can be written in the form

$$da_i = p_i \, dt + q_i \, d\xi , \qquad d\vartheta_i = r_i \, dt + s_i \, d\xi \qquad\qquad (12.4, 5)$$

where

$$p_1 = -\frac{B_1 a_1}{2} + \frac{B_1 a_1}{2} \cos 2\varphi_1 - \frac{Da_1^3}{8} + \frac{Da_1^3}{8} \cos 4\varphi_1 + \frac{Ca_1^3}{4\Omega_1} \sin 2\varphi_1$$

$$+ \frac{Ca_1^3}{8\Omega_1} \sin 4\varphi_1 - \frac{\Omega_1 E a_1^3}{4} \sin 4\varphi_1 + \frac{K}{4\Omega_1} a_1 a_2 \sin (2\varphi_1 + \varphi_2)$$

$$+ \frac{K}{4\Omega_1} a_1 a_2 \sin (2\varphi_1 - \varphi_2) + \frac{\Omega_2^2 L_1 a_2^2}{2\Omega_1} \sin (\varphi_1 + 2\varphi_2)$$

$$+ \frac{\Omega_2^2 L_1 a_2^2}{2\Omega_1} \sin (\varphi_1 - 2\varphi_2) + \frac{3(g - K_1 f)^2 a_1}{16\Omega_1^2} + \frac{(g - K_1 f)^2 a_1}{4\Omega_1^2} \cos 2\varphi_1$$

$$+ \frac{(g - K_1 f)^2 a_1}{16\Omega_1^2} \cos 4\varphi_1 ,$$

$$p_2 = -\frac{B_2 a_2}{2} + \frac{B_2 a_2}{2} \cos 2\varphi_2 - \frac{Ha_1^2}{2\Omega_2} \sin \varphi_2 + \frac{\Omega_1^2 K_2 a_1 a_2}{4\Omega_2} \sin (\varphi_1 + 2\varphi_2)$$

$$- \frac{\Omega_1^2 K_2 a_1 a_2}{4\Omega_2} \sin (\varphi_1 - 2\varphi_2) + \frac{L a_1^2}{4\Omega_2} \sin (2\varphi_1 + \varphi_2)$$

$$- \frac{L a_1^2}{4\Omega_2} \sin (2\varphi_1 - \varphi_2) + \frac{f^2}{4\Omega_2^2 a_2} + \frac{f^2}{4\Omega_2^2 a_2} \cos 2\varphi_2 ,$$

$$q_1 = -\frac{(g - K_1 f) a_1}{2\Omega_1} \sin 2\varphi_1 ,$$

$$q_2 = -\frac{f}{\Omega_2} \sin \varphi_2 ,$$

$$r_1 = -\frac{B_1}{2} \sin 2\varphi_1 - \frac{Da_1^2}{4} \sin 2\varphi_1 - \frac{Da_1^2}{8} \sin 4\varphi_1 + \frac{3Ca_1^2}{8\Omega_1} + \frac{Ca_1^2}{2\Omega_1} \cos 2\varphi_1$$

$$+ \frac{Ca_1^2}{8\Omega_1} \cos 4\varphi_1 - \frac{\Omega_1 E a_1^2}{4} - \frac{\Omega_1 E a_1^2}{2} \cos 2\varphi_1 - \frac{\Omega_1 E a_1^2}{4} \cos 4\varphi_1$$

$$+ \frac{K a_2}{4\Omega_1} \cos (2\varphi_1 + \varphi_2) + \frac{K a_2}{4\Omega_1} \cos (2\varphi_1 - \varphi_2) + \frac{K a_2}{2\Omega_1} \cos \varphi_2$$

$$+ \frac{\Omega_2^2 L_1 a_2^2}{2\Omega_1 a_1} \cos (\varphi_1 + 2\varphi_2) + \frac{\Omega_2^2 L_1 a_2^2}{2\Omega_1 a_1} \cos (\varphi_1 - 2\varphi_2)$$

$$- \frac{(g - K_1 f)^2}{4\Omega_1^2} \sin 2\varphi_1 - \frac{(g - K_1 f)^2}{8\Omega_1^2} \sin 4\varphi_1 ,$$

$$r_2 = -\frac{B_2}{2}\sin 2\varphi_2 - \frac{Ha_1^2}{2\Omega_2 a_2}\cos \varphi_2 + \frac{\Omega_2 K_2 a_1}{2}\cos \varphi_1 + \frac{\Omega_2 K_2 a_1}{4}\cos (\varphi_1 + 2\varphi_2)$$

$$+ \frac{\Omega_2 K_2 a_1}{4}\cos (\varphi_1 - 2\varphi_2) + \frac{La_1^2}{4\Omega_2 a_2}\cos (2\varphi_1 + \varphi_2)$$

$$+ \frac{La_1^2}{4\Omega_2 a_2}\cos (2\varphi_1 - \varphi_2) - \frac{f^2}{2\Omega_2^2 a_2^2}\sin 2\varphi_2 \,,$$

$$s_1 = -\frac{g - K_1 f}{2\Omega_1}(1 + \cos 2\varphi_1)\,,$$

$$s_2 = -\frac{f}{\Omega_2 a_2}\cos \varphi_2$$

when we use abbreviations

$$K = \Omega_2^2 K_1 - J\,, \qquad L = 2\Omega_1^2 L_2 - H\,.$$

Starting from the differential equations (12.4, 5), the corresponding Fokker Planck Kolmogorov equation which determines the probability density can be evaluated. Because the latter differential equation depends in a rather complicated manner on a_ν, ϑ_ν and t, we introduce, for an iterative solution, first a transformation of the partial amplitudes and phases with suitably chosen small correction functions ε_i and δ_i:

$$\begin{aligned} a_i &= A_i + \varepsilon_i(A_\nu, \Phi_\nu)\,, \\ \varphi_i &= \Phi_i + \delta_i(A_\nu, \Phi_\nu)\,. \end{aligned} \qquad i = 1, 2\,, \qquad (12.4, 6)$$

We shall write, for the transformed phases,

$$\Phi_i = \Omega_i t + \Theta_i\,, \qquad (12.4, 7)$$

that is,

$$\vartheta_i = \Theta_i + \delta_i\,.$$

The aim is to determine iteratively, for the four stochastic processes A_i, Θ_i ($i = 1, 2$), the stochastic differentials, which we shall write in the form

$$dA_i = P_i\,dt + Q_i\,d\xi\,, \qquad d\Theta_i = R_i\,dt + S_i\,d\xi\,. \qquad (12.4, 8)$$

Ito's differentiation formula (compare for instance ARNOLD (1973)) leads to the differentials

$$d\varepsilon_i = \sum_{j=1}^{2}\frac{\partial \varepsilon_i}{\partial A_j}\,dA_j + \sum_{j=1}^{2}\frac{\partial \varepsilon_i}{\partial \Phi_j}\,d\Phi_j + \nabla \varepsilon_i\,dt$$

of the stochastic processes $\varepsilon_i(A_\nu, \Phi_\nu)$, $\nu = \{1, 2\}$, if ∇ denotes the operator of the additional Ito terms,

$$\nabla = \frac{1}{2}\sum_{j=1}^{2}\left(Q_j^2\frac{\partial^2}{\partial A_j^2} + S_j^2\frac{\partial^2}{\partial \Phi_j^2}\right) + Q_1 Q_2\frac{\partial^2}{\partial A_1\,\partial A_2}$$

$$+ S_1 S_2\frac{\partial^2}{\partial \Phi_1\,\partial \Phi_2} + \sum_{j, k=1}^{2}Q_j S_k\frac{\partial^2}{\partial A_j\,\partial \Phi_k}\,.$$

Using this and equations (12.4, 7), (12.4, 8), (12.4, 5), we get from (12.4, 6)

$$\mathrm{d}A_i = \left[-\sum_{j=1}^{2} \left(\Omega_j \frac{\partial \varepsilon_i}{\partial \Phi_j} + \frac{\partial \varepsilon_i}{\partial A_j} P_j + \frac{\partial \varepsilon_i}{\partial \Phi_j} R_j \right) - \nabla \varepsilon_i + p_i(A_\nu + \varepsilon_\nu, \Phi_\nu + \delta_\nu) \right] \mathrm{d}t$$

$$+ \left[-\sum_{j=1}^{2} \left(\frac{\partial \varepsilon_i}{\partial A_j} Q_j + \frac{\partial \varepsilon_i}{\partial \Phi_j} S_j \right) + q_i(A_\nu + \varepsilon_\nu, \Phi_\nu + \delta_\nu) \right] \mathrm{d}\xi . \qquad (12.4, 9)$$

Analogously, the equation

$$\mathrm{d}\Theta_i = \left[-\sum_{j=1}^{2} \left(\Omega_j \frac{\partial \delta_i}{\partial \Phi_j} + \frac{\partial \delta_i}{\partial A_j} P_j + \frac{\partial \delta_i}{\partial \Phi_j} R_j \right) - \nabla \delta_i + r_i(A_\nu + \varepsilon_\nu, \Phi_\nu + \delta_\nu) \right] \mathrm{d}t$$

$$+ \left[-\sum_{j=1}^{2} \left(\frac{\partial \delta_i}{\partial A_j} Q_i + \frac{\partial \delta_i}{\partial \Phi_j} S_j \right) + s_i(A_\nu + \varepsilon_\nu, \Phi_\nu + \delta_\nu) \right] \mathrm{d}\xi$$

can be found.

12.5. Fokker Planck Kolmogorov equation

The Fokker Planck Kolmogorov equation of the probability density $w(A_\nu, \Theta_\nu, t)$ is, for a system of differential equations of the form (12.4, 8),

$$\frac{\partial w}{\partial t} + \frac{\partial (P_1 w)}{\partial A_1} + \frac{\partial (P_2 w)}{\partial A_2} + \frac{\partial (R_1 w)}{\partial \Theta_1} + \frac{\partial (R_2 w)}{\partial \Theta_2}$$

$$= \frac{1}{2} \frac{\partial^2 (Q_1^2 w)}{\partial A_1^2} + \frac{1}{2} \frac{\partial^2 (Q_2^2 w)}{\partial A_2^2} + \frac{1}{2} \frac{\partial^2 (S_1^2 w)}{\partial \Theta_1^2} + \frac{1}{2} \frac{\partial^2 (S_2^2 w)}{\partial \Theta_2^2}$$

$$+ \frac{\partial^2 (Q_1 Q_2 w)}{\partial A_1 \, \partial A_2} + \frac{\partial^2 (Q_1 S_1 w)}{\partial A_1 \, \partial \Theta_1} + \frac{\partial^2 (Q_1 S_2 w)}{\partial A_1 \, \partial \Theta_2} + \frac{\partial^2 (Q_2 S_1 w)}{\partial A_2 \, \partial \Theta_1} + \frac{\partial^2 (Q_2 S_2 w)}{\partial A_2 \, \partial \Theta_2} + \frac{\partial^2 (S_1 S_2 w)}{\partial \Theta_1 \, \partial \Theta_2} .$$

$$(12.5, 1)$$

Even in the important stationary case

$$\frac{\partial w}{\partial t} = 0$$

which we assume in what follows, no closed-form solutions of this equations are known. We shall use an approximation method (compare STRATONOVICH (1961)) to find an iterative solution.

In the first approximation, in the sense of the averaging method, we consider in (12.5, 1) only the non-oscillating (that is, not depending on Φ_1, Φ_2) expressions and seek correspondingly also the probability density as a function of A_1 and A_2 only (compare STRATONOVICH (1961)). For such a probability density, (12.5, 1) simplifies to

$$Tw + T_1 \frac{\partial w}{\partial A_1} + T_2 \frac{\partial w}{\partial A_2} = \frac{Q_1^2}{2} \frac{\partial^2 w}{\partial A_1^2} + \frac{Q_2^2}{2} \frac{\partial^2 w}{\partial A_2^2} + Q_1 Q_2 \frac{\partial^2 w}{\partial A_1 \, \partial A_2} \qquad (12.5, 2)$$

where

$$
\left.
\begin{aligned}
T &= \frac{\partial P_1}{\partial A_1} + \frac{\partial P_2}{\partial A_2} + \frac{\partial R_1}{\partial \Theta_1} + \frac{\partial R_2}{\partial \Theta_2} - \frac{1}{2}\frac{\partial^2 Q_1^2}{\partial A_1^2} - \frac{1}{2}\frac{\partial^2 Q_2^2}{\partial A_2^2} \\
&\quad - \frac{1}{2}\frac{\partial^2 S_1^2}{\partial \Theta_1^2} - \frac{1}{2}\frac{\partial^2 S_2^2}{\partial \Theta_2^2} - \frac{\partial^2(Q_1 Q_2)}{\partial A_1\, \partial A_2} - \frac{\partial^2(Q_1 S_1)}{\partial A_1\, \partial \Theta_1} - \frac{\partial^2(Q_1 S_2)}{\partial A_1\, \partial \Theta_2} \\
&\quad - \frac{\partial^2(Q_2 S_1)}{\partial A_2\, \partial \Theta_1} - \frac{\partial^2(Q_2 S_2)}{\partial A_2\, \partial \Theta_2} - \frac{\partial^2(S_1 S_2)}{\partial \Theta_1\, \partial \Theta_2}, \\
T_1 &= P_1 - \frac{\partial Q_1^2}{\partial A_1} - \frac{\partial(Q_1 Q_2)}{\partial A_2} - \frac{\partial(Q_1 S_1)}{\partial \Theta_1} - \frac{\partial(Q_1 S_2)}{\partial \Theta_2}, \\
T_2 &= P_2 - \frac{\partial Q_2^2}{\partial A_2} - \frac{\partial(Q_1 Q_2)}{\partial A_1} - \frac{\partial(Q_2 S_1)}{\partial \Theta_1} - \frac{\partial(Q_2 S_2)}{\partial \Theta_2}.
\end{aligned}
\right\} \qquad (12.5,\ 3)
$$

The correction functions ε_i, δ_i in (12.4, 6) are not yet used (put equal to zero) *in first approximation*. We assume for simplicity $\Omega_2 > \Omega_1$ and exclude an internal resonance $\Omega_2 = 2\Omega_1$. Then we get

$$
\left.
\begin{aligned}
T &= -\frac{B_1 + B_2}{2} - \frac{3DA_1^2}{8} + \frac{(g - K_1 f)^2}{16\Omega_1^2} - \frac{f^2}{4\Omega_2^2 A_2^2}, \\
T_1 &= -\frac{B_1 A_1}{2} - \frac{DA_1^2}{8} - \frac{(g - K_1 f)^2 A_1}{16\Omega_1^2}, \\
T_2 &= -\frac{B_2 A_2}{2} + \frac{f^2}{4\Omega_2^2 A_2}, \\
Q_1^2 &= \frac{(g - K_1 f)^2 A_1^2}{8\Omega_1^2}, \\
Q_2^2 &= \frac{f^2}{2\Omega_2^2}, \\
Q_1 Q_2 &= 0 .
\end{aligned}
\right\} \qquad (12.5,\ 4)
$$

In this approximation, the coefficients (12.5, 4) of the Fokker Planck Kolmogorov equation (12.5, 2) do *not* reveal the dependence on non-linear inertia and restoring forces.

We now use the transformation (12.4, 6) in order to find a *higher-order solution* of the Fokker Planck Kolmogorov equation. For simplicity assume

$$ K_2 = L_1 = 0 $$

which holds in particular for the autoparametric vibration absorber.

Choose the correction function ε_1 so that in the equations (12.4, 8), (12.4, 9) for P_1 the terms of p_1 which are of greatest order of magnitude and independent of phase and which contain neither the square of the excitation coefficients f, g nor ε_i, δ_i multiplied by other small parameters, are cancelled by the expressions

$$ -\sum_{j=1}^{2} \Omega_j \frac{\partial \varepsilon_1}{\partial \Phi_j} $$

in (12.4, 9). We shall succeed in doing so by choosing

$$\varepsilon_1 = \frac{B_1 A_1}{4\Omega_1} \sin 2\Phi_1 + \frac{DA_1^3}{32\Omega_1} \sin 4\Phi_1 - \frac{CA_1^3}{8\Omega_1^2} \cos 2\Phi_1 - \frac{CA_1^3}{32\Omega_1^2} \cos 4\Phi_1$$

$$+ \frac{EA_1^3}{16} \cos 4\Phi_1 - \frac{KA_1 A_2}{4\Omega_1(2\Omega_1 + \Omega_2)} \cos (2\Phi_1 + \Phi_2)$$

$$- \frac{KA_1 A_2}{4\Omega_1(2\Omega_1 - \Omega_2)} \cos (2\Phi_1 - \Phi_2).$$

Correspondingly, cancel the terms of greatest order of magnitude in P_2, R_1 and R_2 which do not depend on phase by choosing

$$\varepsilon_2 = \frac{B_2 A_2}{4\Omega_2} \sin 2\Phi_2 - \frac{LA_1^2}{4\Omega_2(2\Omega_1 + \Omega_2)} \cos (2\Phi_1 + \Phi_2)$$

$$+ \frac{LA_1^2}{4\Omega_2(2\Omega_1 - \Omega_2)} \cos (2\Phi_1 - \Phi_2),$$

$$\delta_1 = \frac{B_1}{4\Omega_1} \cos 2\Phi_1 + \frac{DA_1^2}{8\Omega_1} \cos 2\Phi_1 + \frac{DA_1^2}{32\Omega_1} \cos 4\Phi_1 + \frac{CA_1^2}{4\Omega_1^2} \sin 2\Phi_1$$

$$+ \frac{CA_1^2}{32\Omega_1^2} \sin 4\Phi_1 - \frac{EA_1^2}{4} \sin 2\Phi_1 - \frac{EA_1^2}{16} \sin 4\Phi_1$$

$$+ \frac{KA_2}{4\Omega_1(2\Omega_1 + \Omega_2)} \sin (2\Phi_1 + \Phi_2) + \frac{KA_2}{4\Omega_1(2\Omega_1 - \Omega_2)} \sin (2\Phi_1 - \Phi_2)$$

$$+ \frac{KA_2}{2\Omega_1\Omega_2} \sin \Phi_2$$

and

$$\delta_2 = \frac{B_2}{4\Omega_2} \cos 2\Phi_2 + \frac{LA_1^2}{4\Omega_2(2\Omega_1 + \Omega_2) A_2} \sin (2\Phi_1 + \Phi_2)$$

$$+ \frac{LA_1^2}{4\Omega_2(2\Omega_1 - \Omega_2) A_2} \sin (2\Phi_1 - \Phi_2).$$

These correction functions have to be inserted in (12.4, 8), (12.4, 9). Thus we can find, if we also exclude an internal resonance $\Omega_2 = 4\Omega_1$, the additional terms of second approximation in P_1 independent of phase

$$P_{1\,\mathrm{add}} = \frac{EB_1 A_1^3}{8} + \frac{CDA_1^5}{32\Omega_1^2}. \tag{12.5, 5}$$

By (12.5, 5) we get for the coefficients (12.5, 4) of the Fokker Planck Kolmogorov equation (12.5, 2) — because the other terms are, as terms neglected in second approximation, of power two in the excitation coefficients — the additional terms

$$T_{1\,\mathrm{add}} = P_{1\,\mathrm{add}}. \tag{12.5, 6}$$

and

$$T_{\mathrm{add}} = \frac{\partial P_{1\,\mathrm{add}}}{\partial A_1} = \frac{3EB_1 A_1^2}{8} + \frac{5CDA_1^4}{32\Omega_1^2}. \tag{12.5, 7}$$

If we consider instead of the system (12.4, 1) the corresponding system with one degree of freedom for x_1 only, which reads, by use of (12.4, 3),

$$x_1'' + \Omega_1^2 x_1 = -B_1 x_1' - D x_1^2 x_1' - C x_1^3 - E(x_1'^2 + x_1 x_1'') - (g - K_1 f)\,\dot{\xi}(t)\,x_1$$

with the coefficient $g - K_1 f$ of parametric excitation composed by the direct parametric excitation and the autoparametric excitation in (12.4, 1), the probability density of the amplitude is given by an ordinary differential equation, and thus in second approximation by (11.1, 20):

$$\left.\begin{aligned}
w(A_1) &= c A_1^{1 - \frac{8\Omega_1^2 B_1}{(g - K_1 f)^2}}\, e^{-U A_1^2 - V A_1^4}, \\[2mm]
U &= \frac{\Omega_1^2 D}{(g - K_1 f)^2} - \frac{2\Omega_1^2 E B_1}{(g - K_1 f)^2} + \frac{5 C B_1}{2(g - K_1 f)^2} - \frac{E}{2}, \\[2mm]
V &= \frac{1}{16(g - K_1 f)^2}\,(3 D C - 2\Omega_1^2 D E - 5 C E B_1 + 2\Omega_1^2 E^2 B_1).
\end{aligned}\right\} \quad (12.5,\,8)$$

By analogy with this form of solution, which enables us to separate the behaviour of the probability density for $A_1 \to 0$ where the exponential factor tends to one from the behaviour for greater A_1 where the exponential factor predominates, we shall try to solve the partial differential equation (12.5, 2) iteratively in the form

$$w(A_1, A_2) = c A_1^i A_2^j\, e^{\sum c_{kl} A_1^k A_2^l} \qquad (12.5,\,9)$$

choosing from the beginning $c_{00} = c_{10} = 0$ by analogy with (12.5, 8).

Abbreviating (12.5, 9) by

$$w = c A_1^i A_2^j\, e^{\Phi}$$

we find, for $A_1 \neq 0$, $A_2 \neq 0$, that

$$\left.\begin{aligned}
\frac{\partial w}{\partial A_1} &= c A_1^i A_2^j\, e^{\Phi}\left(\frac{i}{A_1} + \frac{\partial \Phi}{\partial A_1}\right), \\[2mm]
\frac{\partial^2 w}{\partial A_1^2} &= c A_1^i A_2^j\, e^{\Phi}\left[\left(\frac{i}{A_1} + \frac{\partial \Phi}{\partial A_1}\right)^2 - \frac{i}{A_1^2} + \frac{\partial^2 \Phi}{\partial A_1^2}\right], \\[2mm]
\frac{\partial w}{\partial A_2} &= c A_1^i A_2^j\, e^{\Phi}\left(\frac{j}{A_2} + \frac{\partial \Phi}{\partial A_2}\right), \\[2mm]
\frac{\partial^2 w}{\partial A_2^2} &= c A_1^i A_2^j\, e^{\Phi}\left[\left(\frac{j}{A_2} + \frac{\partial \Phi}{\partial A_2}\right)^2 - \frac{j}{A_2^2} + \frac{\partial^2 \Phi}{\partial A_2^2}\right].
\end{aligned}\right\} \quad (12.5,\,10)$$

Consider exponents Φ up to the sixth power in A_1, A_2, assuming $c_{kl} = 0$ for $k + l \geq 7$ in (12.5, 7). Then

$$\begin{aligned}
\frac{\partial \Phi}{\partial A_1} &= 2 c_{20} A_1 + c_{11} A_2 + 3 c_{30} A_1^2 + 2 c_{21} A_1 A_2 + c_{12} A_2^2 + 4 c_{40} A_1^3 + 3 c_{31} A_1^2 A_2 \\
&\quad + 2 c_{22} A_1 A_2^2 + c_{13} A_2^3 + 5 c_{50} A_1^4 + 4 c_{41} A_1^3 A_2 + 3 c_{32} A_1^2 A_2^2 + 2 c_{23} A_1 A_2^3 \\
&\quad + c_{14} A_2^4 + 6 c_{60} A_1^5 + 5 c_{51} A_1^4 A_2 + 4 c_{42} A_1^3 A_2^2 + 3 c_{33} A_1^2 A_2^3 \\
&\quad + 2 c_{24} A_1 A_2^4 + c_{15} A_2^5
\end{aligned}$$

and analogous expressions for $\partial \Phi / \partial A_2$ and the second derivatives will hold.

We have to insert all these expressions into (12.5, 10) and the resulting expressions, as well as (12.5, 4), (12.5, 6), (12.5, 5), (12.5, 7) into (12.5, 2), and divide by w. A comparison of equal powers of A_1, A_2 leads, after a lengthy analysis, to equations for i, j, c_{kl}.

We find for the lowest powers,

$$\frac{1}{A_2^2}: \quad (j-1)\frac{f}{4\Omega_2^2} = 0$$

and

$$\frac{1}{A_2}: \quad \frac{fc_{01}}{4\Omega_2^2} = 0$$

from which it follows because forced excitation does not vanish, that

$$j = 1, \quad c_{01} = 0.$$

Using this, we find

$$\frac{A_1}{A_2}: \quad c_{11} = 0,$$

$$1: \quad \frac{f^2}{\Omega_2^2}c_{02} = -\frac{1+i}{2}B_1 - B_2 + (1-i^2)\frac{(g-K_1f)^2}{16\Omega_1^2} \qquad (12.5, 11)$$

and

$$\frac{A_1^2}{A_2}: \quad c_{21} = 0,$$

$$A_1: \quad c_{12} = 0,$$

$$A_2: \quad c_{03} = 0.$$

In the next step we derive from the coefficients of

$$\frac{A_1^3}{A_2}: \quad c_{31} = 0,$$

$$A_1^2: \quad \frac{f^2}{\Omega_2^2}c_{22} = \frac{3+i}{8}(EB_1 - D) - B_1c_{20} - (1+i)\frac{(g-K_1f)^2 c_{20}}{4\Omega_1^2}, \qquad (12.5, 12)$$

$$A_1A_2: \quad c_{13} = 0,$$

$$A_2^2: \quad 4c_{04} = -\frac{\Omega_2^2 B_2 c_{02}}{f^2} - c_{02}^2, \qquad (12.5, 13)$$

further that

$$\frac{A_1^4}{A_2}: \quad c_{41} = 0,$$

$$A_1^3: \quad \frac{f^2}{\Omega_2^2}c_{32} + \frac{3}{2}\left[B_1 + (3+2i)\frac{(g-K_1f)^2}{8\Omega_1^2}\right]c_{30} = 0,$$

so that we can choose

$$c_{30} = c_{32} = 0,$$

$$A_1^2A_2: \quad c_{23} = 0,$$

$$A_1A_2^2: \quad c_{14} = 0,$$

$$A_2^3: \quad c_{05} = 0.$$

Finally the coefficients yield

$$\frac{A_1^5}{A_2}: \quad c_{51} = 0 \,,$$

$$A_1^4: \quad \frac{f^2}{\Omega_2^2} c_{42} = \left(-\frac{D}{4} + \frac{EB_1}{4} c_{20} - \frac{(g - K_1 f)^2}{4\Omega_1^2} [c_{20}^2 + 2(2 + i)\, c_{40}] \right.$$

$$\left. - 2B_1 c_{40} + \frac{(5 + i)\, DC}{32\Omega_1^2} \,, \right. \tag{12.5, 14}$$

$$A_1^3 A_2: \quad c_{33} = 0 \,,$$

$$A_1^2 A_2^2: \quad \frac{4f^2}{\Omega_2^2} c_{24} = \left[iB_1 + B_2 - (3 + 2i - i^2) \frac{(g - K_1 f)^2}{8\Omega_1^2} \right] c_{22} \,, \tag{12.5, 15}$$

$$A_1 A_2^3: \quad c_{15} = 0 \,,$$

$$A_2^4: \quad \frac{9f^2}{2\Omega_2^2} c_{06} = -\left(B_2 + \frac{2f^2}{\Omega_2^2} c_{02} \right) c_{04} \,. \tag{12.5, 16}$$

It shows that the only eventually non-vanishing coefficients we obtain are $j = 1$, c_{02} given by (12.5, 11), c_{22} given by (12.5, 12), c_{04} given by (12.5, 13), c_{42} given by (12.5, 14), c_{24} given by (12.5, 15), and c_{06} given by (12.5, 16), whereas i, c_{20}, c_{40}, c_{60} remain undetermined. In order to secure the correspondence with the one-degree-of-freedom solution (11.1, 9), choose in *first approximation*

$$i = 1 - \frac{8\Omega_1^2 B_1}{(g - K_1 f)^2} \tag{12.5, 17}$$

and

$$c_{20} = -\frac{\Omega_1^2 D}{(g - K_1 f)^2} \,, \qquad c_{40} = 0 \,. \tag{12.5, 18}$$

Consequently, (12.5, 11) simplifies to

$$c_{02} = -\frac{\Omega_2^2 B_2}{f^2} \,,$$

so that (12.5, 13) reads

$$c_{04} = 0$$

and (12.5, 12) leads to

$$c_{22} = \frac{(3 + i)\, \Omega_2^2 E B_1}{8 f^2} \,.$$

If we use the *second approximation* (12.5, 8) of the one-degree-of-freedom solution, we have to choose (12.5, 17) and instead of (12.5, 18)

$$c_{20} = -\frac{\Omega_1^2 D}{(g - K_1 f)^2} + \frac{2\Omega_1^2 E B_1}{(g - K_1 f)^2} - \frac{5C B_1}{2(g - K_1 f)^2} + \frac{E}{2} \,, \tag{12.5, 21}$$

$$c_{40} = -\frac{1}{16(g - K_1 f)^2} (3DC - 2\Omega_1^2 DE - 5CE B_1 + 2\Omega_1^2 E^2 B_1) \,. \tag{12.5, 22}$$

Using these formulae, we get from (12.5, 12)

$$c_{22} = -\frac{(3+i)\,\Omega_2^2}{16\Omega_1^2 f^2}\left[E(g - K_1 f)^2 + 2\Omega_1^2 E B_1 - 5CB_1\right],$$ (12.5, 23)

from (12.5, 14)

$$c_{42} = -\frac{\Omega_2^2}{32\Omega_1^2 f^2}\left[2i\Omega_1^2 DE - 5(2+i)\,DC + 2E^2(g - K_1 f)^2\right.$$
$$\left. + \frac{50C^2 B_1^2}{(g - K_1 f)^2} + 3(3-i)\,\Omega_1^2 E^2 B_1 - 5(3-2i)\,ECB_1\right]$$

and from (12.5, 15)

$$c_{24} = \frac{\Omega_2^2}{4f^2}\left(B_2 - \frac{3+i}{1-i}\,B_1\right)c_{22}.$$

Formula (12.5, 16) yields

$$c_{06} = 0.$$

The formulae thus found show that c_{02} and c_{20} (in (12.5, 21) because of the first negative term of highest order of magnitude) are negative. In formula (12.5, 22) for c_{40} the first two terms are of highest order of magnitude, they yield a negative coefficient c_{40} if

$$3C > 2\Omega_1^2 E.$$ (12.5, 24)

The first approximation formula (12.5, 20) gives $c_{22} > 0$ for $i > -3$ whereas the second approximation formula (12.5, 23) leads to $c_{22} < 0$ if both (or neither) of the conditions

$$i > -3,$$
$$E(g - K_1 f)^2 + 2\Omega_1^2 E B_1 > 5CB_1$$

hold. If in (12.5, 24) the equality sign holds, so that $c_{40} \approx 0$, the second condition (12.5, 25) simplifies with $E > 0$ to $i > -5$, that is, the only condition for $c_{22} < 0$ is $i > -3$.

The highest order part of c_{42} is negative for

$$2i\Omega_1^2 E > 5(2+i)\,C,$$

whereas c_{24} has the same sign as c_{22} if

$$(1-i)\,B_2 > (3+i)\,B_1.$$

The order of magnitude of c_{20} and c_{40} is

$$\frac{D}{(g - K_1 f)^2},$$

whereas the order of c_{02} and c_{22} is $B_{1,2}/f^2$, the order of c_{42} is D/f^2 and the order of c_{24} is $B_{1,2}^2/f^4$ if we assume $\Omega_{1,2}$, E and C of order one and $(g - K_1 f)^2$ and B_1 of the same order.

We evaluate the coefficients for the example $\Omega_1 = 1$, $\Omega_2 = 2.5$, $C = E = K_1 = 1$, $D = 0.1$, $B_1 = 0.001$, $B_2 = 0.004$, $g = 0$ and different values of the forced excitation parameter f. The first approximation formulae (12.5, 17), (12.5, 18), (12.5, 19)' (12.5, 20) yield

f	i	c_{20}	c_{02}	c_{22}	
0.07	-0.6327	-20.4082	-5.1020	0.3774	(12.5, 25)
0.10	0.2000	-10.0000	-2.5000	0.2500	
0.20	0.8000	-2.5000	-0.6250	0.0742	

The second approximation formulae (12.5, 21), (12.5, 22), (12.5, 23) and the ensuing formulae give

f	c_{20}	c_{40}	c_{22}	c_{42}	c_{24}
0.07	-20.0100	-1.2372	-0.3586	31.9125	-0.2916
0.10	-9.5500	-0.6063	-0.8750	20.3047	0.0000
0.20	-2.0125	-0.1516	-1.3730	5.6599	0.8045

$$(12.5, 26)$$

12.6.   Behaviour of the solution

In order to evaluate the extreme values of the probability density, we have to determine the points of intersection of the curves

$$\frac{\partial w}{\partial A_1} = 0 \,,$$

that is,

$$\sum k c_{kl} A_1^k A_2^l + i = 0$$

and

$$\frac{\partial w}{\partial A_2} = 0 \,,$$

that is,

$$\sum l c_{kl} A_1^k A_2^l + 1 = 0 \,.$$

A numerical evaluation is possible in every case. For $k + l < 6$ we can find a cubic equation for A_1^2,

$$4 c_{40} c_{22} A_1^6 + 2(c_{22} c_{20} + 2 c_{40} c_{02}) A_1^4 + [2 c_{20} c_{02} + (i - 1) c_{22}] A_1^2 + i c_{02} = 0 \,,$$

$$(12.6, 1)$$

and a corresponding formula

$$A_2^2 = - \frac{1}{2 c_{22} A_1^2 + 2 c_{02}} \,.$$

$$(12.6, 2)$$

Neglecting c_{40}, (12.6, 1) yields the solution

$$4 c_{22} c_{20} A_1^2 = -2 c_{20} c_{02} + (1 - i) c_{22}$$

$$\pm \sqrt{4 c_{20}^2 c_{02}^2 + (1 - i)^2 c_{22}^2 - 4(1 + i) c_{22} c_{20} c_{02}} \,.$$

$$(12.6, 3)$$

If we further neglect c_{22}, we get

$$A_1^2 = -\frac{i}{2c_{20}}, \qquad A_2^2 = -\frac{1}{2c_{02}}. \tag{12.6, 4}$$

If forced excitation is so small that $i < 0$ ($f = 0.07$ in the examples above), (12.6, 4) has no real solution A_1, A_2, that is, the probability density diminishes monotonically from $w = \infty$ for $A_1 = 0$. If, as for the other examples, $i > 0$, (12.6, 4) determines a maximum value of the probability density. Using in the first approximation the values (12.5, 25), in the second approximation (12.5, 26) and neglecting c_{22} by (12.6, 4) or taking c_{22} into consideration by (12.6, 3), (12.6, 2), we can find for $f = 0.1$

	$A_{1\,max}$	$A_{2\,max}$
first approximation, without c_{22}	0.1000	0.4472
first approximation, with c_{22}	0.1003	0.4474
second approximation, without c_{22}	0.1023	0.4472
second approximation, with c_{22}	0.1014	0.4464

and for $f = 0.2$

	$A_{1\,max}$	$A_{2\,max}$
first approximation, without c_{22}	0.4000	0.8944
first approximation, with c_{22}	0.4049	0.9253
second approximation, without c_{22}	0.4458	0.8944
second approximation, with c_{22}	0.3756	0.7815

In the first approximation the coefficient c_{22} is positive and shifts the maximum to greater values of A_1 and A_2, whereas in the second approximation c_{22} is negative and has the opposite influence.

As in the first approximation c_{22} is positive, (12.6, 3), (12.6, 2) lead to a second extreme value (a minimum) of the probability density:

$$A_{1\,min} = 3.1544, \qquad A_{2\,min} = 6.3182 \quad \text{for} \quad f = 0.1,$$

$$A_{1\,min} = 2.8670, \qquad A_{2\,min} = 5.7470 \quad \text{for} \quad f = 0.2.$$

For these values the probability density is practically zero, for instance for $f = 0.1$ w_{min}/w_{max} is less than 10^{-41}.

In the second approximation this minimum disappears because then c_{22} is negative.

By means of (12.5, 25), (12.5, 26) we can determine how the c_{kl} terms contribute to the exponent in the probability density formula (12.5, 9) for $A_1 = A_{1\,max}$, $A_2 = A_{2\,max}$ (in second approximation, with c_{22}):

	Contribution of					
f	c_{20}	c_{02}	c_{22}	c_{40}	c_{42}	c_{24}
0.1	-0.09817	-0.49825	-0.00179	-0.00006	0.00043	0
0.2	-0.28376	-0.38175	-0.11825	-0.00301	0.06873	0.04232

This shows that c_{20}, c_{02} and in the second place c_{22}, have the greatest influence on the exponent. The influence of c_{42} is greater than that of c_{40} and c_{24}.

An especially strong influence of c_{42} is revealed in the example $\Omega_1 = 3$, $\Omega_2 = 10$, $C = g = 0$, $E = 6$, $D = 0.1$, $B_1 = 0.001$, $B_2 = 0.004$ and $K_1 f = 0.3$, particularly for $K_1 = 1.5$, $f = 0.2$ or $K_1 = 3$, $f = 0.1$ or $K_1 = 6$, $f = 0.05$. The formulae of second approximation and (12.6, 4) give

$$i = 0.2 , \qquad c_{20} = -5.8 , \qquad c_{40} = 7.05 , \qquad c_{24} = 0$$

and

K_1	f	c_{02}	c_{22}	c_{42}	$A_{1\,max}$	$A_{2\,max}$
1.5	0.05	− 160	− 576	− 1578	0.1313	0.0559
3	0.1	− 40	− 144	− 394.5	0.1313	0.1118
6	0.2	− 10	− 36	− 98.6	0.1313	0.2236

that is, the same ratio

$$c_{42} : c_{02} = 9.8625 .$$

The general formulae of the approximate maximum amplitudes (12.6, 4) on the system parameters are found, by inserting (12.5, 17), (12.5, 21) and (12.5, 19), to be

$$A_{1\,max}^2 = \frac{(g - K_1 f)^2 - 8\Omega_1^2 B_1}{2\Omega_1^2 D - 4\Omega_1^2 E B_1 + 5CB_1 - E(g - K_1 f)^2} ,$$

$$A_{2\,max}^2 = \frac{f^2}{2\Omega_2^2 B_2} .$$

These show that the parameters of the first equation (12.4, 1) and the forced excitation f combined with K_1 determine $A_{1\,max}$ whereas the parameters Ω_2, f and B_2 of the second equation (12.4, 1) determine $A_{2\,max}$. Increasing eigenfrequency Ω_2 and linear damping B_2 diminish $A_{2\,max}$. Increasing non-linear damping D and non-linear restoring force C and in general also increasing eigenfrequency Ω_1 and linear damping B_1 diminish $A_{1\,max}$. The amplitude $A_{2\,max}$ is proportional to forced excitation f, whereas $A_{1\,max}$ increases with the combination $g - K_1 f$ of forced and parametric excitation.

Fig. 12.6, 1 gives the maximum amplitude $A_{1\,max}$ as dependent on B_1 for $\Omega_1 = K_1 = E = C = 1$, $D = 0.1$, $g = 0$ and different values of f. The dependence of $A_{1\,max}$ on f for different parametric excitation g and $B_1 = 0.001$ is shown by the full curves of Fig. 12.6, 2, the dashed line giving the corresponding maximum amplitudes $A_{2\,max}$ for $\Omega_2 = 2.5$ and $B_2 = 0.004$.

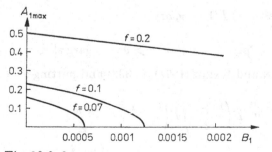

Fig. 12.6, 1

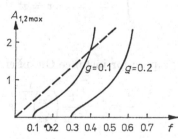

Fig. 12.6, 2

The above discussion of the probability density $w(A_1, A_2)$ was possible without evaluating the integration constant c in this function. In order to determine quantitatively the probility density we have to evaluate c by the normalization condition

$$\iint\limits_0^\infty w \, \mathrm{d}A_1 \, \mathrm{d}A_2 = 1 \qquad (12.6, 5)$$

where w is of the form (12.5, 9). Using all non-vanishing coefficients evaluated above and the abbreviations

$$\left.\begin{array}{l} P = -c_{02} - c_{22}A_1^2 - c_{42}A_1^4 , \\[2mm] Q = -c_{24}A_1^2 \end{array}\right\} \qquad (12.6, 6)$$

as well as the integration formulae (11.2, 1), (11.2, 2), we can reduce (12.6, 5) to the single integral

$$\frac{1}{c} = \int\limits_0^\infty A_1^i \, \mathrm{e}^{c_{20}A_1^2 + c_{40}A_1^4} J(A_1) \, \mathrm{d}A_1 \qquad (12.6, 7)$$

where

$$J(A_1) = \int\limits_0^\infty A_2 \, \mathrm{e}^{-PA_2^2 - QA_2^4} \, \mathrm{d}A_2$$

$$= \begin{cases} \dfrac{1}{\sqrt{8Q}} \, \mathrm{e}^{\frac{P^2}{8Q}} \, D_{-1}\left(\dfrac{P}{\sqrt{2Q}}\right) & \text{for } Q > 0 , \\[4mm] \dfrac{1}{2P} & \text{for } Q = 0 , \quad P > 0 \end{cases} \qquad (12.6, 8)$$

and D_{-1} given by (10.2, 8), (10.2, 10).

The integral (12.6, 7) can be evaluated numerically. If we assume

$$c_{20}, c_{02} \quad \text{and} \quad c_{22} < 0 ,$$

but

$$c_{42} = c_{24} = c_{40} = 0$$

we get

$$\frac{1}{c} = -\frac{1}{2} \int\limits_0^\infty \frac{A_1^i \, \mathrm{e}^{c_{20}A_1^2}}{c_{02} + c_{22}A_1^2} \, \mathrm{d}A_1$$

or, using the integration formula

$$\int\limits_0^\infty \frac{x^{\nu-1} \, \mathrm{e}^{-\varrho x}}{x + \tau} \, \mathrm{d}x = \tau^{\nu-1} \, \mathrm{e}^{\varrho\tau} \, \Gamma(\nu) \, \Gamma(1 - \nu, \varrho\tau)$$

$$\text{for } \mu > 0 , \quad \nu > 0 , \quad |\arg \tau| < \pi$$

(compare for instance GRADŠTEJN and RYŽIK (1971), p. 333) and putting $x = A_1^2$,

$$\frac{1}{c} = -\frac{1}{4c_{22}}\left(\frac{c_{22}}{c_{02}}\right)^{\frac{1-i}{2}} \mathrm{e}^{-\frac{c_{02}c_{20}}{c_{12}}} \, \Gamma\left(\frac{1+i}{2}\right) \Gamma\left(\frac{1-i}{2}, -\frac{c_{02}c_{20}}{c_{22}}\right)$$

$$\text{for } i > -1 .$$

If additionally $c_{22} = 0$ holds, we find that

$$\frac{1}{c} = -\frac{\Gamma\left(\dfrac{1+i}{2}\right)}{4c_{02}(-c_{20})^{\frac{1+i}{2}}} \quad \text{for} \quad i > -1 \, . \tag{12.6, 9}$$

If we introduce instead of (12.6, 6) the abbreviations

$$P' = -c_{20} - c_{22}A_2^2 - c_{24}A_2^4 \, ,$$
$$Q' = -c_{40} - c_{42}A_2^2 \, ,$$

the integration formula (11.2, 1) leads to the single integral

$$\frac{1}{c} = \int\limits_0^\infty A_2 \, e^{c_{02}A_2^2} J'(A_2) \, \mathrm{d}A_2$$

with

$$J'(A_2) = \int\limits_0^\infty A_1^i \, e^{-P'A_1^2 - Q'A_1^4} \, \mathrm{d}A_1$$

$$= \begin{cases} \dfrac{1}{2}(2Q')^{\frac{-1+i}{4}} e^{\frac{P'^2}{8Q'}} D_{\frac{-1+i}{2}}\left(\dfrac{P'}{\sqrt{2Q'}}\right) & \text{for} \quad i > -1, \quad Q' > 0 \, , \\[4mm] \dfrac{1}{2}(P')^{\frac{-1+i}{2}} \Gamma\left(\dfrac{1+i}{2}\right) & \text{for} \quad i > -1, \quad Q' = 0, \quad P' > 0 \, , \end{cases}$$

which has no advantage for the evaluation of c in comparison with (12.6, 7), but shows *generally* that, corresponding to the one-dimendional case of Chapter 10,

$$i > -1 \, ,$$

that is

$$(g - K_1 f)^2 > 4\Omega_1^2 B_1$$

is the integrability condition for the probability density. If this condition does not hold, the probability of positive amplitudes is zero and no vibration arises.

Under the assumption

$$c_{20}, c_{02} \quad \text{and} \quad c_{42} < \varrho \, ,$$

but

$$c_{40} = c_{22} = c_{24} = 0$$

taking into consideration besides the second order coefficients the greatest coefficient c_{42}, which corresponds better with the last example, (12.6, 7) reads

$$\frac{1}{c} = -\frac{1}{2}\int\limits_0^\infty \frac{A_1^i \, e^{c_{20}A_1^2}}{c_{02} + c_{42} A_1^4} \, \mathrm{d}A_1 \, .$$

The transformation

$$\sqrt{\frac{c_{42}}{c_{02}}} A_1^2 = x$$

and the integration formula (compare GRADŠTEJN and RYŽIK (1971), p. 337)

$$\int\limits_0^\infty \frac{x^{\nu-1}\, e^{-\varrho x}}{1 + x^2}\, dx = \frac{\pi V_\nu(2\varrho,0)}{\sin \pi\nu} \qquad (\nu > 0, \varrho > 0)$$

give

$$c = -\frac{4 c_{02}^{\frac{3-i}{4}}\, c_{42}^{\frac{1+i}{4}} \sin\left(\frac{1+i}{2}\pi\right)}{\pi V_{\frac{1+i}{2}}\left(-2 c_{20}\sqrt{\frac{c_{02}}{c_{42}}},\ 0\right)} \tag{12.6, 10}$$

where V_ν are Lommel functions of two variables which can be represented by

$$V_{\frac{1+i}{2}}(2\varrho, 0) = \frac{\sqrt{\varrho}}{\Gamma\left(\frac{1-i}{2}\right)}\, S_{-\frac{i}{2}, \frac{1}{2}}(\varrho) \tag{12.6, 11}$$

with the asymptotic development (compare GRADŠTEJN and RYŽIK (1971), p. 1000)

$$S_{-\frac{i}{2}, \frac{1}{2}}(\varrho) = \varrho^{-\frac{2+i}{2}} \sum_{\nu=0}^{n-1} \frac{(-1)^\nu\, \Gamma\left(\frac{3}{4}+\frac{i}{4}+\nu\right)\Gamma\left(\frac{1}{4}+\frac{i}{4}+\nu\right)}{\left(\frac{\varrho}{2}\right)^\nu \Gamma\left(\frac{3}{4}+\frac{i}{4}\right)\Gamma\left(\frac{1}{4}+\frac{i}{4}\right)} + O\left(\varrho^{-\frac{i}{2}-2n}\right)$$

$$\text{if} \quad i \neq -1, -3, -5, \dots \tag{12.6, 12}$$

The *probability* that the amplitudes A_1 and A_2 are greater than certain *levels* can be readily determined by help of the two-dimensional probability density $w(A_1, A_2)$.

The probability that A_1 is greater than a level α_1, as dependent on A_2, is

$$W(\alpha_1, A_2) = \int\limits_{\alpha_1}^\infty w\, dA_1\,,$$

while the corresponding probability that A_2 exceeds a level α_2 is

$$W(A_1; \alpha_2) = \int\limits_{\alpha_2}^\infty w\, dA_2\,.$$

The probability that a certain amplitude A_1 occurs, independently of A_2, is

$$W A_1) = \int\limits_0^\infty w\, dA_2\,,$$

the probability of a certain amplitude A_2 independently of A_1 is

$$W(A_2) = \int\limits_0^\infty w\, dA_1\,.$$

By combination and use of (12.6, 5) we get the probability that A_1 is greater than a level α_1, independently of A_2,

$$W_1(\alpha_1) = \int\limits_{\alpha_1}^{\infty}\int\limits_{0}^{\infty} w \, \mathrm{d}A_1 \, \mathrm{d}A_2 = 1 - \int\limits_{0}^{\alpha_1}\int\limits_{0}^{\infty} w \, \mathrm{d}A_1 \, \mathrm{d}A_2 \qquad (12.6, 13)$$

(Fig. 12.6, 3), the probability

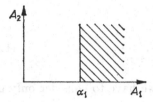

Fig. 12.6, 3

$$W_2(\alpha_2) = \int\limits_{0}^{\infty}\int\limits_{\alpha_2}^{\infty} w \, \mathrm{d}A_1 \, \mathrm{d}A_2 = 1 - \int\limits_{0}^{\infty}\int\limits_{0}^{\alpha_2} w \, \mathrm{d}A_1 \, \mathrm{d}A_2 \qquad (12.6, 14)$$

that A_2 exceeds α_2, independently of A_1 (Fig. 12.6, 4), and the probability

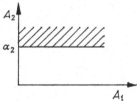

Fig. 12.6, 4

$$W_{12}(\alpha_1, \alpha_2) = \int\limits_{\alpha_1}^{\infty}\int\limits_{0}^{\infty} w \, \mathrm{d}A_1 \, \mathrm{d}A_2 + \int\limits_{0}^{\alpha_1}\int\limits_{\alpha_2}^{\infty} w \, \mathrm{d}A_1 \, \mathrm{d}A_2$$

$$= 1 - \int\limits_{0}^{\alpha_1}\int\limits_{0}^{\alpha_2} w \, \mathrm{d}A_1 \, \mathrm{d}A_2 \qquad (12.6, 15)$$

that A_1 exceeds a level α_1 and A_2 exceeds a level α_2 (Fig. 12.6, 5).

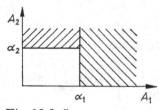

Fig. 12.6. 5

We assume the probability density takes the form

$$w = c A_1^i A_2 \, \mathrm{e}^{c_{20}A_1^2 + c_{02}A_2^2 + c_{40}A_1^4 + c_{22}A_1^2 A_2^2 + c_{42}A_1^4 A_2^2}.$$

By analogy with (12.6, 8) we find that

$$\int\limits_{\alpha_2}^{\infty} A_2 \, \mathrm{e}^{c_{02}A_2^2 + c_{22}A_1^2 A_2^2 + c_{42}A_1^4 A_2^2} \, \mathrm{d}A_2 = - \frac{1}{2} \frac{\mathrm{e}^{(c_{02} + c_{22}A_1^2 + c_{42}A_1^4)\,\alpha_2^2}}{c_{02} + c_{22}A_1^2 + c_{42}A_1^4} \qquad (12.6, 16)$$

if the expression in parentheses is negative, in particular that

$$\int_0^\infty A_2 \, e^{c_{02}A_2^2 + c_{22}A_1^2 A_2^2 + c_{42}A_1^4 A_2^2} \, \mathrm{d}A_2 = -\frac{1}{2(c_{02} + c_{22}A_1^2 + c_{42}A_1^4)} . \tag{12.6, 17}$$

The probability (12.6, 13) becomes, with (12.6, 17), if we neglect $c_{40}A_1^4$ in comparison with $c_{20}A_1^2$,

$$W_1(\alpha_1) = 1 + \frac{c}{2} \int_0^{\alpha_1} \frac{A_1^i \, e^{c_{20}A_1^2}}{c_{02} + c_{22}A_1^2 + c_{42}A_1^4} \, \mathrm{d}A_1 . \tag{12.6, 18}$$

This integration not being possible in closed form, we have to consider only one expression in the denominator.

For not too great values α_1, we can neglect $(c_{22}A_1^2 + c_{42}A_2^2) \, A_2^2$ in comparison with $c_{02}A_2^2$ respectively with $c_{20}A_1^2$ and get, because of (12.2, 5),

$$W_1(\alpha_1) = -\frac{c}{4c_{02}} (-c_{20})^{-\frac{1+i}{2}} \Gamma\left(\frac{1+i}{2}, -c_{20}\alpha_1^2\right) \tag{12.6, 19}$$

or, considering (12.6, 9),

$$W_1(\alpha_1) = \frac{\Gamma\left(\dfrac{1+i}{2}, -c_{20}\alpha_1^2\right)}{\Gamma\left(\dfrac{1+i}{2}\right)} . \tag{12.6, 20}$$

For numerical evaluation the development

$$\Gamma(\nu, \mu) = \Gamma(\nu) - \sum_{n=0}^\infty \frac{(-1)^n \, \mu^{\nu+n}}{n! \, (\nu + n)!} \quad (\nu \neq 0, \ -1, \ -2, \ ...)$$

corresponding to (10.2, 4) can be used. The asymptotic representation of $\Gamma(\nu, \mu)$ for great $|\mu|$

$$\Gamma(\nu, \mu) = \mu^{\nu-1} \, e^{-\mu} \left[1 + O\left(\frac{1}{\mu}\right)\right] \tag{12.6, 21}$$

(compare for instance GRADŠTEJN and RYŽIK (1971), p. 956) yields

$$W_1(\alpha_1) = \frac{e^{c_{20}\alpha_1^2}}{(-c_{20})^{\frac{1-i}{2}} \alpha_1^{1-i} \Gamma\left(\dfrac{1+i}{2}\right)} \left[1 + O\left(\frac{1}{c_{20}\alpha_1^2}\right)\right] \quad \text{for} \quad |c_{20}| \, \alpha_j^2 \gg 1 . \tag{12.6, 22}$$

The probability $W_{12}(\alpha_1, \alpha_2)$ becomes, if we neglect the same terms and use (12.2, 2), (12.2, 5), (12.6, 17),

$$W_{12}(\alpha_1, \alpha_2) = \left(1 - e^{c_{02}\alpha_2^2}\right) \frac{\Gamma\left(\dfrac{1+i}{2}, -c_{20}\alpha_1^2\right)}{\Gamma\left(\dfrac{1+i}{2}\right)} + e^{c_{02}\alpha_2^2} . \tag{12.6, 23}$$

Special cases of this formula are (12.6, 18) for $\alpha_2 = \infty$ and

$$W_2(\alpha_2) = e^{c_{02}\alpha_2^2}$$

for $\alpha_1 = \infty$. Analogously to (12.6, 21), (12.6, 22), formula (12.6, 23) admits an asymptotic representation for $|c_{20}| \alpha_1^2 \gg 1$.

If we write (12.6, 18) in the form

$$W_1(\alpha_1) = -\frac{c}{2} \int\limits_{\alpha_1}^{\infty} \frac{A_1^i \, e^{c_{20}A_1^2}}{c_{02} + c_{22}A_1^2 + c_{42}A_1^4} \, dA_1 \,,$$

we can, for greater values of α_1, neglect $c_{02} + c_{22}A_1^2$ in comparison with $c_{42}A_1^4$ and thus get instead of (12.6, 19)

$$W_1(\alpha_1) = -\frac{c}{4c_{42}} (-c_{20})^{\frac{3-i}{2}} \, \Gamma\left(\frac{i-3}{2}, -c_{20}\alpha_1^2\right).$$

Using (12.6, 10) with (12.6, 11) and the first term of the development (12.6, 12) as well as the asymptotic representation (12.6, 22) for $|c_{20}| \alpha_1^2 \gg 1$, we find that

$$W_1(\alpha_1) = \frac{c_{02}\Gamma\left(\dfrac{1-i}{2}\right)\sin\left(\dfrac{1+i}{2}\,\pi\right) e^{c_{20}\alpha_1^2}}{\pi c_{42}(-c_{20})^{\frac{1-i}{2}} \alpha_1^{5-i}}. \tag{12.6, 24}$$

An example for the probability W_1 is given in Fig. 12.6, 6 where, corresponding to the examples p. 393, $i = 0.2$, $c_{20} = -5.8$ and $c_{42} : c_{02} = 9.8625$. The dashed line is found from formula (12.6, 22), the full line from formula (12.6, 24), taking into consideration c_{42}. It shows that neglecting c_{42} yields probabilities of about the same order of magnitude. For values $\alpha_1 < 0.5$ the asymptotic formulae used do not hold.

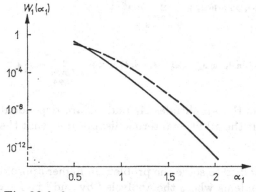

Fig. 12.6, 6

The probability $W_2(\alpha_2)$ is sketched in Fig. 12.6, 7 wirh $\Omega_2 = 7$ and different values of B_2/f^2. It shows how this probability diminishes with increasing damping B_2 respectively augments with increasing forced excitation f depending on the level α_2.

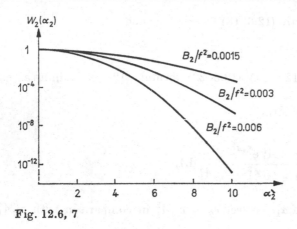

Fig. 12.6, 7

12.7. Application of computer algebra

The computer algebra methods sketched in Section 6.8 have been applied to the problem at hand in order both to check and to generalize the analytical results found. The developed program consists, according to the analytical methods, of four steps. Firstly it transforms the differential equations into the standard form. Secondly it realizes the iterative elimination of the fluctuating terms up to the approximation wanted. For this purpose the program expands amplitudes and phases into series with a small parameter. Thirdly the program realizes the averaging in the excitation terms. At last it provides the ansatzfunction chosen, inserts it into the Fokker Planck equation and determines, by means of coefficient comparison with respect to the powers of the amplitudes, the set of coupled equations for the different coefficients of the ansatzfunction.

The computer algebra program not only verified the above formulae, it also realized the generalized ansatzfunctions

$$w(A_1, A_2) = c\, e^{\, i\ln A_1 + j\ln A_2 + i_1 \ln A_1 \cdot \ln A_2 + \sum\limits_{\substack{k,l\ \text{even} \\ k+l \neq 6}} c_{kl} A_1^k A_2^l}$$

and

$$w(A_1, A_2) = c\, e^{\, i\ln A_1 + j\ln A_2 + i_1 \ln A_1 \cdot \ln A_2 + i_2 A_2^2 \ln A_1 + j_2 A_1^2 \ln A_2 + \sum\limits_{\substack{k,l\ \text{even} \\ k+l \neq 4}} c_{kl} A_1^k A_2}$$

which take into consideration that the amplitudes A_1 and A_2 are dependent also if one of them is small. It showed that the additional terms disappear so that the ansatz (12.5, 9) is confirmed.

The program also gives the possibility to solve the problem in higher approximation or to solve even more complex problems where the analysis "by and" is not possible or not suitable.

Other problems of coupled, especially of autoparametric vibrations as for instance vibrations of beam systems (frames) or vibrations of structures filled with liquid (containers) lead to similar equations of motion which can be investigated analogously (compare BARR (1969), BARR and McWHANNELL (1971), IBRAHIM and BARR (1975)).

Appendix

Method of obtaining phase trajectories in the phase plane

Consider a set of differential equations

$$\left.\begin{aligned} \frac{\mathrm{d}x}{\mathrm{d}t} &= X(x, y)\,, \\[2mm] \frac{\mathrm{d}y}{\mathrm{d}t} &= Y(x, y) \end{aligned}\right\} \tag{A, 1}$$

and assume that $X(x, y)$, $Y(x, y)$ are functions of x and y such that they satisfy the conditions of a unique solution in the given space x, y, t. With the exception of the singular points defined by the equations

$$X(x, y) = 0\,, \qquad Y(x, y) = 0 \tag{A, 2}$$

a single trajectory then passes through each point in the (x, y) plane. These assumptions are, for example, always satisfied if $X(x, y)$ and $Y(x, y)$ are analytical functions of x and y.

The calculation starts at a point of the phase plane whose coordinates are determined by the initial conditions $x(0) = x_0$, $y(0) = y_0$ (assuming that this point is not exactly a singular one); the slope of the trajectory passing through it is given by the equation

$$\frac{\mathrm{d}y}{\mathrm{d}x} = N = Y(x_0, y_0)/X(x_0, y_0)\,. \tag{A, 3}$$

(A, 1) implies

$$\mathrm{d}s = [(\mathrm{d}x)^2 + (\mathrm{d}y)^2]^{1/2} = [X^2 + Y^2]^{1/2}\,\mathrm{d}t\,. \tag{A, 4}$$

The coordinates of the next point on the trajectory are approximately obtained from the coordinates of the point located on the tangent to the trajectory at a distance $\mathrm{d}s$ from the starting point (Fig. A, 1). They are described by the equations

$$\left.\begin{aligned} x_1 &= x_0 + \mathrm{d}x = x_0 + \mathrm{d}s\ \mathrm{sgn}\ [X(x_0, y_0)]/(1 + N^2)^{1/2}\,, \\[2mm] y_1 &= y_0 + \mathrm{d}y = y_0 + \mathrm{d}s\ N\ \mathrm{sgn}\ [X(x_0, y_0)]/(1 + N^2)^{1/2}\,. \end{aligned}\right\} \tag{A, 5}$$

The calculations can be carried out using one of the following three procedures:

(1) A constant step along the trajectory, that is $\mathrm{d}s = M = \text{const.}$
(2) A constant time step, that is $\mathrm{d}t = M = \text{const.}$ We can then substitute in (A, 5)

$$\mathrm{d}s = [(\mathrm{d}x)^2 + (\mathrm{d}y)^2]^{1/2} = (X^2 + Y^2)^{1/2}\,M\,. \tag{A, 6}$$

(3) Combination of (1) and (2). Substituting in (A, 5)

$$\mathrm{d}s = (X^2 + Y^2)^K\,M \tag{A, 7}$$

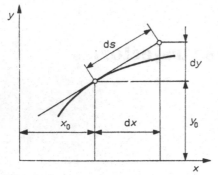

Fig. A, 1

we obtain from (A, 4)

$$dt = (X^2 + Y^2)^{K-1/2} M \qquad\qquad (A, 8)$$

where $0 < K < 0.5$.

The disadvantage of procedure (1) (a constant step along the trajectory, that is motion at a constant velocity) is its inability to provide clear information about the proximity of the singular point. In procedure (2) the proximity of the singular point is readily discerned (the motion of the plotter is slowed down); however, the step along the trajectory ds is very small in the vicinity of, and fairly large at some distance from the singular point. The third procedure, if properly set up, does away with the drawbacks while, at the same time, stressing the merits of the former two.

Below is a schematic program prepared for a Hewlett-Packard (type 9830 A) calculator. Since this calculator uses only capital letters, the following notation is introduced in the program:

$$x = A , \qquad y = B , \qquad dx = C , \qquad dy = D .$$

Note: To define $X(x, y)$ and $Y(x, y)$ which are denoted by symbols in the schematic program, it is necessary to substitute the appropriate expressions.

The procedure outlined below includes all three versions.

```
10   SCALE ...
20   S = ±1 (+ for positive, — for negative time)
30   M =
40   DISP „A=, B=";
50   INPUT A, B
60   PLOT A, B
70   X = S * X(A, B)
80   Y = S * Y(A, B)
90   N = Y/X
      1 C = M * SGNX/(SQR(1 + N * N))
100   2 C = M * SGNX * (SQR(X * X + Y * Y))/(SQR(1 + N * N))
      3 C = M * SGNX * ((X * X * Y * Y) ↑ K)/(SQR(1 + N * N))
110  D = N * C
120  A = A + C
130  B = B + D
140  GO TO 60
150  END
```

(* denotes multiplication, ↑ raising to a power)

The run is terminated by the command STOP.

The example which follows shows a schematic program for obtaining the time development of a dependent variable, for example, $x = x(t)$.

```
10  SCALE
20  S = ±1
30  M =
35  T = 0
40  DISP ,,A=, B=";
50  INPUT A, B
60  PLOT A, B
70  X = S * X(A, B)
80  Y = S * Y(A, B)
90  N = Y/X
    1 C = M * SGNX * (SQR(X * X + Y * Y))/(SQR(1 + N * N))
100 2 C = M * SGNX * (SQR(X * X + Y * Y))/(SQR(1 + N * N))
    3 C = M * SGNX * ((X * X + Y * Y) ↑ K)/(SQR(1 + N * N))
110 D = N * C
120 A = A + C
130 B = B + D
    1 T = T + M/SQR(X * X + Y * Y))
135 2 T = T + M
    3 T = T + M/((X * X + Y * Y) ↑ (0.5 − K)
140 G0 T0 60
150 END
```

The chief merit of the procedure just described is the simplicity of the program. In some cases, for example, when two stable singular points and a saddle point but no limit cycle exist in the phase plane, the procedure can be started from any point and the trajectory will always tend to one of the stable singular points regardless of the size of the step. This circumstance has no qualitative effect; a similar statement can hardly be made of other methods.

The size of the step must usually be reduced when the trajectories are very dense, for example, when the stable cycle lies close to the unstable one. In cases of this sort, reduction of the size of the step alone is frequently not enough and the progress of the solution must be checked by repeating the calculation at both positive and negative time. Difficulties are sometimes encountered with slightly damped systems, especially when the damping is non-linear so that virtually no damping exists in the neighbourhood of the equilibrium position.

The method of solution of trajectories in the phase space in which the system is defined by the equations

$$\frac{dx}{dt} = X_j(x_1, \ldots, x_n) \qquad (j = 1, 2, \ldots, n), \tag{A, 9}$$

which satisfy similar conditions of unique solution in the $(x_1, \ldots, x_n, t)$ space, resembles that described by (A, 1). The resulting relations similar to those obtained for (A, 1) are as follows:

$$ds = [\sum_{j=1}^{n} (dx_j)^2]^{1/2} = dt \, (\sum_{j=1}^{n} X_j)^{1/2}, \tag{A, 10}$$

26*

$$\frac{x_k}{x_1} = N_k, \qquad (N_1 = 1; k = 2, 3, \ldots, n),$$

$$(A, 11)$$

$$dx_k = N_k \, dx_1,$$

$$(A, 12)$$

$$dx_1 = dt \, \text{sgn} \, (X_1) \, (\sum_{j=1}^{n} X_1^2)^{1/2} / (1 + \sum_{k=2}^{n} N_k^2)^{1/2}.$$

$$(A, 13)$$

Bibliography

ANDRONOV, A. A.; VITT, A. A.; CHAJKIN, S. E.
(Андронов, А. А.; Витт, А. А.; Хайкин, С. Е.)
1959 Теория колебаний, второе изд., Гостехиздат, Москва.
(Engl. transl. Theory of oscillations, Pergamon Press, Oxford etc. 1966).

ARNOLD, L.
1973 Stochastische Differentialgleichungen. Theorie und Anwendung, R. Oldenbourg Verlag, München—Wien.

BABICKIJ, V. I. (Бабицкий, В. И.)
1978 Теория виброударных систем, Наука, Москва.

BARR, A. D. S.
1969 Dynamic instabilities in moving beams and beam systems, Proc. 2nd Int. Congr. Theory Mach. Mechanisms, Zakopane, Vol. 1, 365.

BARR, A. D. S.; McWHANNELL, D. C.
1971 Parametric instability in structures under support motion, J. Sound Vib. 14, 491 to 509.

BAXTER, K. G.
1971 The nonlinear response of mechanical systems to parametric random excitation, Ph. D. Thesis, Syracuse Univ., New York.

BECKER, L.
1972 Experimentelle und numerische Untersuchung von Kombinationsresonanzen, Diss. TU Karlsruhe.

BENZ, G.
1962 Schwingungen nichtlinearer gedämpfter Systeme mit pulsierenden Speicherkennwerten, Diss. TH Karlsruhe.
1965 Ein Beitrag zur Berechnung der Resonanzkurven aus nichtlinearen Bewegungsgleichungen, ZAMM 45, T 101—T 104.

BLAQUIERE, A.
1966 Nonlinear system analysis, Academic Press, New York—London.

BLECHMAN, I. I. (Блехман, И. И.)
1971 Синхронизация динамических систем, Наука, Москва.
1981 Синхронизация в природе и технике, Наука, Москва.

BOGOLJUBOV, N. N.; MITROPOL'SKIJ, J. A. (Боголюбов, Н. Н.; Митропольский, Ю. А.)
1963 Асимптотические методы в теории нелинейных колебаний, Изд. физ.-мат. лит., Москва.
(Germ. transl. Asymptotische Methoden in der Theorie der nichtlinearen Schwingungen, Akademie-Verlag, Berlin 1965).

BOGUSZ, W.
1966 Stability of non-linear systems (in Polish), Państwowe Wydawnictwo Naukowe, Warszawa.

BOLOTIN, V. V. (Болотин, В. В.)
1956 Динамическая устойчивость упругих систем, Гостехиздат, Москва.
(German transl. Kinetische Stabilität elastischer Systeme, VEB Deutscher Verlag der Wissenschaften, Berlin 1961;
Engl. transl. The dynamic stability of elastic systems, Holden Day, San Francisco 1964).

1979 Случайные колебания упругих систем, Наука, Москва.

BONDAR', N. G. (Бондарь, Н. Г.)
1978 Нелинейные колебания, возбуждаемые импульсами, Вища школа, Киев, Донецк.

BOSCH, M.
1965 Über das dynamische Verhalten von Stirnradgetrieben unter besonderer Berücksichtigung der Verzahnungsgenauigkeit, Diss. TH Aachen.

BULGAKOV, B. V. (Булгаков, Б. В.)
1954 Колебания, Гостехиздат, Москва.

BUTENIN, N. V. (Бутенин, Н. В.)
1962 Элементы теории нелинейных колебаний, Судпром, Ленинград.

CAUGHEY, T. K.
1971 Nonlinear theory of random vibrations, Advances in Applied Mechanics, Vol. 11.

CESARI, L.
1963 Asymptotic behavior and stability problems in ordinary differential equations, 2nd ed., Springer-Verlag, Berlin—Göttingen—Heidelberg.

ČETAEV, N. G. (Четаев, Н. Г.)
1955 Устойчивость движения, Гостехиздат, Москва.

CHARKEVIČ, A. A. (Харкевич, А. А.)
1953 Автоколебания, Гостехиздат, Москва.
1956 Нелинейные и параметрические явления в радиотехнике, Гостехиздат, Москва.

CHAS'MINSKIJ, R. Z. (Хасьминский, Р. З.)
1969 Устойчивость сустем дифференциальных уравнений при случайных возмущениях их параметров, Наука, Москва.

CRANDALL, S. H.; MARK, W. D.
1963 Random vibrationi n mechanical systems, Academic Press, New York and London.

DIMENTBERG, F. M. (Диментберг, Ф. М.)
1959 Изгибные колебания вращающихся валов, Изд. АН СССР, Москва.
(Eng. transl. Flexural vibrations of rotating shaft, Butterworth and Co., London 1961).

DIMENTBERG, M. F. (Диментберг, М. Ф.)
1980 Нелинейные стохастические задачи механических колебаний, Наука, Москва.

DIMENTBERG, M. F.; GORBUNOV, A. A. (Диментберг, М. Ф,; Горбунов, А. А.)
1975 Некоторые задачи диагностики колебательной системы со случайным параметрическим возбуждением, Прикл. мех., Отд. мат., мех. и киб., АН СССР, 4, 71-75, XI.

DIMENTBERG, M. F.; ISIKOV, N. E.; MODEL, R.
(Диментберг, М. Ф.; Исиков, Н. Е.; Модель, Р.)
1981 Колебания системы с кубически-нелинейным демпфированием при одновременном периодическом и случайном параметрическом возбуждении, Известия АН СССР, Механика твёрдого тела, 36, 22—24.

DOWELL, E. H.
1981 Non-linear oscillator models in bluff body aero-elasticity, J. Sound Vib. 75, 251 to 264.

EBELING, W.; ENGEL-HERBERT, H.
1982 Stochastic theory of kinetic transitions in nonlinear mechanical systems; Успехи Механики (Advances in Mechanics) 5, Number 3/4, 41-60.

EBELING, W; HERZEL, H.; RICHERT, W.; SCHIMANSKY-GEYER, L.
1986 Influence of noise on Duffing—Van der Pol oscillators, ZAMM 66 (to appear).

EICHER, N.
1981 Einführung in die Berechnung parametererregter Schwingungen, Dokumentation Weiterbildung TU Berlin.

EVAN-IWANOWSKI, R. M.
1976 Resonance oscillations in mechanical systems, Elsevier Sc. Publ. Comp., Amsterdam —Oxford—New York.

FIALA, V.
1976 Solution of non-linear vibration systems by means of analogue computers, Monographs and Memoranda, Nat. Res. Inst. for Machine Design, Nr. 19, Běchovice.

FIALA, V.; TONDL, A.
1974 Contribution to the solution of the Duffing equation (in Czech), ARITMA Computing Technique, No. 3, 23—32.

FROLOV, K. V.; FURMAN, F. A. (Фролов, К. В.; Фурман, Ф. А.)
1980 Прикладная теория виброзащитных сустем, Машиностроение, Москва.

GICHMAN, I. I.; SKOROCHOD, A. V. (Гихман, И. И.; Скороход, А. В.)
1968 Стохастические дифференциальные уравнения, Наукова думка, Киев.
1975 Теория случайных процессов, т. 3, Наука, Москва.

GRADŠTEJN, I. S.; RYŽIK, I. M. (Градштейн, И. С.; Рыжик, И. М.)
1971 Таблицы интегралов, сумм, рядов и произведений, Изд. пятое,

GRIGOR'EV, N. V. (Григорьев, Н. В.)
1961 Нелинейные колебания элементов машин и сооружений, Гостехиздат, Москва— Ленинград.

GROBOV, V. A. (Гробов, В. А.)
1961 Асимптотические методы расчёта изгибных колебаний валов турбомашин, Изд. АН СССР, Москва.

GUCKENHEIMER, J.; HOLMES, PH.
1983 Nonlinear oscillations, dynamical systems and bifurcation of vector fields, Springer-Verlag, New York—Berlin—Heidelberg— Tokyo.

HAAG, J.; CHALEAT, R.
1960 Problèmes de théorie générale des oscillations et de chronométrie, Gauthier-Villars, Eyrolles, Paris.

HAGEDORN, P.
1978 Nichtlineare Schwingungen, Akademische Verlagsgesellschaft, Wiesbaden.

HAKEN, H.
1982 Synergetik, Springer-Verlag, Berlin—Heidelberg—New York.

HAXTON, R. S.; BARR, A. D. S.
1972 The autoparametric vibration absorber, Trans. ASME, 119—125.

HAYASHI, CH.
1964 Nonlinear oscillations in physical systems, McGraw-Hill, New York.

HOLMES, P. J. (Ed.)
1980 New approaches to nonlinear problems in dynamics, SIAM, Philadelphia.

HORN, J.
1948 Gewöhnliche Differentialgleichungen, 5. erw. Aufl., Walter de Gruyter u. Co., Berlin (West).

HORTEL, M.
1968 Forced vibrations in weakly non-linear parametric gear systems with several degrees of freedom (in Czech), Strojnicky časopis 19, 414—432.

1969 Analysis of forced vibrations in weakly non-linear parametric gear systems with several degrees of freedom (in Czech), Strojnicky časopis 20, 35—57.

1970 Einfluß der Lagerelastizität auf die Amplituden-Frequenzcharakteristik eines Zahnradsystems, Strojnicky časopis 21, 598—620.

HORTEL, M.; SCHMIDT, G.
1979 Gedämpfte und selbsterregte Schwingungen in einem nichtlinearen parametererregten System mit kinematischen Kopplungen, Proc. 8th Int. Conf. Nonl. Osc., Academia, Prague, Vol. I, 331—340.
1981 Untersuchungen von Parameternichtlinearitäten bei Übersetzungsgetrieben, ZAMM, Vol. I, 331—34 61, 21—28.
1983 Frequenzmitnahme bei zwangs- und selbsterregten mechanischen Schwingungen, ZAMM 64, 23—30.

IBRAHIM, R. A.; BARR, A. D. S.
1975 Autoparametric resonance in a structure containing a liquid, J. Sound Vib. 42, 159—179, 181—200.

IBRAHIM, R. A.; ROBERTS, J. W.
1976 Broad band random excitation of a two-degree-of-freedom system with autoparametric coupling, J. Sound Vib. 44, 335—348.
1977 Stochastic stability of the stationary response of a system with autoparametric coupling, ZAMM 57, 643—649.

KAMKE, E.
1959 Differentialgleichungen, Lösungsmethoden und Lösungen, 6. Aufl., Akademische Verlagsgesellschaft Geest und Portig K.-G., Leipzig.

KARAČAROV, K. A.; PILJUTIK, A. G. (Карачаров, К. А.; Пилютик, А. Г.)
1962 Введение в техническую теорию устойчивости движения, Гос. издат. физ.-мат. литературы, Москва.

KAUDERER, H.
1958 Nichtlineare Mechanik, Springer-Verlag, Berlin—Göttingen—Heidelberg.

KAZAKEVICH, V. V. (Казакевич, В. В.)
1974 Автоколебания в компрессорах, Машиностроение, Москва.

KEL'ZON, A. S.; JAKOVLEV, V. I. (Кельзон, А. С.; Яковлев, В. И.)
1971a Сужение зоны автоколебаний нагруженного вала, вращающегося в подшипниках скольжения, Известия АН СССР, Механика твёрдого тела, 5, 36—43.
1971b Переход через зону автоколебаний вертикального вала с учётом сил инерции смазки, Доклады АН СССР, 2, 289—292.

KLOTTER, K.
1980 Technische Schwingungslehre, Erster Band: Einfache Schwinger, 3., völlig neubearb. u. erw. Aufl., herausgeg. mit Unterstützung durch G. BENZ, Teil B: Nichtlineare Schwingungen, Springer-Verlag, Berlin—Heidelberg—New York.

KLOTTER, K.; KOTOWSKI, G.
1939 Über die Stabilität der Bewegung des Pendels mit oszillierendem Aufhängepunkt, ZAMM 19, 289—296.

KOBRINSKIJ, A. E.; KOBRINSKIJ, A. A. (Кобринский, А. Е.; Кобринский, А. А.)
1973 Виброударные системы, Наука, Москва.

KOLOVSKIJ, M. Z. (Коловский, М. З.)
1966 Нелинейная теория виброзащитных систем, Наука, Москва.

KONONENKO, V. O. (Кононенко, В. О.)
1964 Колебательные системы с ограниченным возбуждением, Наука, Москва.

KOTEK, Z.; KUBÍK, S.
1962 Non-linear circuits (in Czech), SNTL, Praha.

KROPAČ, O.
1972 Relations between distributions of random vibratory processes and distributions of their envelopes, Apl. Matematiky 17, No. 2, 75—112.

KÜHNLENZ, J. (Кюнленц, Ю.)
1979 Об одом асимптотическом методе построения стохастических уравнений колебаний высших приближений и его применение, в кн. Асимптотические методы нелинейной механики, 66—79, Киев.

Kušul, M. Ja. (Кушул, М. Я.)
1963 Автоколебания роторов, Наука, Москва.

Landa, P. S. (Ланда, П. С.)
1980 Автоколебания в системах с конечным числом степеней свободы, Наука, Москва.

Lennox, W. C.; Kuak, Y. C.
1976 Narrow-band excitation of a nonlinear oscillator, Trans. ASME, June 1976, 340—344.

Linke, H.
1970 Untersuchungen zur Ermittlung dynamischer Zahnkräfte von einstufigen Stirnradgetrieben mit Geradeverzahnung, Diss. TU Dresden.

Ljapunov, A. M. (Ляпунов, А. М.)
1950 Общая задача об устойчивостидвижения, Гостехиздат, Москва—Ленинград.

Lur'e, A. I. (Лурье, А. И.)
1951 Некоторые нелинейные задачи теории автоматического регулирования, Гостехиздат, Москва—Ленинград.

Magnus, K.
1961 Schwingungen, B. G. Teubner, Stuttgart.

Magnus, W.; Oberhettinger, F.
1943 Formeln und Sätze für die speziellen Funktionen der mathematischen Physik, Springer-Verlag, Berlin—Göttingen—Heidelberg.

Malkin, I. G. (Малкин, И. Г.)
1952 Теория устойчивости движения, Наука, Москва.
(Germ. transl. Theorie der Stabilität einer Bewegung, Akademie-Verlag, Berlin 1959).
1956 Некоторые задачи теории нелинейных колебаний, Гостехиздат, Москва.

Mansour, W. M.
1972 Quenching of limit cycles of a van der Pol oscillator, J. Sound Vib. 25, 395—405.

Massa, E.
1967 On the instability of parametrically excited two degrees of freedom vibrating systems with viscous damping, Meccanica 2, 243—255.

McLachlan, N. W.
1947 Theory and application of Mathieu functions, Oxford University Press, Oxford.
1950 Ordinary nonlinear differential equations in engineering and physical sciences, Clarendon Press, Oxford.

McQueen, D. H.
1976 On the dynamics of compressor surge, J. Eng. Sc. 18, 234—238.

Merker, H.
1981 Über den nichtlinearen Einfluß von Gleitlagern auf die Schwingungen von Rotoren Fortschritts-Berichte der VDI Zeitschriften, VDI-Verlag, Düsseldorf.

Mettler, E.
1949 Allgemeine Theorie der Stabilität erzwungener Schwingungen elastischer Körper, Ing.-Arch. 17, 418—449.
1965 Schwingungs- und Stabilitätsprobleme bei mechanischen Systemen mit harmonischer Erregung, ZAMM 45, 475—484.
1970 Kinetische Instabilität eines elastischen Trägers unter Parametererregung durch rotierende Unwuchten, Ing.-Arch. 39, 171—186.

Minorsky, N.
1947 Introduction to non-linear mechanics, I. W. Edwards, Ann Arbor.

Mitropol'skij, Ju. A. (Митропольский, Ю. А.)
1971 Метод усреднения в нелинейной механике, Наукова думка, Киев.

MITROPOL'SKIJ, JU. A.; KOLOMIEC, V. G. (Митропольский, Ю. А.; Коломиец, В. Г.)
1976 Применение асимптотических методов в стохастических системах, в кн. Приближённые методы исследования нелинейных систем, Киев.

MODEL, R.
1978a Untersuchung nichtlinearer Zufallsschwingungen, Diss. TH Magdeburg.
1978b Kombinationsresonanz eines stochastisch erregten Schwingungssystems, ZAMM 58, 377—382.

MOLERUS, O.
1963 Laufunruhige Drehzahlbereiche mehrstufiger Stirnradgetriebe, Diss. TH Karlsruhe.

MÜLLER, P. C.; SCHIEHLEN, W. O.
1976 Lineare Schwingungen, Akademische Verlagsgesellschaft, Wiesbaden.

NAYFEH, A. H.; MOOK, D. T.
1979 Nonlinear oscillations, John Wiley and Sons, New York.

NEJMARK, JU. I. (Неймарк, Ю. И.)
1972 Метод точечных отображений в теории нелинейных колебаний, Наука, Москва.

NIKOLAENKO, N. A. (Николаенко, Н. А.)
1967 Вероятностные методы динамического расчета машиностроительных конструкций, Машиностроение, Москва.

NISHIKAWA, Y.
1964 A contribution to the theory of nonlinear oscillations, Nippon Printing and Publishing Comp., Osaka.

NOVÁK, M. (Новак, М.)
1963 Анализ экспериментальных нелинейных резонансных кривых. Труды международного симпозиума по нелинейным колебаниям, III, 305—311, Изд. АН УССР, Киев.

PANOVKO, JA. G.; GUBANOVA, I. I. (Пановко, Я. Г.; Губанова, И. И.)
1979 Устойчивость и колебания упругих систем, Наука, Москва.

PARKS, P. C.; TONDL, A.
1979 Non-linear oscillations in wave power machines, Proc. 7th Int. Conf. Nonlin. Osc., Academia, Praha, Vol. I, 69—86.

PETERKA, F.
1970 Theory of motion of a two-mass mechanical system with impacts and its application to a dynamic impact damper (in Czech), Academia, Praha.
1981 Introduction into vibrations of mechanical systems with internal impacts (in Czech), Academia, Praha.

PHILIPPOW, E.
1963 Nichtlineare Elektrotechnik, Akademische Verlagsgesellschaft Geest und Portig K.-G., Leipzig.

PISARENKO, G. S. (Писаренко, Г. С.)
1955 Колебания упругих систем с учётом рассеяния энергии в материале, Издат. АН УССР, Киев.

POPP, K.
1982 Chaotische Bewegungen beim Duffing-Schwinger, Festschrift Prof. Magnus, München, 269—296.

PŮST, L.; TONDL, A.
1956 Introduction to the theory of non-linear and quasiharmonic vibrations of mechanical systems (in Czech), Academia, Praha.

RAGUL'SKENE, V. L. (Рагульскене, В. Л.)
1974 Виброударные системы, Минтис, Вильнюс.

RAGUL'SKIS, K. M. (Рагульскис, К. М.)
1963 Механизмы на вибрирующем основании, Каунас.

RIEGER, N. F.
1980 Stability of rotors in bearings, Lectures Rotor Dynamics Colloquium CISM, Udine.

RUDOWSKI, J.
1979 Two-frequency limit cycles in self-excited vibration systems (in Polish), PAN, Warszawa.

RUDOWSKI, J.; SZEMPLINSKA-STUPNICKA, W.
1977 Two-frequency limit cycles in self-excited vibration systems (in Polish), PAN, Warszawa.

SCHMIDT, G.
1961 Über die Biegeschwingungen des gelenkig gelagerten axial pulsierend belasteten Stabes, Math. Nachr. 23, 75—132.
1963 Mehrfache Verzweigungen bei gelenkig gelagerten längs pulsierend belasteten Stäben, Math. Nachr. 26, 25—43.

1965a Die Wechselwirkung erzwungener und parametererregter Schwingungen bei flachen Schalen, Rev. Méc. Appl. 10, 47—78.
1965b Das Zusammenwirken von Quer- und Längsresonanzen bei Stäben, Arch. Mech. Stos. 17, 233—247.
1965c Nonlinear parametric vibrations of sandwich plates, Proc. Vibr. Problems 6, 209—228.
1965d Zum Verhältnis von Resonanzbreiten und Maximalamplituden bei parametererregten Schwingungen, Abh. dt. Akad. Wiss., Klasse Math., Phys. u. Techn., 234—243.
1967a On the stability of combination oscillations, Proc. Vibr. Problems 8, 35—45.
1967b Über Schüttelschwingungen bei Motoren, Publ. Inst. Math., Belgrad 7, 111—122.
1967c Instabilitätsbereiche bei rheolinearen Schwingungen, Monatsber. dt. Akad. Wiss. Berlin 9, 405—411.
1967d Über nichtlineare Drehschwingungen rotierender Wellen, Rev. Méc. Appl. 12, 527—541.
1969a Resonanzlösungen nichtlinearer Schwingungsgleichungen, Nova Acta Leopoldina 34, Nr. 188, 1—60.
1969b Zur Stabilität der Resonanzlösungen nichtlinearer Schwingungsgleichungen, Nova Acta Leopoldina 34, Nr. 188, 61—68.
1973 On the dynamic stability of systems with a finite number of degrees of freedom, Eq. diff. et fonctionelles non linéaires, Paris, 489—505.
1974 Parametrically excited nonlinear vibrations, Lectures Int. Centre Mech. Sciences (CISM), Udine, 1—100.
1975 Parametererregte Schwingungen, VEB Deutscher Verlag der Wissenschaften, Berlin (Russ. Transl. Мир, Москва 1978).
1977a Probability densities of parametrically excited random vibrations, Proc. IUTAM-Symp. Stochastic problems in dynamics, 1976, Pitman, London—San Francisco—Melbourne.
1977b Vibrating mechanical systems with random parametric excitation, Proc. 14th IUTAM Congress, Delft 1976, Theor. and Appl. Mech., ed. by W. T. KOITER, Amsterdam—New York—Oxford, 439—450 (Russ. transl. Moscow 1979, 684—694).
1978 Nonlinear systems under random and periodic parametric excitation, Ber. 7. Int. Konf. Nichtlin. Schwing., Berlin 1975, Abh. Akad. Wiss. DDR, Nr. 6, 341—359.
1979 Forced and parametrically excited nonlinear random vibrations, Proc. 8th Int. Conf. Nonlin. Oscill., Prague 1978, Academia, Praha, Vol. II, 633—638.
1981a Schwingungen unter gleichzeitiger zufälliger Zwangs- und Parametererregung, ZAMM 61, 409—419.
1981b Параметрически возбуждаемые случайные колебания в нелинейных механических системах, Успехи механики 4, вып. 2, 63—88.
1984 Interaction of self-excited, forced and parametrically excited vibrations, Proc. 9th Int. Conf. Nonlin. Osc., Kiev 1981, Vol. 3, 310—315.

1986 Onset of chaos and global analytical solutions for Duffing's oscillator, ZAMM 66 (to appear).

SCHMIDT, G.; SCHULZ, R.
1982 Nonlinear random vibrations of systems with several degrees of freedom, Proc. IUTAM Symp. Random Vibr. and Reliab., Frankfurt 1982,
1983 Analytische und numerische Bestimmung der Schwingungen in Zahnradgetrieben, Tagung Zahnradgetriebe Dresden, 239—245.

SCHMIDT, G.; WEIDENHAMMER, F.
1961 Instabilitäten gedämpfter rheolinearer Schwingungen, Math. Nachr. 23, 301—318.

SCHMIDT, G.; WENZEL, V.
1984 Zufallsschwingungen unter schmalbandiger Parameter- und Zwangserregung, ZAMM 64 (1984).

SCHMIEG, H.
1976 Kombinationsresonanz bei Systemen mit allgemeiner harmonischer Erregermatrix, Tagung Zahnradgetriebe Dresden, 239—245. Diss. Univ. Karlsruhe.

SCHULZ, R.
1983 Verwendung der Formelmanipulation zur Aufstellung und Lösung bestimmter Schwingungsgleichungen der Mechanik. In: Künstliche Intelligenz/analytische Arbeitstechniken, Weiterbildungszentrum für mathematische Kybernetik und Rechentechnik der TU Dresden, Heft 65/83.

1985 Ein Programmsystem zur automatisierten Schwingungsberechnung, Forschungsbericht, Institut für Mechanik der AdW der DDR.

1986 Analytische Berechnung der Drehschwingungen mehrstufiger Stirnradgetriebe, ZAMM 66 (to appear).

SCHULZ, R.; FRIEDRICH, G.
1985 Innere dynamische Zahnkräfte in hochtourigen Planetengetrieben, Maschinenbautechnik (to appear).

SERGEEV, S. I. (Сергеев, С. И.)
1959 Демпфирование механических колебаний, Гос. издат. физ.-мат. литературы, Москва.

SKALICKÝ, A.
1979 Nichtlineare Schwingungen des Fördermediums von Kreiselpumpen im Leitungs-, system, Proc. 8th Int. Conf. Nonlin. Osc. 1978, Academia, Prague 1979, Vol. II 633—658.

SPARROW, C.
1982 The Lorenz equations: Bifurcation, chaos, and strange attractors, Springer-Verlag, Berlin—New York—Heidelberg.

STARŽINSKIJ, V. M. (Старжинский, В. М.)
1977 Прикладные методы нелинейных колебаний, Наука, Москва.

STOKER, J. J.
1950 Nonlinear vibrations in mechanical and electrical systems, Interscience publishers, New York, London.

STRATONOVICH, R. L. (Стратонович, Р. Л.)
1961 Избранные вопросы теории флуктуаций в радиотехнике, Советское радио, Москва.
(Engl. transl. Topics in the theory of random noise, Gordon and Breach, New York—London—Paris 1967).

SVAČINA, J.; FIALA, V.
1980 Verification of limit envelopes method for damping identification (in Czech), Strojnicky časopis 31, 319—330.

TONDL, A.

1961 Experimental investigation of self-excited vibrations of rotors due to the action of lubricating oil film in journal bearings, Monographs and Memoranda No. 1, Nat. Res. Inst. for Machines Design, Prague.

1965 Some problems of rotor dynamics, Academia, Praha/Chapman and Hall Ltd., London.

1967a The effect of the out-of-roundness of journal on rotor and bearings dynamics, Acta Technica ČSAV, 12, No. 1, 62—79.

1967b To the dynamics of mining chain conveyors (in Czech), Strojnicky časopis 18, 197—206.

1967c On resonance vibrations of one-mass non-linear systems with two degrees of freedom (in Czech), Rozpravy ČSAV, ř. techn. věd, 77, No. 4.

1970a Domains of attraction for non-linear systems, Monographs and Memoranda, Nat. Res. Inst. for Machine Design, No. 8, Běchovice.

1970b Self-excited vibrations, Monographs and Memoranda, Nat. Res. Inst. for Machine Design, No. 9, Běchovice.

1970c The effect of elastic foundation and damping on the limit of onset of self-excited vibrations of rotors mounted in externally pressurized bearings, Proc. 6th Conf. on Machine Dynamics, Ústav mechaniky strojov SAV, Bratislava.

1971a The effect of an elastically-suspended foundation mass and its damping on the initiation of self-excited vibrations of a rotor mounted in airpressurized bearings, Gas Bearing Symposium proc., Univ. of Southampton, paper 1, 1—15.

1971b Notes on the paper "Effects of non-linearity due to large deflections in the resonance testing of structures", J. Sound Vib. 17, 429—436.

1973a Analysis of stability of steady — locally stable — solutions for not fully determined disturbances, Monographs und Memoranda, Nat. Res. Inst. for Machine Design, No. 15, Běchovice.

1973b Нелинейные колебания механических систем, Мир, Москва.

1973c Vibration of rigid rotors with vertical shaft mounted in aerostatic bearings, Monographs and Memoranda No. 14, Nat. Res. Inst. for Machine Design, Běchovice.

1973d Some properties of non-linear system characteristics and their application to damping identification. Acta technica ČSAV 18, 166—179.

1974a Some problems of self-excited vibration of rotors, Monographs and Memoranda No. 17, Nat. Res. Inst. for Machine Design, Běchovice.

1974b Notes on the solution of forced oscillations of a third-order non-linear system, J. Sound Vib. 37, 273—279.

1975a Quenching of self-excited vibrations: Equilibrium aspects, J. Sound Vib. 42, 251 to 260.

1975b Quenching of self-excited vibrations: One- and two-frequency vibrations, J. Sound Vib. 42, 261—271.

1975c The application of skeleton curves and limit envelopes to analysis of non-linear vibration, The Shock and Vibration Digest 7, July 1975, No. 7, 3—20.

1976a On the interaction between self-excited and forced vibrations, Monographs and Memoranda No. 20, Nat. Res. Inst. for Machine Design, Praha.

1976b Parametric vibration of a non-linear system, Ing.-Arch. 45, 317—324.

1976c On the stability of steady vibration of Duffing system with degressive characteristic (in Czech), Strojnicky časopis 27, Part I — Stability investigation for small disturbances, 344—352; Part II — Stability investigation for not small disturbances, 577 to 589.

1976d Quenching of self-excited vibrations: Effect of dry friction, J. Sound Vib. 45, 285—294.

1977a Excited vibration of fourth-order non-linear systems, Acta technica ČSAV 22, 480—499.

1977b Application of tuned absorbers to self-excited systems with several masses, Proc. 9th Conf. on Machine Dynamics, Czechoslovak. Acad. Sc., Inst. of Thermomechanics, September 1977.

1978a Skeleton curves and limit envelopes for non-linear systems with several degrees of freedom (in Czech), Strojnicky časopis 29, 633—641.

1978b On the interaction between self-excited and parametric vibrations, Monographs and Memoranda of the Nat. Res. Inst for Machine Design No. 25, Praha.

1978c Solution of phase trajectories (in Czech), Strojnicky časopis 29, 46—56.

1979a Limit envelopes of systems with non-symmetric characteristics (in Czech), Strojnicky časopis 30, 658—665.

1979b On the dynamics of compressor surge, Int. J. of Non-Linear Mechanics 14, 259—266.

1979c Автоколебания механических систем, Мир, Москва.

1980a Determination of the limit of initiation of self-excited vibration of rotors, Int. J. Non-Linear Mechanics 15, 417—428.

1980b Quenching of self-excited vibrations (in Czech), Academia, Praha.

1980c A dynamic model of a system with viscoelastic restoring force (in Czech), Strojnicky časopis 31, 141—145.

1980c Beitrag zur Theorie des Pumpens, Ing.-Arch. 49, 255—260.

1981a On the dynamics of a compressor or centrifugal pump system, Monographs and Memoranda No. 32, Nat. Res. Inst. for Machine Design, Praha.

1982 Zum Problem des gegenseitigen Einflusses von selbsterregten und fremderregten Schwingungen, ZAMM 62, 103—113.

1984 On a special case of inertial excitation of nonlinear systems, Proc. 9th Int. Conf. Nonlin. Oscill., Kiev 1981, Vol. 3, 246—250.

TONDL, A.; BACKOVÁ, I. (ТОНДЛ, А.; БАЦКОВА, И.)
1968 К анализу вынужденных колебаний системы с периодически изменяющимися коэффициентами,
Recue Roumaine des Sc. Techn. — Mécanique Appliquée 13, 113—119.

TROGER, H.
1982 Über chaotisches Verhalten einfacher mechanischer Systeme, ZAMM 62, T 18—T 27.

1984 A simple nonlinear mechanical oscillator with strange behaviour, Proc. 9th Int. Conf. Nonlin. Oscill., Kiev 1981, Vol. 3, 481—483.

UEDA, Y.
1968 Some problems in the theory of nonlinear oscillations, Nippon Printing and Publishing Comp., Osaka.

VALEEV, K. G. (ВАЛЕЕВ, К. Г.)
1963 Об опасности комбинационных резонансов, Прикл. мат. мех. 27, 1134—1142.

Vibracii v technike (Вибрации в технике)
1978ff. In 6 volumes, edited by V. N. ČELOMEJ, E. J. GRIGOLJUK, K. V. FROLOV and others, especially vol. 1: Колебания линейных систем, ed. by V. V. Bolotin, 1978, vol. 2: Колебания нелинейных механических систем, ed. by I. I. BLECHMAN, 1979 and vol. 3: Колебания машин, конструкций и их элементов; ed. by F. M. DIMENTBERG and K. S. KOLESNIKOV, 1980, Mašinostroenie Moskva.

VOJTÍŠEK, S.; JANÁČ, K.
1969 Solution of non-linear systems, Academia, Praha.

WEDIG, W.
1978 Moments and probability densities of parametrically excited systems and continuous systems, Ber. 7. Int. Konf. Nichtlin. Schwing., Berlin 1975, Abh. AdW DDR, N6, 469—492.

WEIDENHAMMER, F.
1955 Rheolineare Drehschwingungen in Kolbenmotoren, Ing.-Arch. 23, 262—269.

1966 Parametererregte Schwingungen ausgewuchteter Rotoren, ZAMM 46, T 145—T 148.

WENZEL, V.
1978 Zur Theorie parametererregter Schwingungen und ihre Anwendung auf Platten- und Schalenprobleme, Diss. TH Magdeburg.

YUNG-CHEN LU
1976 Singularity theory and an introduction to catastrophe theory, Springer-Verlag, New York—Heidelberg—Berlin.

ZAKRŽEVSKIJ, M. V. (ЗАКРЖЕВСКИЙ, М. В.)
1980 Колебания существенно-нелинейных механических систем, Зинатне, Рига.

Index